19th International Conference on Cold Fusion 2015 (ICCF19)

Published in the Journal of Condensed Matter Nuclear Science (JCMNS) Volume 19

Padua, Italy
13-17 April 2015

Editors:

Jean-Paul Biberian
George Miley
Michael McKubre
Peter Hagelstein
Xing Zhong Li
Edmund Storms

ISBN: 978-1-5108-2588-8

Printed from e-media with permission by:

Curran Associates, Inc.
57 Morehouse Lane
Red Hook, NY 12571

Some format issues inherent in the e-media version may also appear in this print version.

Printed by Curran Associates, Inc. (2016)

For permission requests, please contact
International Society for Condensed Matter Nuclear Science (ISCMNS)
at the address below.

International Society for Condensed Matter Nuclear Science (ISCMNS)
The Willows, Hobro,
Wolverley, Kidderminster
Worcs DY11 5ST
England

www.iscmns.org
webmaster@iscmns.org

Additional copies of this publication are available from:

Curran Associates, Inc.
57 Morehouse Lane
Red Hook, NY 12571 USA
Phone: 845-758-0400
Fax: 845-758-2633
Email: curran@proceedings.com
Web: www.proceedings.com

J. Condensed Matter Nucl. Sci. 19 (2016) 1–324
 ISSN 2227-3123

JOURNAL OF CONDENSED MATTER NUCLEAR SCIENCE

Volume 19 2016

CONTENTS

PREFACE

RESEARCH ARTICLES

JOURNAL OF CONDENSED MATTER NUCLEAR SCIENCE

Experiments and Methods in Cold Fusion

Proceedings of the ICCF 19 Conference, April 13–17, 2015, Padua, Italy

VOLUME 19, June 2016

JOURNAL OF CONDENSED MATTER NUCLEAR SCIENCE

Experiments and Methods in Cold Fusion

Preface

It has been a huge privilege and a source of pride to organize the 19th edition of the International Conference on Condensed Matter Nuclear Science. Held at the historic "Palazzo della Ragione" in Padua, April 13–17, 2015, ICCF-19 included over 500 registered participants from more than 29 countries, being one of the most successful conferences of this series ever.

Being the first one organized by a private company it saw the creation of "The Engineering Applications Committee" with the express purpose to encourage the generation of products derived from years of scientific research.

We had the honor to receive the High Patronage of the Office of the Prime Minister of the Republic of Italy as well as the privilege of support from the following:

- Minister of the Economic Development of the Republic of Italy.
- Italian Agency for New Technologies, Energy and Sustainable Development (ENEA).
- National Research Council of Italy (CNR).
- The International Society for Condensed Matter Nuclear Science.
- The Italian Industrial Company Association (Confindustria).
- Padua Chamber of Commerce.

We are very grateful to them and to our sponsors, the New Energy Foundation and the Sidney Kimmel Institute for Nuclear Renaissance for their help and support.

With our great pleasure and gratitude the opening day has seen as special speakers:

- Massimo Bitonci
  Mayor of Padova
- Rodolfo Cetera
  Vice President of Confindustria Padova
- Vincenzo Vespri
  Valuator of Research & Development Project – Ministry of Instruction, University e Research
- Francesco Maria Avato
  University of Ferrara
- Eugenio Grippo
  Gianni, Origoni, Grippo, Cappelli & Partners Law Firm
- Tom Darden
  CEO Cherokee Investment Fund
- Cerivaldo da Silva Lira
  Ministry of Brasil Economic Development
- Robert V. Duncan
  President of the ICCF-19 Scientific Committee – Dean of Texas Tech University

A heartfelt thanks goes to the International Advisory Committee to Michael McKubre, Vittorio Violante, Graham Hubler, and to everyone who has contributed with their own works at the research and development of this fascinating science. We hope we have given at least as much as the ICCF has given to us. The scientific results of this conference are collected in this book, a treasure of the latest research coming from all over the world that clearly exemplifies the progress made in the past decade, and that set new ambitious goals for the coming years. The target of a new, clear, safe, and sustainable source of energy is increasingly visible to all of us. The collaborations between research centers, private companies, researchers, and universities as well as the competition between different ideas, theorems and ways of thinking are the evidence of a consolidated will to succeed.

Sincerely,

Antonio La Gatta
(*General Chair, ICCF-19, Engineer and Professor, TSEM Founder*)
May 2016

J. Condensed Matter Nucl. Sci. 19 (2016) 1–9

Research Article

Effect of Cathode Pretreatment and Chemical Additives on H/D Absorption into Palladium via Electrochemical Permeation

Orchideh Azizi*, Jinghao He, Dan T. Paterson, Arik El-Boher, Dennis Pease and Graham Hubler†

The Sidney Kimmel Institute for Nuclear Renaissance (SKINR), Department of Physics and Astronomy, University of Missouri, Columbia, Missouri, MO 65211, USA

Abstract

In this study, we applied a variety of palladium pretreatment methods, including nitric acid etching, Aqua regia etching and electrochemical oxidation/reduction to determine the effect of cathode initial condition on hydrogen absorption into palladium during the loading process. The electrochemical behavior of palladium was characterized using electrochemical permeation, electrochemical desorption, and electrochemical impedance spectroscopy (EIS). The hydrogen diffusion coefficient in palladium membrane was obtained using the time-lag method. The lattice diffusivity of hydrogen as well as the amounts and distribution of the diffused and trapped hydrogen were analyzed using the desorption rate of hydrogen. We also investigated the effect of pyridine, a permeation promoter on hydrogen absorption in palladium membranes. Electrochemical parameters of double layer capacitance, charge transfer resistance and surface roughness were characterized using electrochemical impedance spectroscopy for both treated and untreated cathodes. This study shows that the ability to load Pd with hydrogen is only mildly affected by surface pretreatment and the chemical additives, and therefore, we consider the pretreatments a second order effect regarding the ability to load.

Keywords: Absorption, Deuterium, Impedance spectroscopy, Palladium, Permeation

1. Introduction

Palladium hydrogen system has been intensively investigated due to the exceptional ability of palladium to adsorb and absorb hydrogen. Achieving consistent high loading of Hydrogen (H) and Deuterium (D) in Palladium (Pd) is one of the most challenging issues in the field of Anomalous Heat Effect (AHE). Numerous researchers have suggested that the concentration of deuterium loaded into palladium may have to exceed a critical value before the generation of excess heat [1–3]. Studies show that the hydrogen permeation and loading behavior is sensitive to the surface treatment of the sample [4,5] and chemical impurities [6–8]. The Sidney Kimmel Institute for Nuclear Renaissance (SKINR) has

*Corresponding author. E-mail: oazizi@missouri.edu
†E-mail: hublerg@missouri.edu

© 2016 ISCMNS. All rights reserved. ISSN 2227-3123

conducted loading studies of 50 μm thick Pd foils in an attempt to enhance the electrochemical activity of palladium and increase loading by using various pretreatment methods and chemical additives. The objectives of this work is to study hydrogen absorption and loading in palladium affected by various surface treatment and chemical additives as permeation promoters using an electrochemical permeation technique.

2. Experimental

The electrochemical permeation measurements of hydrogen were performed in a two-compartment electrochemical cell [9]. In each glass half-cell a platinum plate and Ag–AgCl electrode were used as a counter and reference electrode, respectively. The electrolyte solution was 0.1 M LiOH made using 18.2 MΩ-cm Millipore water.

Palladium foils of 50 μm thickness were prepared using cold rolling of a 1 mm thick palladium (Holland Moran) followed by thermal annealing at 850°C for 60 min in a protocol to ensure all samples having the same initial conditions. Such a Pd foil with surface area of 3.14 cm^2 was used as a membrane (working electrode). To study the effect of initial conditions on hydrogen permeation and hydrogen loading into palladium, four types of palladium samples were used as a membrane. They are Pd foil etched in concentrated nitric acid for three minutes (type 1), Pd foil etched in Aqua regia for 3 min (type 2), pretreated Pd by electrochemical oxidation/reduction (type 3), and un-etched palladium (type 4).

In the electrochemical permeation cell the palladium surface at which water molecules were reduced is referred to as the entry surface and the corresponding surface at which the diffusing hydrogen was oxidized is labeled the exit surface.

During the permeation measurements, the entry side of the membrane was subjected to galvanostatic cathodic charging. The exit side of the membrane was continuously polarized at a constant potential close to open circuit potential. Under this potential, hydrogen atoms escaping from the membrane were oxidized. The current measured at the exit side corresponds to the permeation rate of hydrogen through the membrane. Figure 1 shows schematic of the electrochemical permeation cell.

3. Results and Discussion

3.1. Effect of chemical etching on hydrogen permeation and loading

Figure 2 shows the hydrogen permeation currents measured at the exit side for different charging currents at the entry side and for un-etched Pd and Pd foils etched in nitric acid and in Aqua regia. After 10 min of permeation, the hydrogen charging current from the entry side was interrupted and the entry side of the membrane was polarized immediately at the same potential of the exit side. At charging currents less than 7 mA, after few seconds, the permeation rate reaches the steady-state value and remains constant with time. At charging currents higher than 7 mA, the permeation current after reaching the maximum value decreases with increasing of time. Figure 2(d) shows that permeation current of palladium etched in Aqua regia decreased slightly faster than those of Pd etched in nitric acid and un-etched palladium.

Figure 3(a) shows the permeation efficiency (steady-state permeation current/charging current) vs. charging current for palladium membranes without etching and etched in nitric acid and Aqua regia. At charging currents less than 7 mA, the permeation efficiency for all three electrodes was close to 90%. At charging current higher than 7 mA the permeation efficiency dropped to 60% and decreased further to 20% by increasing the charging current up to 40 mA. These results show that the permeation efficiency on Pd etched in Aqua regia at charging current less than 7 mA is slightly lower than those of other two electrodes. To pinpoint the underlying mechanism, we plotted the steady-state permeation current as a function of the square root of the charging current in Fig. 3(b), and found that permeation current is linearly dependent upon the square root of the charging current at charging currents lower than 7 mA. Furthermore, this linear relationship is independent on the surfaces of un-etched or etched Pd, suggesting the

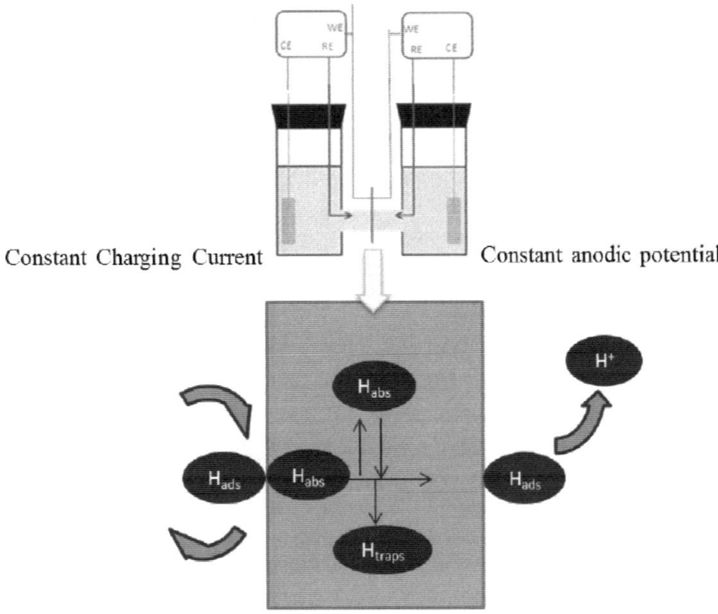

Figure 1. Schematic of electrochemical permeation test.

permeation process is fairly well determined by hydrogen diffusion rate [10]. In other words, water discharge and hydrogen absorption in Pd are controlled processes at low charging currents and pretreatment of palladium has no effect on the surface kinetics. However, at charging current larger than 7 mA the steady-state permeation current no longer follows the linear relationship and appears to be not affected by charging current much, especially for strongly etched Pd (in Aqua regia). In this case, hydrogen recombination flux increase and absorption flux decreases.

For hydrogen permeation measurements under galvanostatic charging, the diffusion coefficient through palladium membrane is conventionally calculated from the lag time of current transient [11] according to $D = L^2/2t_1$, where L is the thickness of the palladium membrane, and t_1 is the time at which the permeation current reaches 0.63 of the steady-state value. Using the results of permeation current versus time in Fig. 1, we determined the diffusion coefficient of hydrogen in both etched and un-etched palladium is about $3.5 \times 10^{-7}\text{cm}^2\text{s}^{-1}$, which is in good agreement with the literature [12,13].

Figure 4(a) presents the steady-state polarization behavior of palladium in 0.1 M LiOH obtained by sweeping the potential at a sweep rate of 0.16 mV/s. The steady-state polarization curve reveals that at low over-potentials the Tafel slope is about –0.18, but it changes to –0.25 at over-potentials higher than –1.1 V.

The entry potential versus hydrogen permeation current in Fig. 4(b) show particularly sharp changes at a potential of –1.2 (V) above which the permeation current is independent of potential. Steady-state polarization data in conjunction with permeation current data reveals that the mechanism of hydrogen evolution from Pd in alkaline solution at low charging current starts with water discharge and the formation of adsorbed hydrogen followed by hydrogen absorption into Pd. At higher charging current when the concentration of hydrogen increases near the surface of palladium entry side the mechanism changes to a coupled discharge and desorption reactions, in which the rate of hydrogen desorption increases significantly and the absorption flux decreases [14].

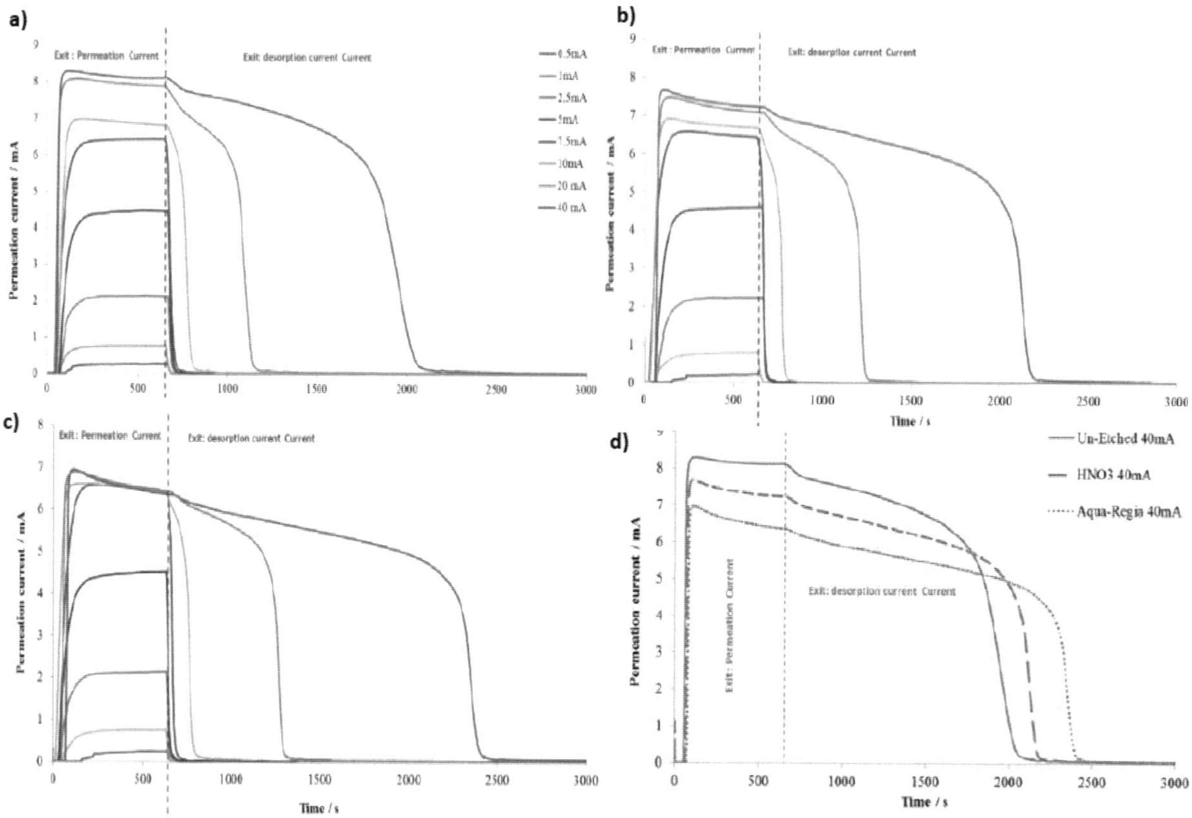

Figure 2. Permeation currents on the exit side of palladium membrane during applying different cathodic currents from the entry side for 10 min in 0.1 M LiOH solution. (a) un-etched palladium; (b) Pd foil etched in concentrated nitric acid for 3 min; (c) Pd foil etched in Aqua regia for 3 min and (d) comparing the permeation current vs. time for all three electrodes at 40 mA charging current. Data after vertical dashed line represents hydrogen discharge process under no charging current from the entry side of the membrane.

To study the hydrogen loading into Pd membrane the electrochemical desorption method was used. As shown in Fig. 1, after 10 min, the cathodic current was switched off and entry side was polarized immediately at the same potential of the exit side. From this moment the anodic currents were recorded at the both sides of the membrane and they correspond to hydrogen desorption rate from the Pd membrane. Integration of the desorption current – time curves at both sides of the membrane leads to the total amounts of diffusible and absorbed hydrogen in the palladium membrane (see inset of Fig. 5).

The results show that for all palladium membranes, etched and un-etched, the hydrogen concentration at the entry side is higher than that at the exit side.

The H/Pd loading ratio, calculated using hydrogen desorption data for un-etched and etched palladium, is presented in Fig. 5. The results reveal that the hydrogen loading increases with increasing charging current as expected, and H/Pd loading ratio is slightly higher for palladium etched in Aqua regia in comparison to Pd etched in nitric acid and un-etched palladium.

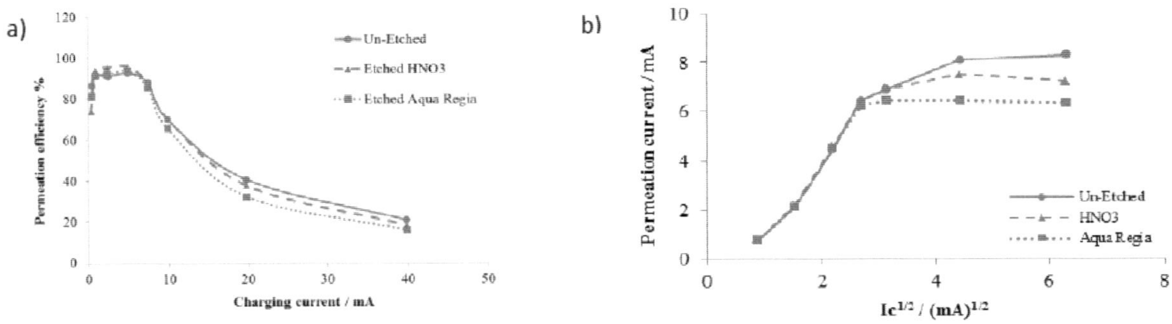

Figure 3. (a) Permeation efficiency as a function of charging current; (b) Permeation current as a function of the square root of the charging current (I_c) for un-etched Pd, etched Pd in nitric acid and etched Pd in Aqua regia.

Figure 6 shows the electrochemical impedance spectroscopy data (Nyquest plots) obtained at different potentials on etched and un-etched palladium membranes in 0.1 M LiOH solution. The data shows that the charge transfer resistance deceased by increasing the potentials which is in good agreement with steady-state polarization data.

Also fitted data showed that the double layer capacitance is slightly higher for etched Pd in Aqua regia probably due to higher surface area on roughed surfaces.

3.2. Effect of high current

Figure 7 shows the discharge current vs. time at entry side (Fig. 7(a)), the permeation current vs. time at exit sides after hydrogen charging for 10, 20 and 40 min at 40 mA charging current (Fig. 7(b)). The atomic ratio of H/Pd vs. charging time obtained from electrochemical desorption technique at both sides of the membrane (Fig. 7(c)) reveals that hydrogen loading into Pd increased linearly with time for the first 40 minutes and hydrogen loading reaches close to 25%. For longer charging time at 40 mA the hydrogen loading increased to about 89% and remained almost constant for charging times higher than 3 h.

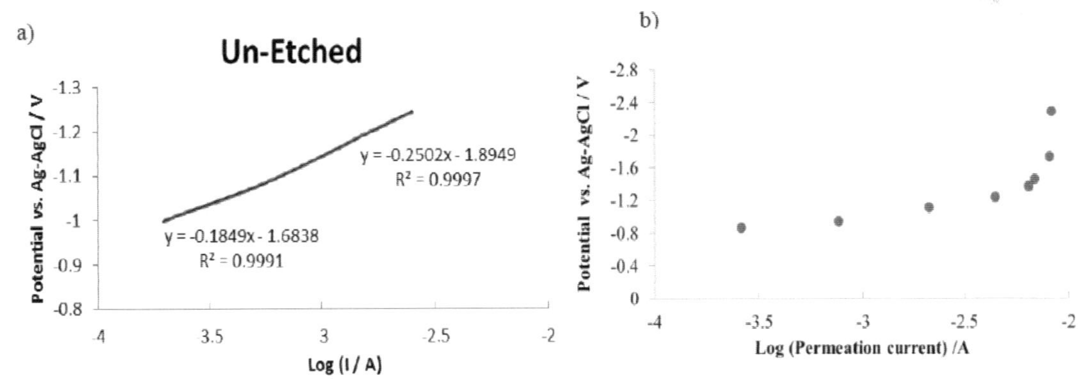

Figure 4. (a) Steady-state polarization curve of un-etched palladium in 0.1 M LiOH solution; (b) Dependence of the log of hydrogen permeation current on potential for un-etched 50 μm thick Pd in 0.1 M LiOH solution.

3.3. Electrochemical pretreatment

The effect of specific electrochemical oxidation reduction procedure was studied in attempt to improve the electrochemical activity of palladium and increase the hydrogen loading into palladium. In this procedure a symmetric square-wave potential ± 1.5 V was applied to the Pd electrode for about 20–30 s, with frequency of 800 Hz. Figure 8(a) shows the cyclic voltammetry (CV) obtained from untreated and electrochemically treated palladium. The treated electrode presents almost the same features as the untreated electrode, but the treated cathode gives higher currents during the whole potential sweep, suggesting the cathode is more active after electrochemical treatment.

Figure 8(b) shows the permeation current vs. time at several different charging currents at the entry side for unetched, Aqua regia etched, and electrochemically treated palladium membranes, among which the electrochemically treated membrane has the highest permeation current.

3.4. Chemical impurities

Alternative to chemical etching and electrochemical pretreatment, we also studied the effect of Pyridine, a permeation promotor on hydrogen permeation current and loading behavior of palladium. As shown in Fig. 9(a), pyridine improves the hydrogen permeation after adding 1 mM pyridine to the entry side at a 10 mA charging current. The permeation current increased during the initial 10 min charging, but it soon reached the steady-state value. Adding pyridine to the both sides of the permeation cell results in a decrease of the steady-state permeation current in comparison to adding pyridine at only one side. However the permeation current is still higher than that in pure electrolyte.

The desorption data from both sides of the membrane shows less hydrogen loading into the palladium membrane in the presence of pyridine. In the presence of pyridine the hydrogen coverage on the entry side is less than in pure electrolyte. As a result the permeation efficiency is slightly higher than that in pure electrolyte under the same condition.

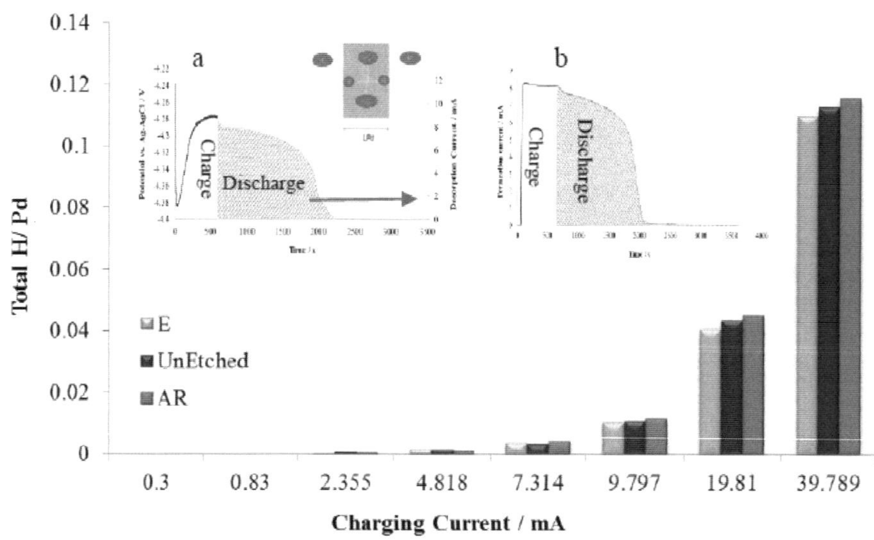

Figure 5. Total atomic ratio of H/Pd calculated using hydrogen discharge data as a function of charging current, in 0.1 M LiOH solution. Inset: Desorption rate of hydrogen from palladium membrane after 10 min charging at 10 mA current; (a) entry side and (b) exit side .

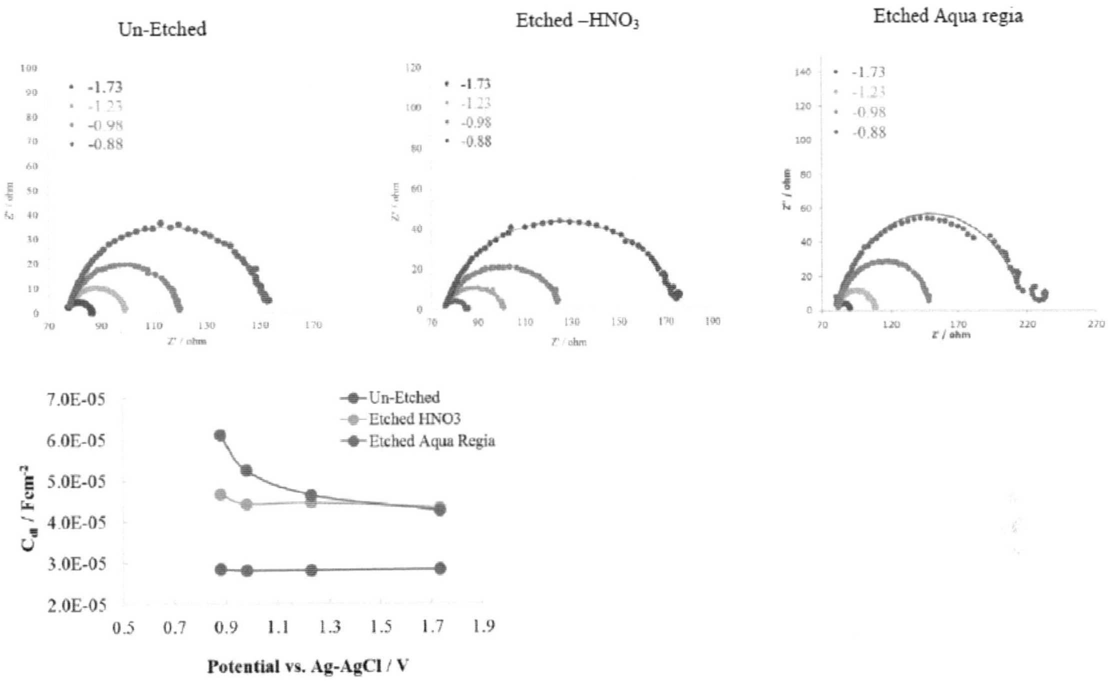

Figure 6. Potential vs. Ag–AgCl/V.

Figure 9(b) shows the effect of pyridine concentrations on hydrogen permeation current. For the concentration up to 5 mM the increasing pyridine concentrations resulted in a high permeation current. But for higher pyridine concentration, the permeation current decreased but it was still higher than that in pure electrolyte as pyridine may block the active sites for water reduction at the entry side.

4. Conclusion

At low charging currents, the permeation efficiency for all un-etched and etched palladium was close to 90% and they are determined by hydrogen diffusion in Pd membranes. Hydrogen diffusion coefficient around $3.5 \times 10^{-7} cm^2/s$ was obtained using the time-lag method. The kinetics of hydrogen is controlled by water discharge followed by, hydrogen absorption into Pd. At intermediate charging currents the permeation efficiency was about 60%. At high charging currents, it further decreased to 20%, suggesting the hydrogen recombination flux increases significantly and the absorption flux decreases.

Hydrogen loading in Pd etched in Aqua regia is slightly higher than that of un-etched Pd mainly due to higher surface area and more available active sites. Square-wave electrochemical oxidation reduction pretreatment of palladium appears to improve H loading in palladium. Presence of pyridine affects the mechanism of hydrogen reactions and slightly enhances the permeation current, but it does not have significant effects on improving H loading in palladium. Our study shows that the ability to load Pd with hydrogen is only mildly affected by surface pretreatment and the chemical additives studied, and therefore, we consider the pretreatments a second order effect regarding the ability

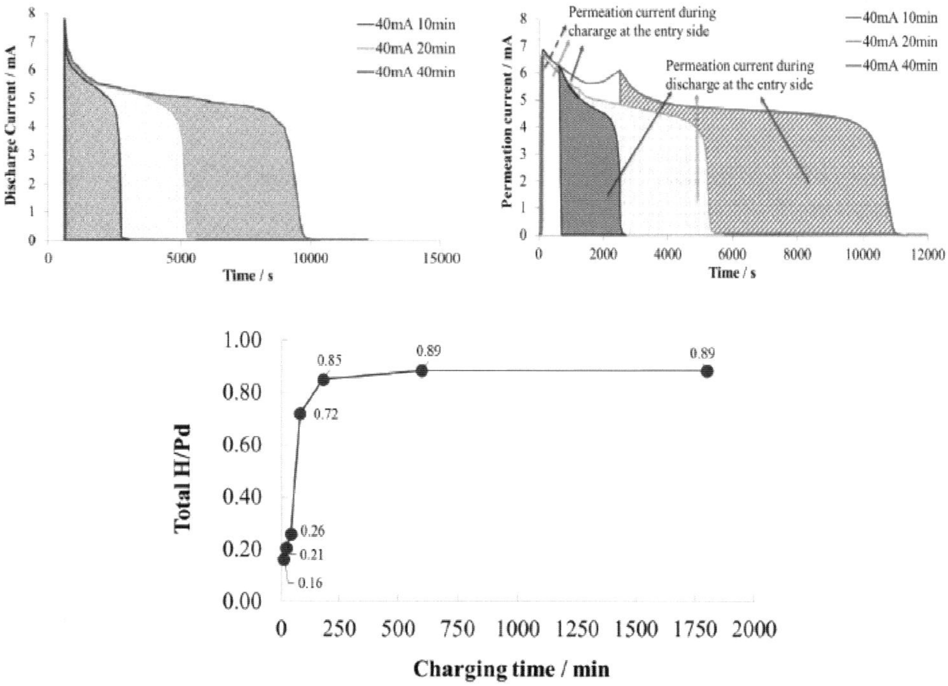

Figure 7. (a) Hydrogen discharge current vs. time at 50 mV at the entry side of the Pd foil; (b) Hydrogen permeation vs. time at the exit side. All data obtained at charging current of 40 mA for 10, 20 and 40 min; (c) Atomic ratio of H/Pd loading as a function of time at charging current of 40 mA.

to load which suggests that subtle alterations of the kinetics of surface processes like Tafel recombination, probably caused by surface impurities, govern the de-loading.

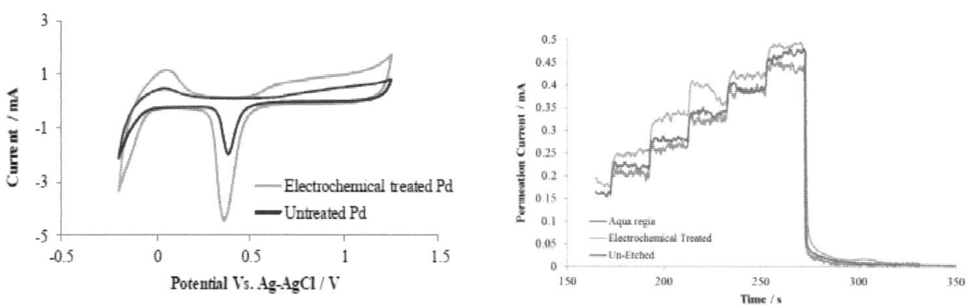

Figure 8. (a) Cyclic voltammograms of untreated palladium and electrochemically treated palladium in 0.1 M LiOH, scan rate of 100 mV/s. (b) Permeation current as a function time for electrochemically treated and untreated palladium in 0.1 M LiOH solution.

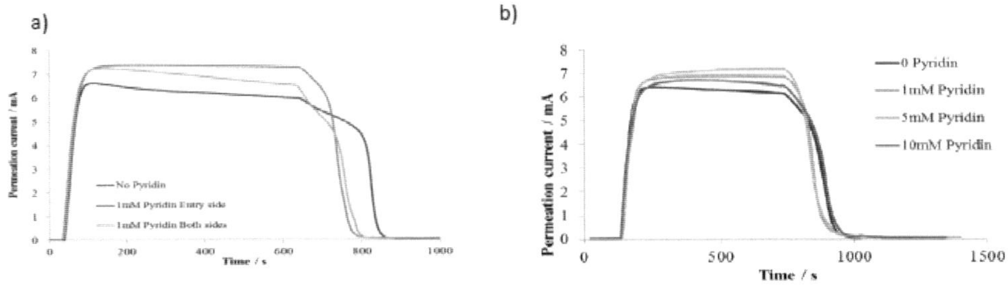

Figure 9. (a) Permeation current versus time obtained in the absence and presence of pyridin; (b) Permeation current versus time obtained in the absence and presence of 1, 5 and 10 mM pyridin.

Result of loading at high charging current is somewhat surprising in that a loading ratio of 0.89 was readily achieved in 3 h in a 50 μm thick foil, and, that the fact that hydrogen was constantly being removed at the exit side. For foils loaded from both sides, it generally takes much longer than 3 h to reach H/Pd of 0.89 and often, such high loading cannot be reached at all. This may indicate that internal stresses caused by the large volume change with increasing H/Pd ratio are more easily accommodated by the foil in one-side charging

Acknowledgment

This work is fully supported by Mr. Sidney Kimmel

References

[1] G.K. Hubler, U.S. Surface and Coatings Technology **201** (2007) 8568.
[2] M.C.H. McKubre et al., *Fifth Int. Conf. on Cold Fusion*, IMRA Europe, Sophia Antipolis Cedex, France, Monte-Carlo, Monaco, 1995.
[3] R.A. Oriani, *Fourth Int. Conf. on Cold Fusion*, Electric Power Research Institute, Lahaina, Maui, 1993, 312 Hillview Ave., Palo Alto, CA, USA.
[4] T. Harris and R.M. Latanislon, *Int. J. Hydrogen Energy* **14** (1989) 623.
[5] T.A. Green and T.I. Quickenden, *J. Electroanaly. Chem.* **368** (1994) 121.
[6] T. Maoka and M. Enyo, *Surface Technol.* **9** (1979) 147.
[7] L. Birry and A. Lasia, *Electrochim. Acta* **51** (2006) 3356.
[8] J.-N. Han, S.-I. Pyun and T.-H. Yang, *J. Electrochem. Soc.* **144** (1997) 4266.
[9] M.A.V. Devanathan and Z. Stachurski, *Proc. Roy. Soc. A: Mathematical, Physical and Eng. Sci.* **270** (1962) 90.
[10] M. Zamanzadeh, A. Allam, H.W. Pickering and G.K. Hubler, *J. Electrochem. Soc.* **127** (1980) 1688.
[11] S. Schuldiner and J.P. Hoare, *J. Electrochem. Soc.* **103** (1956) 178.
[12] J.G. Early, *Acta Metall.* **26** (1978) 1215.
[13] J. Bowker and G.R. Piercy, *Metall. Trans.* **16A** (1985) 715.
[14] J. O'M Bockris, R. Sundaresan, Z. Minevski and D. Letts, *Fourth Int. Conf. on Cold Fusion*, Electric Power Research Institute, Lahaina, Maui, 1993, 312 Hillview Ave., Palo Alto, CA, USA.

J. Condensed Matter Nucl. Sci. 19 (2016) 10–16

Research Article

Calorimetric and Radiation Diagnostics of Water Solutions Under Intense Light Irradiation

Yu.N. Bazhutov*

Terrestrial Magnetism, Ionosphere and Radiowave Propagation Institute RAS (IZMIRAN), Moscow, Russia

A.I. Gerasimova

Russian State Agricultural Correspondence University, Moscow region, Balashikha, Russia

V.V. Evmenenko

Scientific Research Center of Electronics and Computing, Moscow, Russia

V.P. Koretskiy†

Scientific Research Center of Engineering Physical Problems "ERZION" (ERZION Center), Moscow, Russia

A.G. Parkhomov and Yu.A. Sapozhnikov

Lomonosov Moscow State University, Moscow, Russia

Abstract

A series of experiments with light irradiation of saltwater solutions and alkalis in special vessels was performed. The sources of irradiation were: a Light Emitting Diode (LED) lamp, a matrix of LEDs, and a laser. All of them were in the red wavelength. Possible calorimetric diagnostics and nuclear radiation (gamma rays, X-rays, and neutrons) were monitored. Liquid scintillation diagnostics of tritium were used. Gamma-ray radiation detection with an NaI scintillation detector, and X-ray radiation detection with Geiger counters was performed. No such radiation was observed. Excess heat was not detected in any experiment. Neutrons were measured with help of ^{3}He counters placed in a paraffin barrel. Small neutron emissions (up to 100 neutrons) was observed in the form of a series of short bursts (lasting a few milliseconds) during some minutes at the background level. The generation of tritium after LED lamp and matrix of LEDs irradiation of water solutions LiOH and Na_2CO_3has been demonstrated in some experiments.

Keywords: Calorimetry, Erzion model, Gamma-ray, Neutrons, Red light irradiation, Tritium, Water solutions, X-ray

*E-mail: erzion@mail.ru
†E-mail: tarhanval@gmail.com

© 2016 ISCMNS. All rights reserved. ISSN 2227-3123

1. Introduction

This work is an extension of the research [1–4] on the nuclear effects during irradiation of various water solutions. However in these previous tests, a careful diagnosis of the products after light exposure of electrolytes was not carried out. In our present experiments, we performed not only continuous monitoring of possible nuclear emanations (neutrons, gamma rays, and X-rays), but also the calorimetric diagnostics and scintillation diagnosis of tritium after LED and laser irradiation of water solutions of salts and alkalis.

We used the predictions of the Erzion model [5] to define what chemical elements in water solutions can produce nuclear radiation. We reviewed all reactions of Erzion-nuclear exchange for isotopes of light chemical elements from $H^1$to O^{18} and selected those predicted to produce neutrons (1–3) and tritium (8) (with the percentage composition of the parent nucleus in the natural isotope mixture):

$$H^1(E^-, E^0)\ \mathbf{n} + 1{,}65\ \text{MeV} \qquad (100\ \%) \qquad (1)$$

$$H^2(E^-, E_N)\ \mathbf{n} + 5{,}6\ \text{MeV} \qquad (0.016\ \%) \qquad (2)$$

$$Li^6\ (E^-, E_N)\ He^5 + 3{,}2\ \text{MeV} \qquad (7.5\ \%)$$
$$He^5 \rightarrow He^4 + \mathbf{n} + 1{,}36\ \text{MeV} \qquad (3)$$

The reactions (1)–(3) runs with the the negatively charged Erzions E^-, which could generate in the reactions:

$$Li^7\ (E_N, E^-)\ Be^8 + 9{,}5\ \text{MeV} \qquad (92.5\ \%) \qquad (4)$$
$$Be^8 \rightarrow 2\ He^4 + 4{,}8\ \text{MeV}$$

$$C^{14}(E_N, E^-)\ N^{15} + 2{,}4\ \text{MeV} \qquad (-) \qquad (5)$$

$$O^{18}\ (E_N, E^-)\ F^{19} + 0{,}2\ \text{MeV} \qquad (0{,}2\ \%) \qquad (6)$$

$$N^{15}\ (E_N, E^-)\ O^{16} + 4{,}3\ \text{MeV} \qquad (0{,}37\ \%) \qquad (7)$$

To return the Enions E_N into reactions (5)–(8) the closed chains $E_N \leftrightarrow E^0$ can runs on the isotopes H^2, C^{13} and O^{17} in the next reaction pairs:

On the basis of these predictions we decided to use an alkaline solution of lithium (LiOH) with the addition of heavy water (D_2O) to increase the length of the Erzion-nuclear reaction chains. From the Erzion model, deuterium is the best of the nuclei converters ($E_0 \rightarrow E_N$) with maximum cross-section reaction of the Erzion-nuclear exchange [6].

We decided also to perform experiments with solutions of Na (NaOH and Na_2CO_3), which monoisotope Na^{23} effectively turns E_N in the negatively charged E^- like nuclei Li^7 (4).

$H^2(E_N, E^0)\mathbf{H}^3 + 0.1$ MeV (0.016 %) (8)

$H^2(E^0, E_N)H^1 + 3.9$ MeV (0.016 %) (9)

$C^{13}(E_N, E^0)C^{14} + 2.0$ MeV (1.1 %) (10)

$C^{13}(E^0, E_N)\,C^{12} + 1.2$ MeV (1.1 %) (11)

$O^{17}(E_N, E^0)\,O^{18} + 1.9$ MeV (0.038%) (12)

$O^{17}\,(E^0, E_N)\,O^{16} + 2.0$ MeV (0.038%) (13)

2. Experimental Configuration

A series of experiments were conducted with red LED lamp, matrix of LEDs, and laser irradiation of various water solutions including LiOH, NaOH and Na_2CO_3, with various additives of D_2O. To do this we created a special setup with options for working cells made from plastic, porcelain, or a Dewar vessel, along with the devices for detecting neutrons, gamma rays and X-rays. A schematic of these configurations is shown in Fig. 1.

The main features of the experiments are shown in Table 1.

An alkaline solution of LiOH with molarity 2.5 M and (5–20) M NaOH, and a solution of salt 2 M Na_2CO_3 were used as the working liquids. The solvent was water from various sources – city water (Exp. Nos. 1–13); water treated with a carbon filter (Exp. Nos. 14,15); and snow melt water (Exp. Nos. 16,17). It was predicted from the Erzion model that the water from open-air sources may be saturated with the Erzions from cosmic rays. In aqueous solutions the heavy water D_2O was added to intensify the possible nuclear reactions.

Table 1. The main features of the experiments.

Exp. no.	Solution type	Solution Volume (l)	Additive D_2O	Source of light	Exposure time (min)	Solution temp. rise (°C)	Emitted energy (J)
1	2.5 M LiOH	1.0	–	Lamp	7	1.0	714
2	2.5 M LiOH	1.0	–	Lamp	9	1.0	918
3	2.5 M LiOH	0.5	10%	Laser	61	1.0	18
4	2.5 M LiOH	0.5	10%	Lamp	60	1.0	6120
5	2.5 M LiOH	0.5	–	URC-LED	72	1.0	8640
6	2.5 M LiOH	0.5	–	LED-4	87	1.0	10 440
7	2.5 M LiOH	0.5	–	LED-15	50	1.0	9000
8	2.5 M LiOH	0.5	–	LED-20	70	2.0	14 400
9	2.5 M LiOH	0.5	5%	LED-20	60	2.5	15 120
10	10 M NaOH	0.5	5%	LED-20	63	1.5	14 400
11	5 M NaOH	0.5	6%	LED-20	60	2.5	14 400
12	20 M NaOH	0.5	10%	LED-20	57	1.0	13 680
13	2 M Na_2CO_3	0.5	2%	LED-16	60	1.0	11 520
14	H_2O	0.5	50%	LED-9	60	1.5	6480
15	2 M Na_2CO_3	5.0	4.5%	LED-9	71	1.0	7668
16	2 M Na_2CO_3	10.0	1.1%	LED-9	80	1.0	8640
17	2 M Na_2CO_3	0.8	1.1%	LED-9	62	1.0	6696

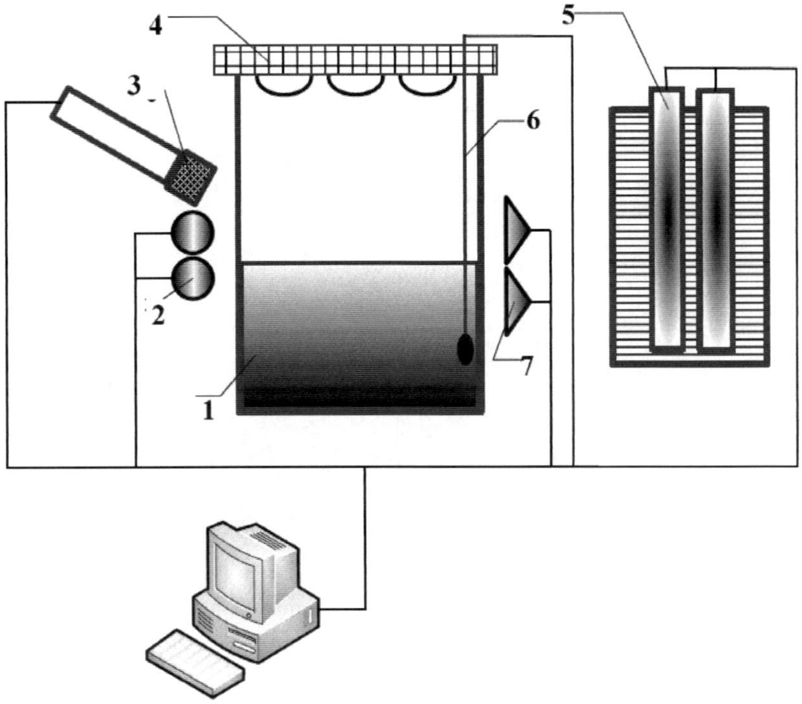

Figure 1. Schematic of the experimental configuration.1–working cell with water solution; 2–Geiger counters type CBM-20; 3–NaI scintillation detector; 4– plate with the sources of red light; 5–He^3 gas neutron counters (Ø3.5 × 35 cm^2) placed in paraffin cylinder (Ø22 × 40 cm^2) or in vessel with water; 6–thermocouple; 7–Geiger counters type CHM-18 with thin window.

The working cells with liquids were placed in the center of a paraffin cylinder with two He^3-counters. These cells were made from plastic vessel (V=1l; Exp. Nos. 1, 2), a porcelain cup (V=0.9l; Exp. Nos. 3–13), or a glass Dewar vessel (V=1l; Exp. Nos. 14, 15, 17). In Experiment No. 16 the working cell was the metal Dewar (V=17.5l) and in this case two He^3-counters were placed in the vessel with water. The efficiency of neutron detectors measured with Cf^{252} – neutron source was 9±0.5%.

The sources of red light were the red LED lamp (with wave length λ=625 ± 25 nm and light power W=1.7 W), the red laser pointer (λ=650 ± 5 nm; W=0.005 W), a matrix of 19 URC-LEDs (λ=665 ± 5 nm; W=0.4 W) and one of the five matrix of LEDs (λ=665 ± 5 nm; W=0.8 – 4.0 W with 4–20 LEDs).

All readings of radiation detectors and thermocouples were recorded continuously with the PC. Tritium registration was conducted with the standard scintillator diagnosis of the 5–10 ml volume solution samples taken before and after each experiment.

The last column in Table 1 is the calculated value of the absorbed (equal to emitted) energy, determined from the source light power, solution volume and exposure time.

A general view of some experimental configurations is shown in Figs. 2 and 3.

Figure 2. The experimental setup with an array of 20 LEDs and neutron counters placed in a paraffin cylinder.

3. Results, Discussion, and Conclusions

Calorimetric diagnostics. No excess heat was detected in any experiment.

From Table 1, we see that the temperature rise during light irradiation of solutions was in the range of 1.5–2.5°C. There is some correlation between the rise of the temperature in the reactor and the emitted heat energy during light

Table 2. Tritium activity for solution samples taken before ("Background") and after ("Effect").

Experiment number	Solution sample	A (Bq/g)	ΔA (Bq/g)	Solution Volume (l)	Exposure time (s)	N_γ ($\times 10^{19}$)	N_T ($\times 10^7$)	N_γ/N_T ($\times 10^{12}$)
11	"Background" "Effect"	235±1 1158±1	923±1	0.5	3600	5445	170	32
12	"Background" "Effect"	1323±1 1340±1	17±1	0.5	3420	518	3	173
13	"Background" "Effect"	280±1 947±1.5	667±2	0.5	3600	3464	120	36
14	"Background" "Effect"	5.3±0.1 13.6±0.1	8.3±0,2	0.5	3600	2455	2	1227
15	"Background" "Effect"	31.1±0.3 52.5±0.3	24±0,4	5.0	4260	2905	5	581
17	"Background" "Effect"	550±10 626±16	76±19	0.8	3720	2536	22	115

A–activity of tritium, ΔA–the difference between the activity "Effect" and "Background"; N_γ–the number of radiating photons from source of light; N_T–the number of produced tritium in all over the irradiated volume; N_γ/N_T–the number of photons to produce a single nucleus of tritium.

Figure 3. The experimental setup with an array of 9 LEDs and solution in the Dewar vessel.

irradiation of solutions. At the same time, the energetic effect is influenced not only by the power of the lamp, but also the duration of the experiment. Thus, in this series of experiments, we were able to establish an upper bound of 16 kJ of possible energy production. If we take the sensitivity of the thermocouple equal to 0.5°C, we can get an upper bound on the potential to generate additional energy at the level 5 mW, which is comparable with the laser power output $\sim$1 mW. Thus, we did not reliably detect excess heat in our experiment.

Nuclear radiations (neutrons, gamma rays, and X-rays) were monitored, but none was detected. In all experiments, the measurements of neutron fluence, dose rate and X-rays fluence were at the background level.

Diagnostics of tritium showed the greatest effect.

Liquid scintillation diagnostics of tritium were carried out on the Chemical Faculty (Department of Radiochemistry) of the Lomonosov Moscow State University. The diagnostics took into account the possible impact of alkaline solution ekzoelectrons. The results obtained for tritium activity for solution samples taken before ("Background") and after ("Effect") are given in Table 2.

The tritium generation depends on the composition and concentration of the water solution and the concentration of D_2O. So, the maximum of tritium quality was formed in Exp. Nos. 11 and 13 using alkaline solutions NaOH and

Na_2CO_3 with D_2O. The same result was in the Exp. Nos. 12 and 17. In other experiments with solution LiOH the "Effect" measurements were significantly smaller.

Thus, experiments showed that tritium can be generated in water solutions NaOH and Na_2CO_3 with small additions of D_2O after irradiation by red light with a wavelength of 650 nm. These results correlate with the Erzion model. Additional investigations are necessary to explain the lack of tritium after irradiation of water solutions of LiOH, and the lack of neutrons in the experiments.

References

[1] Yu.N. Bazhutov, S.Yu. Bazhutova, A.P. Dyad'kin et al., Calorimetric and neutron diagnostics of liquids during laser irradiation, *Proc. 12th Russian Conference Cold Nuclear Transmutation and Ball Lighting*, Moscow, 2005, pp. 41–47.

[2] Yu.N. Bazhutov, S.Yu. Bazhutova, A.P. Dyad'kin, V.V. Nekrasov and V.F. Sharkov, Calorimetric and neutron diagnostics of liquids during laser irradiation, *Proc. 11th Int. Conf. on Cold Fusion, Condensed Matter Nuclear Science,* Marseilles, France, 31 October–5 November, World Scientific, New Jersy, 2006, pp. 374–378.

[3] Dennis Letts and Dennis Cravens, Laser stimulation of deuterated palladium, *ICCF-10*, Boston, USA, 2003.

[4] Mitchell R. Swartz, Photoinduced excess heat from laser-irradiated electrically-polarized palladium cathodes in heavy water, *ICCF-10*, 20 03.

[5] Yu.N. Bazhutov, Influence of spin and parity preservation lows on erzion model predictions in cold fusion experiments, *Proc. 4th Russian Conf. on Cold Fusion and Nuclear Transmutation*, Moscow, 1997, pp. 146–149.

[6] Y.N. Bazhutov, Influence of spin and parity preservation lows on erzion model predictions in cold fusion experiments, *The 7th Int. Conf. on Cold Fusion*, Vancouver, Canada: ENECO Inc., Salt Lake City, UT, 1998, pp. 437–440.

J. Condensed Matter Nucl. Sci. 19 (2016) 17–28

Research Article

Yet Another LENR Theory: Electron-mediated Nuclear Reactions (EMNR)

Andrea Calaon*

Independent Researcher, Monza, Italy

Abstract

An attempt is made to build an LENR theory that does not contradict any basic principle of physics and gives a relatively simple explanation to the plethora of experimental results. A single unconventional assumption is made, namely that nuclei are kept together by a magnetic attraction mechanism, as proposed in the 1980s of the past century by Valerio Dallacasa and Norman Cook. This assumption contradicts a non-proven detail of the standard model, which instead attributes the nuclear force to a residual effect of the strong interaction. The theory is based also on a property of the electron which has been known for long, but has rarely been used: the Zitterbewegung (ZB). This property should allow the magnetic attraction mechanism that binds nucleons together, to manifest also between the electron and any isotope of hydrogen, leading to the formation of three neutral pseudo-particles (the component particles remain separate entities), collectively named here Hydronions (or Hyd). These pseudo-particles can then couple with other nuclei and lead to a fusion reaction "inside" the electron. The Coulomb barrier is not overcome kinetically, but through what could be interpreted as a range extension of the nuclear force itself, realized by the electron when some specific conditions are satisfied. The most important of these necessary conditions is that the electron has to "orbit" the hydrogen nucleus at a frequency of 2.055×10^{16}Hz. This frequency corresponds to photons with an energy of about 85 eV or equivalently a wavelength of 14.6 nm in the Extreme Ultra Violet (EUV). So the large quanta of nuclear energy fractionate into EUV photons during the formation of the Hydronions and during the coupling of Hydronions to other nuclei. The formation of Hydronions requires the so called Nuclear Active Environment (NAE), which is what makes LENR so rare and difficult to reproduce. The numbers suggest that the NAE forms when an unshielded atomic core electron orbital that has an "orbital frequency" near to the coupling frequency is stricken by a naked Hydrogen Nucleus (HNu). This theory therefore implies that the NAE is not inside the metal matrix, but in its immediate neighbourhood. The best candidate atoms for a NAE are listed, based on the energy of their ionization energies. The coincidence with the most common LENR materials appears noteworthy. The Electron Mediated Nuclear Reactions (EMNR) theory can explain also very rapid runaway conditions, radio emissions, biological NAE, and the so called "strange radiation".

Keywords: EMNR theory, Extreme ultra violet, Hydronion, Magnetic attraction, Strange radiation, Zitterbewegung

*E-mail: a.calaon@libero.it

© 2016 ISCMNS. All rights reserved. ISSN 2227-3123

1. Background

The fact that LENR phenomena have been discussed for 26 years and no widely accepted explanation has emerged, suggests strongly that the phenomenon is based on some new or not commonly accepted fact of physics. Any theory trying to explain LENR must therefore have such an "unconventional ingredient".

2. Uncommon Assumption and Electron Nuclear Interaction

The theory proposed here is based on a single "uncommon" assumption, namely that nuclei are kept together by the electromagnetic interaction through the magnetic attraction mechanism proposed by Dallacasa and Cook, e.g. [1,2]. The approach assumes that the magnetic moment of nucleons comes from the rotation of charges (quarks), and not from gluons or intrinsic properties of the quarks. This contradicts a part of the standard model that is so far unproven (e.g. [3]).

The attraction mechanism depends only on the rotation of a point charge. Therefore the electron, with its Zitterbewegung (not different from the supposed internal charge rotation of the nucleons), should be subject to the same magnetic attraction. In other words the electron, if a series of conditions are satisfied, could be affected by what everyone calls the nuclear force. It will be shown that the necessary conditions are not commonly present in condensed matter.

3. Magnetic Attraction Mechanism

The description of the attraction mechanism done by Dallacasa and Cook in [1] is repeated here with some minor changes to the original proposal. If a particle has a magnetic moment it must generate an oscillating magnetic field. Here it is assumed that the magnetic moment of all particles is due to the rotation of a single charge travelling along a circular orbit at the speed of light (c). Whereas this assumption is essentially in accordance with what we know about the electron [4], it is not so for the nucleons, which have three fractional charges of opposite sign. However, for this simple case, one gets:

- Rotation radius: $r = 2\frac{m_{\text{mag}}}{gqc}$.
- Circular frequency: $\nu = \frac{c}{2\pi r}$,

where m_{mag} is the magnetic moment, g the gyromagnetic ratio and q the particle charge.

Let us now evaluate the oscillating magnetic field. Since non-accelerated particles do not emit electromagnetic radiation, the associated magnetic field can not have a radiative component, and can only correspond to the so-called "static" part of the Liénard–Wiechert potential. Moreover, since the speed of the charge is assumed to be equal to c, the only expression for B that could make sense is the simple Biot–Savart law:

$$B_1 = \frac{\mu_0}{4\pi}\frac{q_1\nu_1 \wedge |R_{12}|}{R_{12}^2}, \tag{1}$$

where the subscript 1 is for the emitting charge, and 2 for the place where the magnetic field is evaluated. μ_0 is the vacuum permeability, v_1 the particle speed (in modulus equal to c), R_{12} the radius vector at which the magnetic field has to be evaluated, and $|R_{12}|$ the unit vector in the direction of the vector R_{12}.

When another massive particle with a magnetic moment is immersed into this magnetic field, the magnetic part of the Lorentz force ($F_2 = qv_2 \wedge B_1$) generates a "strong" (oscillating) attractive force. The formula is

$$F_2 = \frac{\mu_0}{\pi}\frac{m_{\text{mag}_2}}{g_2 r_2}\frac{m_{\text{mag}_1}}{g_1 r_1}\frac{|\nu_1| \wedge |\nu_2| \wedge |R_{12}|}{R_{12}^2}, \tag{2}$$

where the m_{mag} are the magnetic moments of the two particles, v the charge speeds, r the charge rotation radius and the notation $|.|$ means a unit vector.

The attractive force reaches high values only if some necessary conditions are satisfied:

- *Alignment*: of the two magnetic moments (i.e. spin): parallel or anti-parallel, apart from precession.
- *Frequency*: the two particles must have the same frequency (synchronous rotation) or one frequency must be an exact integer multiple of the other.
- *Phase*: zero for parallel spins or half-cycle for anti-parallel spins.

In the case of the proton:

- $r_{\text{p}} = 0.105$ fm (much smaller than its charge radius, which is around 0.87 fm),
- $\nu_{\text{p}} = 4.54 \times 10^{23}$ Hz.

At typical nucleon separation distances ($\sim$2 fm), the potential of the attractive force obtained with these data is of the same order of magnitude of the nuclear force, i.e. a few MeV. So, if both nucleons have the same internal charge rotation frequency this kind of force could well be responsible for the so-called "nuclear force". This is what Cook and Dallacasa suggest. Many more considerations could be added regarding this attraction mechanism.

4. Electron Zitterbewegung

The electron manifests as a point charge with an intrinsic and very rapid circular rotation at 2.47×10^{20} Hz, the so-called Zitterbewegung [4]. The rotational motion takes place around the "trajectory" of the electron's centre of mass. The apparent radius of the ZB, as seen by an external (inertial) observer shrinks as the electron speed approaches c (the mass is inversely proportional to the ZB radius). The "static" radius is about 193 fm, much larger than the nucleons, but only about 0.36% of the Bohr radius. The only value that makes sense for the module of the charge speed along the ZB trajectory is the speed of light. The spin, the magnetic dipole moment and the nature of the electron orbitals can be interpreted as consequences of this very rapid charge rotation.

5. Coupling Between Electron and Hydrogen Nucleus

One of the necessary conditions for a net attraction between an electron and a hydrogen nucleus is about the ratio of their intrinsic charge rotation frequencies. This ratio is $\nu_{\text{p}}/\nu_{\text{e}} = 1836.15267389$, and corresponds exactly to the mass ratio of the two particles ($m_{\text{proton}}/m_{\text{electron}}$), or to the inverse ratio of their charge radii ($r_{\text{e}}/r_{\text{p}}$). This ratio is not an integer number and gives no net magnetic attraction.

For the coupling an orbital component of the electron frequency has to be provided. Reaching a synchronous intrinsic charge rotation would require to add to the electron an orbital frequency corresponding to an energy near to 2 GeV, which is far more than any nuclear binding energy, not to mention chemical energies. So the only way to have an attraction between the electron and a hydrogen nucleus is for the electron to "rotate around the nucleus" at a frequency that, added to its intrinsic frequency, reaches 1/1836 of the proton intrinsic frequency (1836 is the nearest integer to the mass ratio). So the necessary orbital contribution $\Delta\nu$ to the electron intrinsic frequency is

$$\Delta\nu = \left(\frac{\nu_{\text{p}}}{1836}\right) - \nu_{\text{e}} = \nu_{\text{p}}\left(\frac{\frac{\nu_{\text{p}}}{\nu_{\text{e}}} - 1836}{1836\frac{\nu_{\text{p}}}{\nu_{\text{e}}}}\right) = 2.055 \times 10^{16}\ \text{Hz}. \tag{3}$$

This frequency corresponds to a photon energy of about 85 eV, or equivalently to a wavelength of 14.6 nm, in the EUV range.

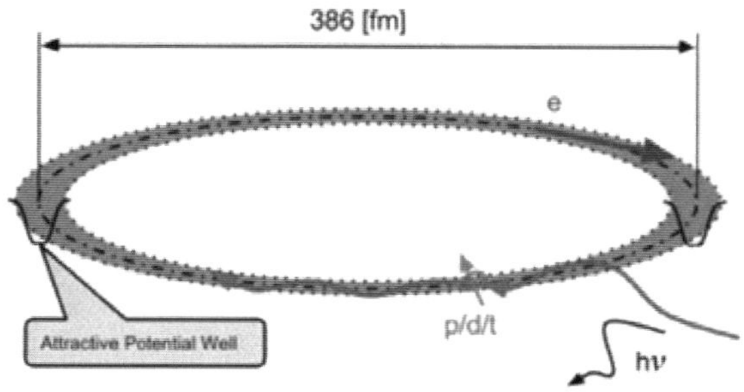

Figure 1. Schemc view of a hydronion with the HNu bound to move along the ZB trajectory.

5.1. Electron orbitals as NAE

Electron orbitals can provide a significant part of the necessary orbital frequency. The rest could come from the energy of the HNu striking the orbital. Valence orbitals have energies always lower than 85 eV and only core orbitals can reach it. Such core orbitals share always their space with other inter-penetrated orbitals, which should prevent the coupling. So in common conditions the electron does not feel any (nuclear) net force towards nuclei.

5.2. Formation of the hydronions: extreme ultraviolet emissions

When an electron and a proton start being attracted towards each other through the magnetic mechanism described above the electron will spiral around and towards the proton. The coupling mechanism is attractive only when the mentioned orbital frequency contribution is present, therefore the electron will spiral towards the HNu at a frequency near to the coupling frequency. This causes the photon emissions to be all in the EUV wavelength range. When the hydrogen nucleus crosses the ZB radius (actually it is the electron that moves towards the HNu, because of the mass ratio) the attractive magnetic force becomes repulsive, and the hydrogen nucleus is captured along the ZB trajectory, as schematically shown in Fig. 1. So the approach produces a bound state between the electron and the HNu. The result is not a new particle because electron and HNu remain separate entities, but it is *a sort of huge and flat neutral nucleus*.

Due to the mass ratio the electron trajectory would look more like a circle fixed at the hydrogen nucleus location, turning around it at about 2.055×10^{16} Hz.

Randell Mills measures in his plasmas [5] an unusual and intense EUV radiation with a wavelength roughly between 10 and 30 nm (the distribution looks roughly like a Cauchy with $E_0 = 65$ eV and $\Gamma/2 = 13$ eV), right across the coupling energy. In light of the present theory Mills should be measuring the emissions due to the formation of Hydronions. EUV radiation is not too easy to measure, while it is absorbed very efficiently in matter and readily transforms into heat. This should be the reason why it was not directly detected in most experiments.

5.3. Three hydronions

As the stable isotopes of hydrogen are three, there are three possible reactions that generate three different Hydronions:

0p:	p+e	$\rightarrow$	pe (hydronius)	+	G_p (MeV)
0d:	d+e	$\rightarrow$	de (deuteronius)	+	G_d (MeV)
0t:	t+e	$\rightarrow$	te (tritionius)	+	G_t (MeV)

G_p, G_d, and G_t stand for the binding energies of the Hydronions.

6. EMNR Take Place in Two Stages

Hydronions have a maximum charge displacement vector of 386 fm and a pulsation frequency around 10^{20} Hz (half of the "reduced" Compton length and frequency), therefore they should travel much more freely than electrons and nuclei inside matter. However, they have a very large magnetic dipole moment if compared to neutrons in fact their magnetic moment is practically equal to the electron's magnetic moment. So their trajectory in free space should be strongly influenced by the presence of a varying magnetic field.

Hydronions can be attracted by other nuclei towards the ZB trajectory though the magnetic attraction. Once the HNu and the second nucleus are both moving along the ZB (like on a race track), they will attract each other through the magnetic attraction because now their magnetic moments are aligned and probably will phase. Figure 2 tries to depict this. As it is well known, the nuclear reaction between the two nuclei takes place only when their distance is down to a few femtometres. In the Quantum Nucleodynamics theory of Cook [1], a nuclear reaction corresponds simply to a rearrangement of the fcc nucleon structure.

The electron provides very special conditions for a nuclear reaction, which is the reason why LENR do no generate the same products of hot fusion; in particular they avoid unstable nuclides. This is one of the most astounding features of LENR the one that make them so interesting, both for energy production and for nuclear waste deactivation. The qualitative physical reason for the preference for stable nuclides should be the fact that the electron of the Hyd keeps perturbing the reacting nuclei, so that the most unstable nuclei can not assemble completely. When the new nuclei being assembled inside the Hyd survive the perturbing action of the electron, the system releases the excess energy and emits photons. The Hyd probably breaks apart after any nuclear reaction that has an energy release higher than its binding energy.

The nuclear reaction that takes place along the electron's ZB can be both a fusion of the HNu (possibly together with the electron) or a fission. Depending on the hydrogen isotope of the Hyd, there are a series of possibilities for the

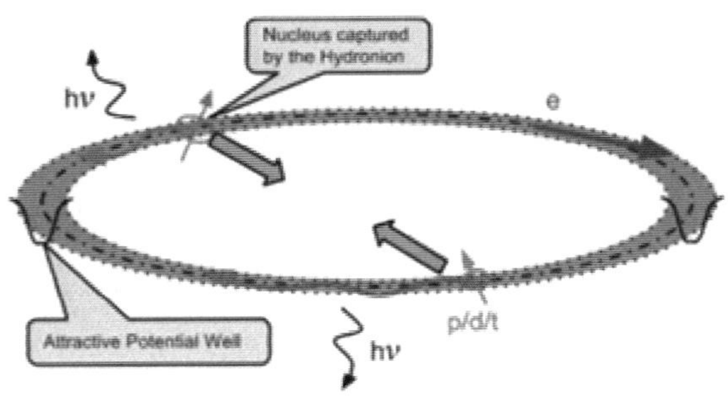

Figure 2. A Hyd has "captured" a nucleus and the two nuclei attract each other emitting photons.

possible *fusion* reactions:

With hydronius (ep)

(p1)	$\mathrm{Nu}(Z,A)$	+	ep	→	$\mathrm{Nu}(Z,A+1)$	+	neutrino	+	photons
(p2)	$\mathrm{Nu}(Z,A)$	+	ep	→	$\mathrm{Nu}(Z+1,A+1)$	+	e	+	photons

With deuteronius (ed)

(d1)	$\mathrm{Nu}(Z,A)$	+	ed	→	$\mathrm{Nu}(Z,A+3)$	+	neutrino	+	photons
(d2)	$\mathrm{Nu}(Z,A)$	+	ed	→	$\mathrm{Nu}(Z+1,A+2)$	+	e	+	photons
(d3)	$\mathrm{Nu}(Z,A)$	+	ed	→	$\mathrm{Nu}(Z,A+1)$	+	ep	+	photons
?(d4)	$\mathrm{Nu}(Z,A)$	+	ed	→	$\mathrm{Nu}(Z+1,A+1)$	+	e + n	+	photons

With tritionius (et)

(t1)	$\mathrm{Nu}(Z,A)$	+	et	→	$\mathrm{Nu}(Z,A+3)$	+	neutrino	+	photons
(t2)	$\mathrm{Nu}(Z,A)$	+	et	→	$\mathrm{Nu}(Z+1,A+2)$	+	e	+	photons
(t3)	$\mathrm{Nu}(Z,A)$	+	et	→	$\mathrm{Nu}(Z,A+2)$	+	ep	+	photons
(t4)	$\mathrm{Nu}(Z,A)$	+	et	→	$\mathrm{Nu}(Z,A+1)$	+	ed	+	photons
?(t5)	$\mathrm{Nu}(Z,A)$	+	et	→	$\mathrm{Nu}(Z+1,A+1)$	+	e + 2n	+	photons

The reactions with the question mark probably are impossible. For the possible *fission* reaction the combinations are many more than those of a fusion.

LENR, which the present theory rename EMNR, take place in two stages:

- *First stage*: Generation of Hydronions (this stage needs a NAE).
- *Second stage*: The Hyd are captured by other nuclei and host nuclear reactions inside the electron.

The flow of Hyd is the "strange radiation" detected in many LENR experiments, as thoroughly summarized in [6]. The second stage should be responsible for the meta-chronous thermal effects and the double optimal operating power measured by Mitchell Swartz (see [7]) and other experimenters. The non-thermal Near-IR radiation measured by Swartz and Verner [8] is due to the not completely thermalized EUV emission; the authors of the paper attributed it to Bremsstrahlung. The strong EUV emission could be part of the reason for the sudden change in resistance of the Nanors that has caused so many electrical equipment failures.

7. Second Stage Reactions with Hydrogen Nuclei

When Hyd react with hydrogen nuclei the possible reactions are shown below.

7.1. Remarks

- All reactions involving the weak interaction (where a neutrino is emitted and a proton combines with an electron to give a neutron) are kinetically much slower than those which do not involve the weak interaction

1e	p+ep	$\rightarrow$	d	+	Neutrino+ max)	1.442 (MeV)	-	G_p
2e	p+ed	$\rightarrow$	t	+	Neutrino+(max)	5.475 (MeV)	-	G_d
3e	d+ep	$\rightarrow$	t	+	neutrino+ max)	5.475 (MeV)	-	G_p
3	d+ep	$\rightarrow$	He^3	+	e +	4.472 (MeV)	-	G_p
4e	d+ed+0.141 (MeV)+G_d	$\rightarrow$	H^4	+	neutrino +	0.00 (MeV)		
	H^4	$\rightarrow$	t	+	n +	3.391 (MeV)		
4.1	d+ed	$\rightarrow$	He^4	+	e +	22.825 (MeV)	-	G_d
4.2	d+ed	$\rightarrow$	t	+	ep +	4.033 (MeV)	-	$G_d + G_p$
5e	t+ep+4.174 (MeV)+ G_p	$\rightarrow$	H^4	+	neutrino +	0.00 (MeV)		
	H^4	$\rightarrow$	t	+	n +	3.391 (MeV)		
5	t+ep	$\rightarrow$	He^4	+	e +	18.792 (MeV)	-	G_p
6e	t+ed+5.318 (MeV)+G_d	$\rightarrow$	H^5	+	neutrino +	0.00 (MeV)		
	H^5	$\rightarrow$	t	+	2n +	2.311 (MeV)		
6.1	t+ed	$\rightarrow$	He^5	+	e +	15.832 (MeV)	-	G_d
	He^5	$\rightarrow$	He^4	+	n +	0.735 (MeV)		
6.2	t+ed+5.616 (MeV)+G_d	$\rightarrow$	H^4	+	ep	0.00 (MeV)	+	G_p
	H^4	$\rightarrow$	t	+	n+	3.391 (MeV)		
7	t(beta decay)	$\rightarrow$	He^3	+	e+antineut. (aver)	5.7 (keV)		
8e	He^3+ep	$\rightarrow$	He^4	+	neutrino+(max)	19.80 (MeV)		
9	He^3+ed	$\rightarrow$	He^4	+	ep + (max)	20.58 (MeV)		

(which is slower than the other interactions). These reactions will practically be relevant only when there are no alternatives that do not require the weak interaction.

- The reactions producing neutrons are all endothermic, apart form reaction 6.1, which requires tritium and deuteronius.
- If $G_p > 1.442$ MeV reaction 1e is endothermic and no deuterium is produced without stimulation. This is very important because it would strongly differentiate between what happens with hydrogen and with deuterium loadings. Deuterium-loading allows the production of He^4 and tritium, and a high energy density if compared to the transmutations and the fissions to which a hydrogen-loaded system is limited. With hydrogen loading there would be no tritium production. The absence of deuterium and tritium in the ash of the tested E-Cat [9] seems to suggest that G_p is in fact higher than 1.442 keV, so that reaction 1e is actually endothermic.
- When He^4 is produced there is always a large energy release, as confirmed by experiments,
- Tritium is generated without the production of free neutrons: this explains the so-called "branching ratio anomaly".
- As already mentioned, when tritium is destroyed in presence of deuteronius (reaction 6.1), neutrons are actu-

ally produced, together with a large quantity of energy,

- If hydronius (ep) decomposes/decays, reaction 4.2 would be a source of protons/molecular hydrogen in experiments with deuterium loading.

8. The Nuclear Active Environment

If an External Core Orbital (which is the first orbital "below" the valence orbitals) with an energy not too far from the coupling energy gets exposed, it can turn into a NAE if a proton strikes it at the right energy. Typically ECOs can get more exposed by the presence of ionic bonds, which displace most of the charge density of the binding orbitals towards the electron acceptor.

In order to have a general idea of the energies of the ECOs of atoms it is possible to look at the ionization energies of the isolated atoms. The real energy of the ECO of a chemically bound atom will always be lower than the charted energy of the ECO of a free atom (ionization energy), because of the partial charge shielding of the valence electrons. However the energy difference between these two cases will decrease with the increasing "ionic" nature of the chemical bond.

By calculating the energy difference between the known ionization energies of all atoms and the coupling energy of 85 eV, it is possible to evaluate which atoms, if ionized (once or more), can offer naked orbitals that could more easily become NAE. Table 1 shows a summary of the atoms that have the energy of their ECO nearest to the coupling energy. The column "Complementary Energy" shows the difference between the ionization energies and the coupling energy.

From Table 1 it is interesting to note a series of points.

- The atomic ion with the ECO energy nearest to the coupling energy is Ca(IV). The energy of its ECO is only 0.5 (eV) from the coupling energy. However in common chemical conditions the maximum oxidation state of Ca is +2 so that Ca(IV) is not present in chemical compounds. The ECO energy of Ca(II) is at 34.09 (eV) distance from the coupling, therefore too much for any contribution from the diffusion of deuterons of Iwamura uses [11]. Instead a fraction of (eV) is an energy that the diffusing deuterium could reach. This suggests that in the multilayer of Iwamura there could be some calcium atom in a bound state that leaves the fifth orbital somehow "exposed".
- Zirconium, in common and stable chemical conditions, exposes an ECO which has the nearest energy to the coupling energy. The atom of zirconium in ZrO_2, while being well ionized (the Pauling electronegativity difference between oxygen and zirconium is 2.11), has an ECO energy of 80.35 (eV), only 4.65 (eV) from the coupling energy. Mitchell Swartz [7] uses ZrO_2 in his Nanor device. In that device the flow of deuterium is helped by a potential difference, which could account for the missing energy of 4.65 (eV). The protons could be in fact accelerated from one hydride grain to another crossing the ZrO_2 matrix.
- Magnesium should be similar to Zr because the ECO energy of Mg(II) is at only 4.9 (eV) from the coupling energy. Iwamura [11] tried this oxide, but did not obtain transmutations. One possible reason for this failure could be that his system cannot provide the deuterons with enough energy to reach the coupling.
- Yttrium was successfully used by Iwamura. This element has two non-ordinary oxidation states at ± 8 (eV) distance from the coupling energy. Their balanced linear combination would require a very low energy contribution from the moment of the deuteron.
- N(III) should have an ECO energy at 7.5 (eV) form the coupling, therefore it is another good candidate for a NAE.
- Li(I) has an ECO at about 9.4 (eV) from the coupling energy and is commonly stably "ionized" in this oxidation state. The missing energy is higher than in the Ca(IV), Zr(IV) and N(III) cases: 9.36 (eV). Lithium is a very good NAE because it can be a gas at the operating temperature of a hypothetical LENR reactor,

Table 1. Summary of the atoms that could have exposed ECO energies nearer to the coupling energy.

Reference experiment	Element	Ionization state	Ionization energy (eV)	Complementary energy (eV)	Ordinary oxidation state
Y. Iwamura	Ca	+4	84.50	0.50	no
	Rb	+5	84.4	0.60	no
	Ga	+4	87.00	–2.00	no
	F	+3	87.14	–2.14	no
Leif Holmlid [10]	K	+4	82.66	2.34	no
	Zn	+4	82.60	2.40	no
	S	+5	88.05	–3.05	"yes"
	Bi	+5	88.30	–3.30	"yes"
	Se	+5	81.70	3.30	"yes"
	Br	+5	88.60	–3.60	"yes"
Mitchell Swartz - NANOR	Zr	+4	80.35	4.65	yes
Y. Iwamura (not working)	Mg	+2	80.14	4.86	yes
	Co	+4	79.50	5.50	"yes"
	Cr	+5	90.63	–5.63	yes
	Kr	+5	78.50	6.50	no
	Sc	+4	91.65	–6.65	no
	N	+3	77.47	7.53	yes
	O	+3	77.41	7.59	no
Y. Iwamura	Y	+5	93.00	–8.00	no
Y. Iwamura	Y	+4	77.00	8.00	no
	Ge	+4	93.5	–8.50	yes
A. Rossi	Ni	+4	76.06	8.94	yes
A. Rossi, H. Mills, Lipinsky, electrolysis	Li	+1	75.64	9.36	yes
	Mn	+5	95.60	–10.60	yes
	Sc	+3	73.49	11.51	yes

whereas all other NAEs are high melting point solids, in which only very thin surfaces near to the HNu source participate to the power generation. A gaseous NAE instead fills an entire volume and can undoubtedly offer higher power densities. Lithium has almost always been present in electrochemical LENR experiments. The cathode surface, before becoming "LENR active", has to develop an oxidized layer with lithium, sufficiently thick as to build a potential difference that accelerates hydrogen nuclei allowing them to gain the missing energy.

9. Hydronion Accumulation inside Condensed Matter

The radius of the Hyd is 193 (fm), about 220 times the proton charge radius. This means that the maximum Hyd cross sectional area is about 50,000 times that of the proton. Moreover Hydronions have a large magnetic moment, essentially equal to that of the electron, which is about 960 times that of the neutron, the only other known long-lived neutral massive particle. Therefore inside condensed matter Hyd should be intensely magnetically scattered because the large magnetic moment and size of the Hyd should force it to interact with the orbital currents and with the intrinsic spin the electrons in the chemical structures it crosses much more intensely than the neutron. So Hyd produced inside condensed matter, if not enough energetic, could rapidly lose momentum and possibly end up trapped in the locations where the gradient of the magnetic field is higher (magnetic traps).

Hyd are neutral, so they interact only minimally with the chemical structure. However probably an extremely high density of Hyd would be able to modify the chemical properties of a substance. So far there are no reports showing change in the chemical properties of "LENR active" materials.

9.1. LENR explosions

Many LENR experimentalists have reported sudden melt down of working LENR active materials, and a number of explosions. As a matter of fact the whole Cold Fusion saga begun with an explosion in the laboratory of Fleischmann and Pons (see [12]). Explosions require a reaction that can involve the whole volume of the exploding material. In all experiments LENR seem to be happening on thin surfaces and not in the entire volume. This suggests that the origin of the volumetric reaction is the volumetric accumulation of a potentially nuclear-active particle. Evidences seem to suggest that this particle can be turned active by thermal and mechanical stimuli: phonons.

A possibility is that phonons, thanks to the magneto-phonon coupling, produce magnetic waves that push the Hyd out of the magnetic traps and force some of them to react. The reactions produce heat that generates a positive feedback that in turn can lead to an explosive outcome. So explosions and some types of runaways should be due to the accumulation of Hyd inside condensed matter followed by phonic stimulation.

9.2. Radio frequency emissions

A number of researchers reported radio frequency emissions (e.g. [13,14]). If the Hyd can end up trapped in condensed matter, they could be stimulated at the Larmor frequency (which is in the radio range for the Hyd) and undergo spin reversal. The following relaxation could be the origin of the radio frequency emissions, similarly to what happens in the Nuclear Magnetic Resonance technique.

10. Some General Remarks

10.1. Susceptibility to a magnetic field

Since the proposed EMNR mechanism is essentially magnetic, it should be susceptible to the presence of a magnetic field. The nature of the NAE could be an additional source of experimental susceptibility to a magnetic field.

10.2. Some nuclei do not react

As the Hot-Cat isotopic measurements show [9], some nuclei seem not react at all. The magnetic coupling in fact has a chance to take place only if a nucleus has a magnetic moment of first or higher order. Probably Ni^{62} does not have magnetic moments, due to its high symmetry. Ni^{62} has the highest binding energy per nucleon, and this should be due to a particularly high symmetry. If one accepts the isospin layered fcc model of the nucleus proposed by Norman Cook [1], the symmetric structure of Ni^{62} could be as shown in Fig. 3.

An additional reason for the inert nature of Ni^{62} is the fact that nuclear reactions inside the electron ZB take place at almost no excess kinetic energy, so that Ni^{62} cannot transform in to any other stable nucleus because there are no nuclei with a higher binding energy per nucleon.

10.3. No tritium accumulation

The magnetic moments of protium and tritium are higher than the magnetic moment of deuterium, in fact $m_\mathrm{d}/m_\mathrm{p} = 0.31$, $m_\mathrm{t}/m_\mathrm{p} = 1.07$. So the reactions with deuterium should be less favourite than those with protium and tritium. Here are summarized the reactions that generate tritium and that destroy it:

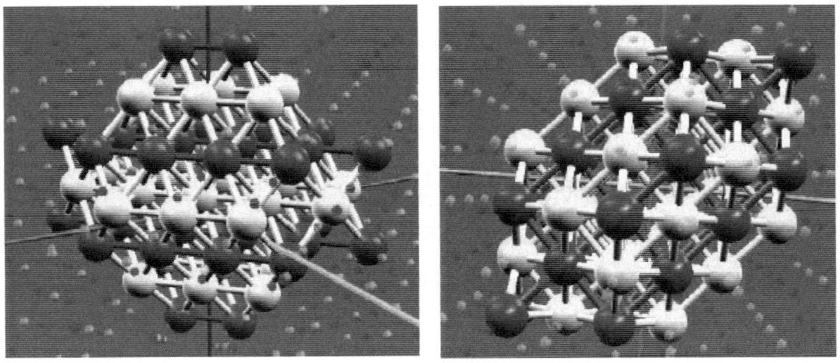

Figure 3. Possible symmetric structure of Ni^{62}.

Main source of tritium

$$4.2:\ \text{d+ed} \rightarrow \text{t + ep} + 4.033\ (\text{MeV}) - G_\text{d} + G_\text{p}.$$

Tritium sink

$$5:\ \text{t+ep} \rightarrow \text{He}^4 + \text{e} + 18.792\,(\text{MeV}) - G_\text{p}.$$

Since all reactions producing tritium (not only those above) involve deuterium, while the elimination of tritium involves protium, tritium should not accumulate. This should be the basic reason for the lack of accumulation of tritium and for its reported elimination apparently correlated to bursts of energy.

11. Summary of Relevant Features of the Proposed EMNR Theory

The proposed Electron-mediated Nuclear Reactions theory has the following fundamental features.

- The Coulomb barrier is not overcome kinetically.
- The nuclear binding energy is electromagnetic and, thanks to the mediation of the electron, a neutral pseudo-particle can form.
- EMNR take place in two stages:
  - *First stage:* Formation of the Hydronions.
  - *Second stage:* the "neutral" Hyd is captured by nuclei and fusion or fission take place "inside the electron".
- The reactions release nuclear energy in form of intense EUV radiation that readily transforms into thermal energy.
- When a Hyd couples to another nucleus it can cause both fusion and fission reactions.
- The preference for stable nuclei is qualitatively explained with the perturbation that the electron charge causes on the forming new nuclei.
- Tritium can be produced without neutrons (no branching ratio problem).
- The NAE is essentially an Electron Core Orbital with a relatively high energy near to 85 (eV) stricken by protons at energies in the (eV) range.

- The NAE is not inside the metal matrix.
- The most promising ECOs are those offered by Zr(IV) and Li(I). Probably, Iwamura managed somehow to use the ECO of Ca(IV), which is the nearest of all atoms to the coupling frequency, but it probably cannot offer a high NAE density.
- The measured radio frequencies emissions could be the "NMR" frequencies of the Hydronions trapped inside metal matrices.
- Biological transmutations are not impossible thanks to a series of ECO that are present in organic matter.
- The strange radiation has the features of the Hydronions.

References

[1] V. Dallacasa and N.D. Cook, *Models of the Atomic Nucleus*, 2nd Edition, Springer, Berlin, 2010, ISBN-10: 3540285695.

[2] N.D. Cook and V. Dallacasa, LENR and nuclear structure theory, *J. Condensed Matter Nucl. Sci.* **13** (2014) 68âĂŞ-7.

[3] M.J. Savage, Nuclear forces from lattice quantum chromodynamics, Presentation at *the Int. Conf. on Nuclear Theory in the Supercomputing Era*, 2013, Iowa State University, arXiv:1309.4752.

[4] D. Hestenes, Zitterbewegung in quantum mechanics, *Found. Phys.* **40**(1) (2010) 1–54.

[5] R. Mills and R. Booker, Soft X-ray continuum radiation from low-energy pinch discharges of hydrogen, *J. Plasma Phys.* **79**(5) (2013) 489–507.

[6] Keith A. Fredericks, Possible detection of tachyon monopoles in photographic emulsions, *Int. Conf. on Condensed Matter Nuclear Science*, Columbia, MO, 2013.

[7] M. Swartz, Amplification and restoration of energy gain using fractionated magnetic fields on ZrO_2–PdD nanostructured CF/LANR quantum electronic component, *J. Condensed Matter Nucl. Sci.* **15** (2015) 66–80.

[8] M. Swartz and G. Verner,Bremsstrahlung in hot and cold fusion, *J. New Energy* **3**(4) (1999) 90–101.

[9] Giuseppe Levi, Evelyn Foschi, Bo Höistad, Roland Pettersson, Lars Tegnér and Hanno Essén, Observation of abundant heat production from a reactor device and of isotopic changes in the fuel, Observation of abundant heat production from a reactor device and of isotopic changes in the fuel, AMS Acta ©ALMA MATER STUDIORUM, Università di Bologna, ISSN: 2038-7954, 2014.

[10] Leif Holmlid and Sveinn Olafsson, Spontaneous ejection of high-energy particles from ultra-dense deuterium D(0), *Int. J. Hydrogen Energy* **40**(33) (2015) 10559–1056.

[11] Y. Iwamura, T. Itoh, N. Yamazaki, H. Yonemura, K. Fukutani and D. Sekiba, Recent advances in deuterium permeation transmutation experiments, *J. Condensed Matter Nucl. Sci.* **10** (2013) 63–67.

[12] M. Fleischmann, S. Pons and M. Hawkins, Electrochemically induced nuclear fusion of deuterium, *J. Electroanal. Chem.* **261** (1989) 301 and errata in Vol. 263.

[13] V. Violante, E. Castagna, S. Lecci, G. Pagano, M. Sansovini and F. Sarto, RF detection and anomalous heat production during electrochemical loading of deuterium in palladium, *Energia, Ambiente e Innovazione* **2–3** (2014) 63–77, DOI 10.12910/EAI2014-6, 2014.

[14] D.A. Kidwell, D.D. Dominguez, K.S. Grabowski and L.F. DeChiaro Jr., Observation of radio frequency emissions from electrochemical loading experiments, *Current Sci.* **108**(4) (2015) 578–581.

J. Condensed Matter Nucl. Sci. 19 (2016) 29–45

Research Article

Observation of Macroscopic Current and Thermal Anomalies, at High Temperature, by Hetero-structures in Thin and Long Constantan Wires Under H_2 Gas

Francesco Celani[*, †], A. Spallone[‡], B. Ortenzi and S. Pella

INFN-LNF, Via E. Fermi 40, 00044 Frascati, Italy

E. Purchi, F. Santandrea, S. Fiorilla, A. Nuvoli, M. Nakamura, P. Cirilli and P. Boccanera

ISCMNS, Latium#1 Group, Via Cavour 26, 03013 Ferentino, Italy

L. Notargiacomo[§]

Dip. Chimica Università "La Sapienza", Pz. A. Moro 5, 00185 Roma, Italy

Abstract

Since 2011, we introduced into LENR field, the use of a Constantan (Cnst) alloy to absorb/adsorb proper amounts of H_2 or D_2 and to generate thermal anomalies even at low temperatures (>200°C). We developed a reactor with a core of sub-micrometric layered Cnst wires that produced measurable excess power (almost reproducible). Subsequently, we used fiberglass sheaths as electrical insulation and found out that this material actually improves reactor performance. In the most recent configuration, we studied the effects of adding Fe nanolayers to the Cnst wires and of including several small knots along their extension, actions that resulted in a larger excess power that grew with increasing wire temperature. We detected a new electric effect: the generation of spontaneous voltage between the ends of a floating wire in the reactor. We performed tests to study results in agreement with Inverse Rydberg Matter model by L. Holmlid.

Keywords: Constantan sub-micrometric surfaces, H_2 ad-/absorption, Iron, Rydberg Matter, Thermodynamic second law

*E-mail: Francesco.Celani@lnf.infn.it

†Also at: ISCMNS, Latium #1 Group, Via Cavour 26, 03013 Ferentino, Italy.

‡Also at: ISCMNS, Latium#1 Group, Via Cavour 26, 03013 Ferentino, Italy.

§Also at: ISCMNS, Latium#1 Group, Via Cavour 26, 03013 Ferentino, Italy.

© 2016 ISCMNS. All rights reserved. ISSN 2227-3123

1. Constantan alloys as H_2-Dissociation Catalyzers

Our investigations concerning the ability of metals such as palladium (Pd) and nickel (Ni) to absorb D_2 and H_2, and to generate anomalous heat at high temperatures, gained new momentum following the introduction of Constantan alloys to LENR research in 2011.

Our original idea was to create a low-cost material able to replace the very expensive (and mechanically weak) palladium in LENR experiments.

We focused our interest on the family of copper-nickel Constantan alloys as materials that fit our purposes because of their ability to dissociate molecular hydrogen [1]. In particular, we selected a low-cost commercial material called ISOTAN44, with atomic composition $Cu_{55}Ni_{44}Mn_1$ (Isabellenhütte Heusler, Germany). Together with a high H_2-diffusion coefficient at high temperature, this material offers good mechanical resistance against the aging effects of the thermal cycles and H_2 absorption/desorption. Moreover, it has very large values of (calculated) catalytic power with respect to hydrogen dissociation, as shown in Table 1.

We demonstrated experimentally that Constantan at nano-/micrometric size and at low temperatures ($T > 120°$C, in comparison with about 2000°C for tungsten) is able to catalyze the dissociation reaction $H_2 \rightarrow 2H$ and absorb/adsorb atomic hydrogen even inside the bulk of the lattice, as well as at the surface. The demonstration was reported in Ref. [1], chapter IV, points 18 and 19, as follows: 18) To get deloading we put the cell under a dynamic vacuum and increased the temperatures. 19) After several hours, we got the original starting value of $R/R_0 = 1$, meaning that the test was fully successful. The fact that H_2 decreases the resistivity of Constantan was first reported in Ref. [2].

Table 1. Catalytic power of different metals and alloys with respect to the reaction $H_2 \rightarrow 2H$, computed in Density Functional Theory [20].

	ΔE (eV)
Ni0.3750 – Cu0.6250	+3.16
Ni0.6250 – Cu0.3750	+2.86
Ni0.8125 – Cu0.1875	+2.10
Ni	+1.74
Ni0.1825 – Cu0.8175	+1.57
Ag0.8125 – Pd0.1875	+0.57
Ag0.625 – Pd0.375	+0.51
Ag0.1875 – Pd0.8125	+0.51
Pd	+0.42
Cu	–1.11
Ag	–1.42

In our experiment we employed Constantan wires of length 100 cm and diameter 0.1– 0.2 mm (Fig. 1). To increase their effective surface available for catalytic processes, wires were subjected to specific thermal/electric treatments that created sub-micrometric and multilayered structures at the surface and deeper in the bulk (Fig. 2). The sub-micrometric structures were simply created by oxidation, with a threshold temperature of 600°C in free air. Such structures are somewhat similar to hetero-structures.

The treatment includes electric high peak power pulses (20 kVA/g of material) with a rise time $T_r < 1\,\mu$s, corresponding to a current density $J > 50$ kA/cm^2, even neglecting skin effects. Such pulses induce extremely fast thermal treatments (warming→cooling) and shock waves. A rough evaluation by fast photo-camera of the color of light emitted from the wire revealed a surface temperature even larger than 1000°C in some tests. At the end of the process we observed glassy materials formed on the wire surface.

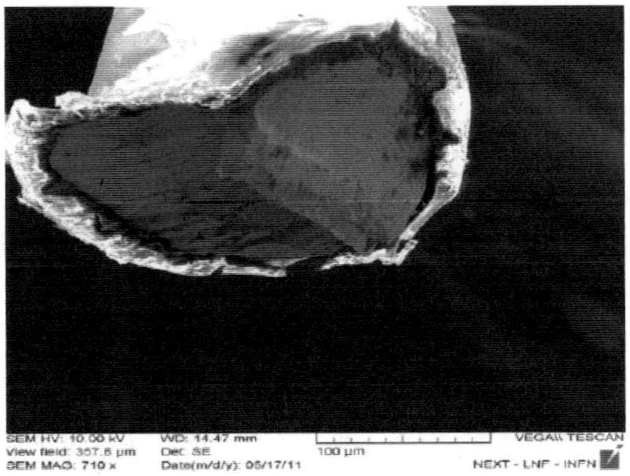

Figure 1. SEM image of the cross section of the Constantan wire as provided by Isabellenhütte Heusler. The plastic cover of the wire is visible as the light area rounding the wire.

This treatment produces sub-micrometric geometries, a sort of chaotic mixture of Ni, NiO, Cu, CuO, $Ni_xCu_yO_z$, reducing/avoiding, at the same time, the usual deleterious self-sintering processes due to high temperatures. SEM observations revealed that the wires so treated had a large number (up to 700 in some samples) of multilayered structures with thickness of 20–100 nm.

The number of layers was roughly characterized by preliminary experiments using 10, 20, 50 and 100 pulses, by SEM and "oblique" cross section of the wire. Among other parameters, we observed that the distance between layers decreases with an increase in the number of pulses. Sadly, the original documents (papers and CD) of this preliminary testing, located in another laboratory, were destroyed by a "third party" in February 2015, together with other materials and documents.

Our treatment was inspired by the Melt Spinning and Quenching metallurgical process, extensively used by Prof. Yoshiaki Arata and his collaborators (Osaka and Tohoku Universities, Japan) to produce nanomaterials (palladium, or Pd_xNi_y, both dispersed into a matrix of ZrO_2 at 65(%) concentration) for his Solid State Fusion devices under

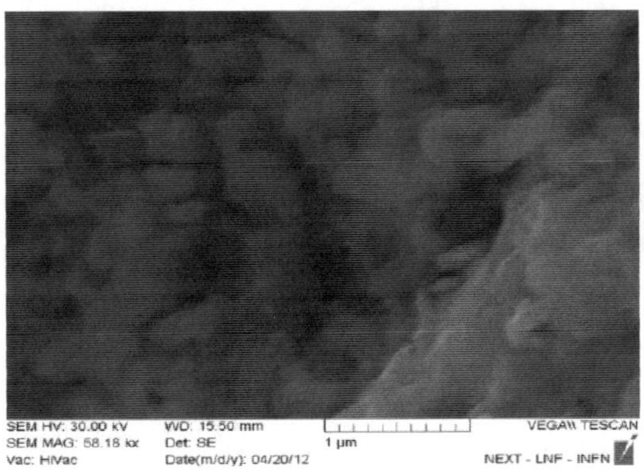

Figure 2. SEM image of the Constantan wire surface at micrometric scale after repeated high-power pulse treatments.

interaction with D_2/H_2 at 150–300°C, pressurized up to 60 atm [3]. In the quenching process their cooling rate was over 100,000 K/s. The important feature of Arata's technique was this "ultra-fast quenching" which yields glassy-like materials. Arata used fast rotating Cu disk where the mixture of palladium (35%) and zirconium (65%) was dropped in a melted state (about 1600°C). Instead of using this kind of disk, we employed ultra-fast powerful electric pulses with fall time <1μs. We needed only cooling from about 900°C down to 650°C. With very long and thin wires it is easy to perform this in a short time, just by natural cooling, because of the dependence of emitted power as T^4 (in Kelvin), i.e. the Stefan–Boltzmann law.

2. First Generation Experiments: the Introduction of Constantan wires

The dissipation reactor we designed to test the new generation Constantan wires consisted of mica sheets (of the type used in electrical heating elements) supported on a central stainless steel (SS) tube ensuring electrical insulation, on which two wires, one active (the surface-modified constantan wire) and the other inert (Ni–Cr control wire), arranged in parallel and helicoidally wound. Because the Ni–Cr wire is intrinsically stable against oxidation or other stresses, it was used to heat the reactor itself (via "indirect" resistive heating, i.e. by radiation and conduction by gas). The core of the reactor was contained inside a borosilicate Schott Duran glass tube with a wall thickness of 3 mm. Temperatures at the external glass wall and inside the reactor were detected by means of several Type K, SS-screened, MgO insulated, thermocouples (diameter 1.5 mm) [1].

Calibrations were made in noble gases (helium, argon) with different power levels applied to the inert wire. The first test was conducted in an atmosphere of H_2/Ar mixture in the ratio 75/25 at 7 bar of total pressure. The power input was 48 W. Figure 3 shows the behavior of the measured quantities over time. The green color line represents the temperature of the external borosilicate wall, while red shows the temperature close to the mica inside the reactor. The key monitor parameter is the ratio R/R_0 between the resistance of the wire at a given temperature T and at room

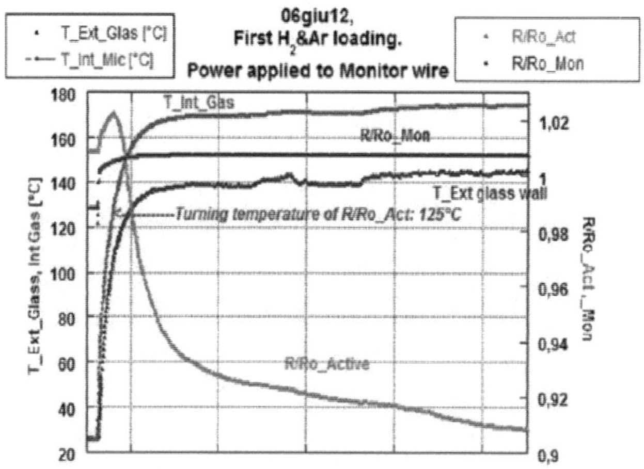

Figure 3. Time behavior of the measured quantities during the first loading by H_2/Ar mixture: temperature of the external glass (*green*), temperature of the internal mica support (*red*), R/R_0 of the inert wire (*blue*), R/R_0 of the active wire (*orange*). Ar inert gas introduced also to reduce H+H probability of uncontrolled recombination to H_2 in the reactor atmosphere.

temperature T_0. For Constantan, this ratio, because of hydrogen interaction, decreases with increasing temperature and time. In the blue line we indicate R/R_0 for the inert (Ni–Cr) wire and in the orange line R/R_0 for the active (Constantan) wire. Recall that one of the key characteristics of Constantan is its normally excellent resistivity stability ($\pm 1\%$) from –75 up to 500°C.

We observed that when the temperature inside the reactor reached 120°C, the R/R_0 of the active wire fell to 0.92 in about 2500 s and to 0.88 in 100,000 s. Correlated with this decrease in resistance, we also observed an increase in excess output power of the reactor. Coinciding with this, all temperatures in the reactor increased. The external glass temperature increased from about 140°C to 150°C; the SS tube temperature from 168°C to 190°C, and the gas temperature (close to the mica) from 170°C to 195°C (Fig. 4). These variations are unrelated to changes in room temperature. Instead, room temperature instabilities somehow helped the anomalous heat generation, perhaps by introducing non-equilibrium conditions. In fact, after a long time, room temperature went back to its initial value, while heat production continued to increase.

During several tests performed in Frascati with mainly Agilent instrumentation (in addition to many ancillary home-made instruments), the excess power ranged between 2 and 12 W (reaching 12 W a few times) at the reference input power value of 48 W. When excess heat was produced, it was typically in the range of 5–10 W. We found no difference in operating the reactor at constant current and constant voltage. Most of the time it was in the constant current regime.

After the National Instruments Annual Meeting (NI-Week, August 2012, Austin, TX, USA), the reactor at Frascati Laboratories, Italy, was disassembled, shipped and reassembled in the USA where it was operated in a public demo for three days. During this event, we observed the highest values of maximum excess power for this experimental set-up: about 21 W with indirect heating (power applied on the inert wire) and about 25 W with direct heating of the active wire with input power of 48 W. Just after NI Week, the reactor was again shipped to South Korea for our participation at the ICCF17 conference at Daejeon. Even there, after overcoming difficulties with the new strict safety rules and conditions of air travel from USA (no gas apart from air inside and no vacuum or pressure conditions allowed), the reactor functioned normally with reasonably reproducible results: 5–15 W of anomalous excess heat with 50 W of electric input power. The values seen were lower than in the test performed during NI-Week. Moreover, for safety reasons, the reactor was operated for most of the time in Constant Voltage regime, both at NI-Week and ICCF17.

After returning to Frascati and performing several cross checks, we realized that the new NI system (the hardware and software set-up since August 2012) in our hands had underestimated the value of electric input power in the regime of Constant Voltage by 11.11% in the range of input power of 40–60 W, i.e. that used in the experiments. Considering this unexpected problem, the real values of AHE power would change to: 8–20 W at Ni-Week, 0–10 W at the ICCF17 Conference (assuming similar hardware arrangements).

We note that all the data were and are fully correct in the constant current regime. The reported lower values are however closer both to our starting values in Frascati and to the results of the subsequent experiments performed by the Martin Fleischmann Memorial Project group which used a batch of Constantan wire with a different starting composition from the old pre-1970s batch we used in the first series of experiments. See Section 3 for further details.

Regarding the overall better performance during NI-Week, in the absence of any systematic study, we can only suspect a positive effect from an accident that happened during the preparation of the set-up: some vapor of silicon oil from the rotary pump went inside the reactor chamber. We did try to remove the possible residual deposits by heating the wires. Anyway, we feel that such an accident was a further strong indication of the catalytic origin of the AHE effect, at least with our materials and operating parameters.

3. Second Generation Experiments: the Addition of Glass

The successive series of experiments exhibited unsatisfactory overall reproducibility. Using SEM/EDS/ICPMS analyses, we found out that the first batches of raw material we used in our experiments, which were produced before 1970, had a composition different from those we used later. Analyses revealed iron contamination in the order of 1000–5000 ppm, and locally up to 10,000 ppm.

Because of a severe budget cut in late 2013, we were forced to redesign and reschedule our experiments with the purpose to study again more deeply some of the most interesting effects obtained in the past and, if possible, to increase the AHE.

In the new experimental set-up we modified the geometrical arrangement of the wires inside the reactor [4]. We used three wires instead of two: a 500-layer Constantan wire, a 2-layer Constantan wire and a platinum wire for control and monitoring purposes. They were inserted inside fiberglass sheaths ($L = 100$ cm, $\Phi_{ext} = 1$ mm, produced by SIGI–Favier, Italy–France Company). Each fiber was porous, of 5 μm mean diameter, and closely braided together. The braid was then twisted around the central SS support that was likewise covered with a fiberglass sleeve with internal diameter of 12 mm. As before, the core was inserted into the thick-wall borosilicate glass tube previously used.

Since February 2013, all the sheaths were embedded in a $Sr(NO_3)_2$ diluted solution and further decomposed in SrO by thermal treatment. Strontium is a material with a low work function for electron emission ($W = 2.59$ eV), similar to the calcium oxide used by Yasuhiro Iwamura at Mitsubishi Heavy Industries Laboratories (Yokohama, Japan) since 1999. Electron emitter materials are empirically recognized to have "beneficial effects" on LENR reaction. We have studied the effects of calcium, barium, strontium and magnesium in electrolytic environment, since about 1995. Iwamura, independently, adopted CaO in his famous transmutation device (2002, JJAP) by flowing deuterium gas. Moreover, he showed that MgO is not effective, similarly to our electrolytic experiments (barium worked best, but it is toxic).

We calculated the emitted power using Stefan–Boltzmann law (emission proportional to T^4, in Kelvin) for the

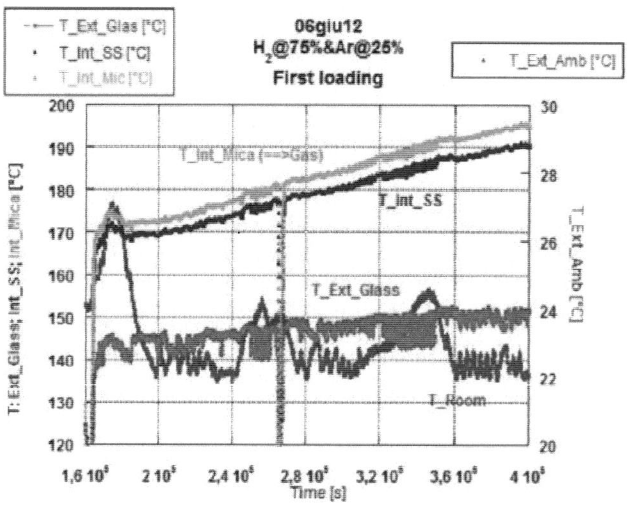

Figure 4. Temperature of the external glass wall (*red*), the internal SS tube (*blue*), the internal mica support (*orange*) and the ambient (*green*) as a function of time.

radiated energy from the glass wall and Newton's law (proportional to temperature) for the energy dissipated by convection.

During the tests several uncontrolled and unexpected phenomena occurred – including spontaneous overheating that damaged the 500-layer wire – making the results unclear. Therefore, we decided that the only way to clarify the role of the fiberglass in the process was to add a larger amount of glass. Even the SS-tube, where the glass-covered wires were twisted around, was inserted inside two more glass hoses.

To calculate excess heat, we decided that tests carried out with a smaller amount of glass would be considered the threshold, i.e. the blank test. Even with such a conservative constraint, the comparison between the two experiments showed that glass has intrinsic co-effects in the generation of anomalous heat. In Fig. 5 we compare the temperatures at the external wall and at the internal SS-tube in the two cases of a small and a large amount of glass. We note that, under the same conditions of pressure and input power, the temperature in the second case is clearly larger than the corresponding values for the experiments with lower glass content in the reactor.

4. Effect of Fiberglass on Hydrogen Storage

The experiments we carried out clarified that the borosilicate fiberglass (type E) we used as electrical insulator plays a role in the generation of excess heat. We signal out the mechanism responsible for this in the adsorption property of (borosilicate) glass, which was observed by Prof. Irving Langmuir since 1920 during his studies on hydrogen dissociation at high temperatures by Tungsten (W) wires (>2000°C, at H_2 low pressures). The hydrogen atoms,

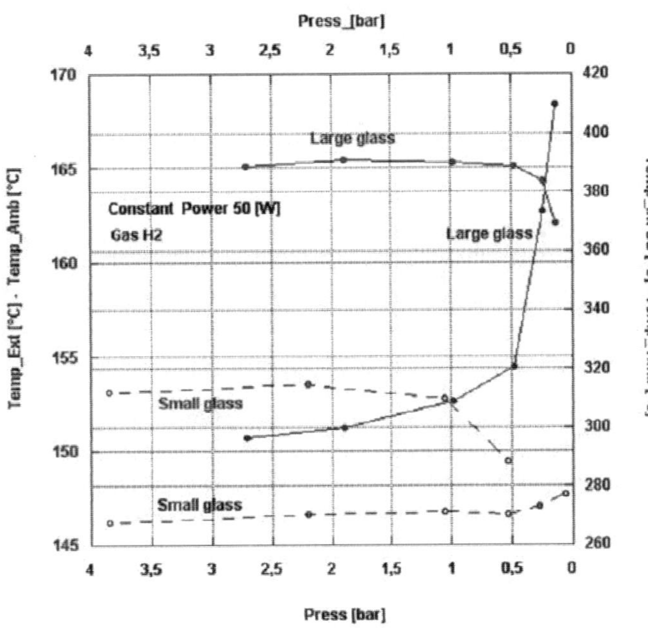

Figure 5. Data comparison between tests conducted with a small and a large amount of glass in the reactor. In red we display external glass temperature, in blue internal SS tube temperature after subtracting ambient temperature. Under the same conditions, temperature is higher when more glass is employed.

produced by the dissociation of molecular hydrogen by means of the catalytic action of our sub-micrometric structured Constantan wires on the reaction (at quite low temperatures, about 200°C), are largely adsorbed onto the surface of the micrometric glass fibers, forming a thin film. Moreover, according to one of our colleagues (Cesare Lorenzetti, private communication), a long-time expert in Fischer–Tropsch type catalysts, the oxides (in the specific case made by iron, potassium, strontium, boron) "locked" inside glassy materials (like fiberglass) are quite stable for long time (months) even in reducing environments and at high temperatures. The stability effect against the "evaporation" of light elements (like potassium) is further enhanced by manganese (present, in the type of Constantan that we used, at 1% concentration).

Langmuir measured the density of hydrogen atoms adsorbed onto the surface of a glass bulb (presumably of type borosilicate) kept at low temperature (about 90 K). The density is of the order of 10^{15} atoms/cm^2. In his experiments, the dissociation was obtained through a hot tungsten filament in H_2 atmosphere and the fraction of dissociated molecules depended on pressure and temperature: the lower the pressure and higher the temperature, the larger the fraction [5,6].

In our experiments the effective surface of each small sheath is larger than 1 m^2. The total surface of the used fibers could be larger than 50 m^2, corresponding at about $10^{20} - 10^{21}$ adsorbed atoms according to Langmuir's measurement. The hydrogen atoms concentrated onto the glass surface by adsorption are much closer to each other than the hydrogen atoms moving in the bulb as a gas, increasing so their probability of recombination in the molecular form. This reaction is largely exothermic with the release of 4.52 eV to the environment.

Moreover, the recombination time is in the time regime of so-called femtochemistry (some-several 10^{-15} s). The experimental demonstration that several chemical reactions have extremely short reaction time ($10^{-12} - 10^{-15}$ s) has been performed since 1980 by Prof. Ahmed H. Zewail using ultra-short laser pulses [7]. Obviously, with such short times, the peak power of recombination is extremely large, $10^{12} - 10^{13}$ times larger than the mean power at steady-state conditions.

With this new equipment configuration, the net effect of the introduction of glass in the reactor is the increase of the rate of the chemical reaction of hydrogen recombination. Therefore, the adsorption property of glass appears to be a co-factor in the generation of heat excess, in addition to the main unknown LENR process of non-chemical origin. We note that nearly all of the chemical reactions produce at most 4.5 eV of energy. In our experiment, the effect lasted for several weeks and its integrated value is very much higher than the product of chemical energy times the amount of material involved. We also suppose that adsorption may enhance absorption of atomic hydrogen in the Constantan lattice, supporting in this way the main LENR process of heat generation. Actually, the possibility that the presence of a hydrogen film closely surrounding the wire favors the diffusion of atomic hydrogen inside the metal should not be excluded. In conclusion, further investigation aimed at clarifying the effects and role of glass in such phenomena is necessary.

5. The Hypothesis of Energy Localization

In 1954 Enrico Fermi and his coworkers J. Pasta and S. Ulam with the help of M. Tsingou performed, at LANL-USA, the numerical simulation of a discrete nonlinear system using the computer MANIAC I [8]. The dynamic system they studied consisted of 64 one-dimensional oscillators coupled to the nearest-neighbors by nonlinear force terms. Fermi's expectation was that the excitation energy associated with a normal node would be distributed along all the normal modes after a sufficiently long time interval; that is to say, the system would have evolved toward a state of energy equipartition. Surprisingly, energy remained localized to few modes and showed a sort of periodicity, returning most of it to its initial mode.

Dr. Brian Ahern (previously at DARPA, USA) has recently revisited the Fermi–Pasta–Ulam problem/paradox, proposing it as a candidate to explain some of the complex phenomena occurring in LENR experiments [9]. While in most solids the atoms reside in parabolic potential wells and therefore undergo simple harmonic motion, generally at high frequency and low amplitude, the atoms of some materials are subject to a non-parabolic potential corresponding to nonlinear force terms. This leads to nonlinear vibrational modes of large amplitude and low frequency. Hence, Ahern proposes that energy localization may take place in structures with a small number of atoms. Both properties – a relatively small number of atoms and nonlinear coupling – are possessed by nanoparticles of size 3–12 nm. This is the "right" dimension suggested by Y. Arata; experimentally reconfirmed by Takahashi–Kitamura, Univ. Osaka, Kobe, Japan, in experiments of LENR by interaction of D_2 and nanoparticles at high temperatures. In nanoparticles, a large fraction of atoms is located at the surface and is subject to nonlinear binding forces with the internal atoms.

The transition to the new regime would be triggered by a pulse of energy that the cluster can receive from thermo-mechanical oscillations of the surrounding medium, or from electrical pulses, even by radiations. As a consequence, a very small number of atoms in the cluster acquire a significantly greater amount of energy than they would in conditions of thermal equilibrium. Locally, the large oscillations manifest as hot regions. This is closely similar to the phenomenon of oscillons in granular media [10]. We note that, in our experimental conditions (i.e. especially the effect of glass fiber sheaths), the local pulse of energy could be provided by the $H + H \rightarrow H_2$ recombination reaction.

Energy localization could clarify the catalysis processes, usually associated to the lowering of the activation energy of a reaction. In this scenario, a few excited atoms, corresponding to a locally hot region, make their energy available for the activation of chemical reactions that could take place only at higher temperatures if the cluster was absent. Ahern sustains that energy localization could also explain the initiation of LENR reactions in H_2/D_2 saturated nanostructures. As a consequence, energy localization circumvents the second law of thermodynamics, since nanoparticles act as Maxwell's demons able to convert part of their thermal energy into valuable chemical energy potential.

The violation of the second law is apparent, because it is not applicable to ensembles with a small number of particles. We quote Maxwell's words [11]: "The truth of the second law is ... a statistical, not a mathematical, truth, for it depends on the fact that the bodies we deal with consist of millions of molecules ... Hence the second law of thermodynamics is continually being violated and that to a considerable extent, in any sufficiently small group of molecules belonging to a real body." Violations of the second law at nanoscales have also been observed experimentally as occasional short-time (less than 2 s) fluctuations around the thermal equilibrium state [12,13].

The catalytic action of nanostructures at the surface of our surface-modified Constantan wires can partially explain the thermal anomalies we found in our experiments, as they enhance H_2 dissociation even at temperatures $T \ll 800°C$. If further confirmed, the energy localization would be the mechanism for coordinating the normal thermal motion of particles in nanometric regions of matter into localized high-energy oscillations.

6. Spontaneous Voltage Generation

In our experimental set-up, we can control only two wires at the same time. For this reason one of the Constantan wires is not connected to the Data Acquisition System (PIXIe, NI) and left unconnected ("floating"). However, we measure periodically its resistance by means of a high sensitivity multimeter (Fluke 187) to evaluate the amount of adsorbed hydrogen. When the wire is heated in the presence of hydrogen, its resistance ratio can decrease up to 0.7. Anyway, from what we know from "open literature", any systematic study on the relationship between the amount of hydrogen absorbed and the decrease of constant resistance is not available. We have just qualitative data about the correlation of the two quantities.

On June 25, 2014, we noted, just by chance, that the floating Constantan wire generated by itself a macroscopic voltage of the order of hundreds of microvolts that resulted as a function of many parameters: temperature, type of gas, pressure, resistance ratio. The highest measured values were about 1400 μV for the voltage and 120 μA for the current with duration of only few hours, while stable outputs were about half. We note that the spontaneous tension cannot be ascribed to the usual Seebeck effect, because we considered only one single wire and not a junction of two

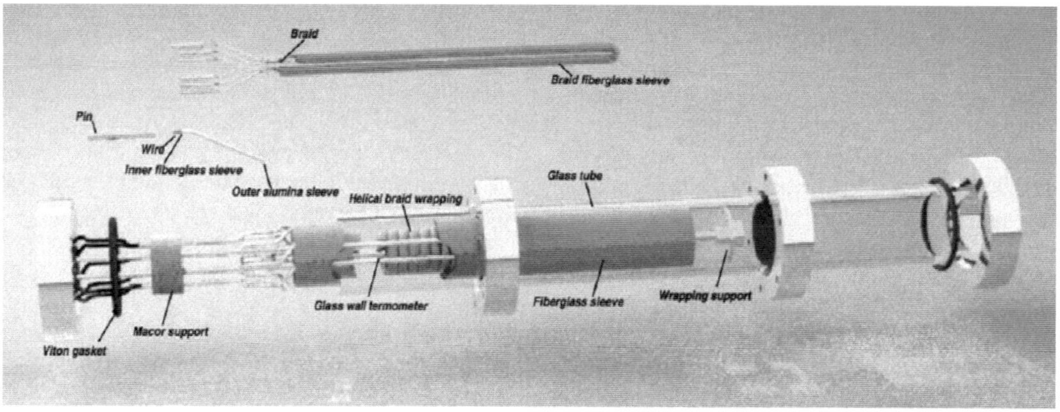

Figure 6. Schematic of the third generation experimental set-up.

different materials as in the thermocouples.

7. New Third Generation Experiments: the Effect of Iron

In previous experiments we discovered that iron contamination in the wires had a positive influence in terms of stability and excess power of the reaction. Hence, we decided to develop a procedure to add iron of nanometric sizes to the surface of Constantan wires and even some microns deep into the bulk during our thermal/electric treatments for the preparation of the nanolayers. From a chemical point of view, iron is characterized by a solubility of hydrogen

Figure 7. Photo of the third generation reactor.

into the lattice that increases largely with temperature: in 100 g of material, from 0.37 cm^3 at 400°C to 7 cm^3 at about 1000°C. To fully exploit this property of iron, the reactor had to reach higher temperatures than we previously obtained. For this reason, we added another sheath over the usual borosilicate glass one, made of alumina (Al_2O_3), with a higher melting temperature ($T_{max} = 1200°C$). Figure 6 shows a complete schematic of the reactor and in Fig. 7 we show a photograph of it, taken during the execution of one of the several experiments we conducted. To increase wire temperature at constant input power, we used a mixture of H_2 with low thermal conductivity noble gases (argon, xenon). The effectiveness of such mixtures was verified by comparing the corresponding excess heat results.

With the new preparation procedure, fully developed at INFN-LNF, we produced 20–40 iron thin layers, similar to multiple hetero-structures, at the wire surface. Moreover, we introduced further non-equilibrium conditions in the wire by making several (up to 20) knots with hole diameter 0.3–2 mm. These knots represent local geometrical variations of the path of the current flowing in the wire and are crossed by the lines of the magnetic field they generate. Because of the large currents (up to 2.3 A) we inject in the wire, they become the locus of large thermal non-homogeneities. The tests we carried out with all different combinations of elements (wire with/without iron/knots) showed that the new configuration (iron plus knots) offers measurable advantages in terms of hydrogen absorption and AHE amount.

As usual, we calculated very conservative estimations of the excess power, using as a blank the corresponding result obtained with the configuration with no iron in the wire. The best result (with Xe/H_2 mixture) was 15 W against an input power of 100 W. The local temperature of the wire was estimated, through the platinum wire, at around 800°C. In Fig. 8, we show the results of the tests carried out in hydrogen atmosphere. $T_{ext} - T_{amb}$ is plotted as a function of the input power in the case of standard Constantan wire (*blue*) and Fe-added Constantan wire (*red*). In green we indicate the excess power of the configuration with iron compared to that without it. Clearly, excess power increases with temperature. This behavior was confirmed by the experiments with Xe/H_2 atmosphere, where larger wire temperatures were reached (Fig. 9). Wire temperature was estimated through SS-support temperature measured by the thermocouples (Fig. 10). For comparison, while at an input power of 90 W, excess power is about 7.5 W in a pure H_2 atmosphere, this raises up to 15 W in a gas mixture.

We also verified the presence of a spontaneous voltage in the non-powered wire in the reactor. In wires having several knots with hole diameter smaller than 1 mm and filled with iron at nanometric size, we observed in gas mixture atmosphere currents up to 150 μA and voltages up to 1900 μV, stable over long times. We compared current generation for different wires with the same number of knots and measured values about 2.2 times larger with the addition of iron. In particular, we ascertained that the local presence of iron in knots was extremely important for maximizing the effect.

8. Discussion on the Indetermination of Temperature Measurements

With regard to the thermometric experiments and the relative indeterminate results, we used three different approaches/typologies/analyses:

(A) Experiments performed in 2012–2013, i.e. the so-called first series.
(B) Experiments performed during 2014, in which we studied in some detail the effect of fiberglass addition, as presented at MIT March 21–23, 2014 Colloquium on Cold Fusion effects.
(C) Experiments concerning the effect of iron addition together with a wire geometry with several knots (main results at ICCF19, April 2015.

First of all, we note that the AHE are considered real by us if, and only if, there are simultaneous increases of both the internal and external temperatures of the reactor, at a given constant input power used as reference. For the experiments of type A we adopted the approach of performing calibrations in inert gases, mainly helium or argon before H_2 addition. Helium has a thermal conductivity similar to H_2. Argon has about seven times lower thermal conductivity than H_2.

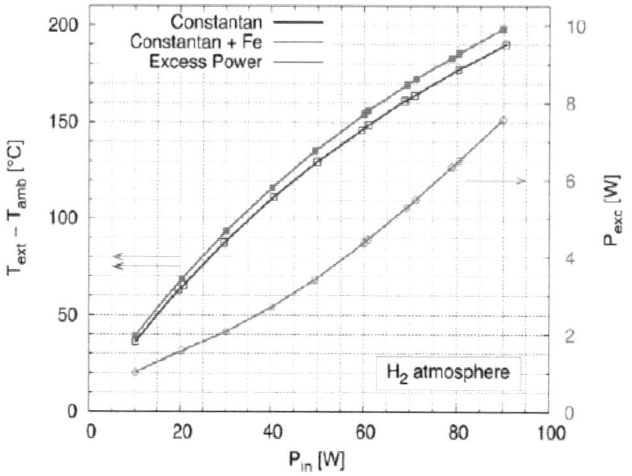

Figure 8. Tests performed in H_2 atmosphere for the configurations with knots with (*red*) and without (*blue*) iron addition. The difference of power between the two configurations is displayed in green.

The improvement of the overall performance using a low thermal conductivity gas, i.e. increasing the wire temperature at constant input power, was discovered by us after 2010, employing both palladium and/or nickel wires. At that time we used flow-calorimetry and the effects were no longer in doubt. The main drawback was the low response time of the system: meaning we could do only two tests per day at most. With isoperibolic-type measurements we can do up to 5–6 experiments/day.

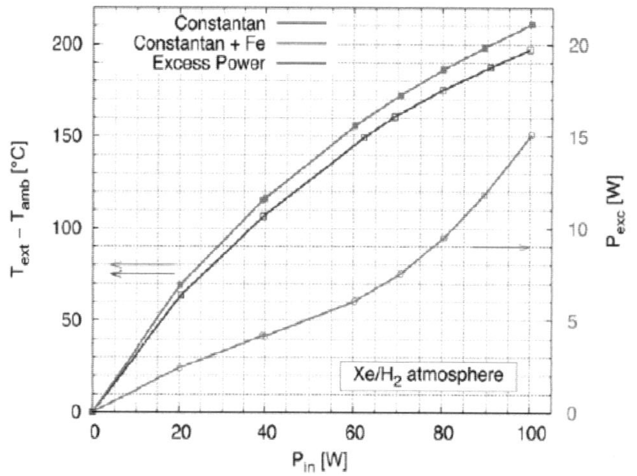

Figure 9. Tests performed in gas mixture atmosphere of Xe (1.4 bar) and H_2 (1.7 bar). The meaning of the colored curves is the same of Fig. 8.

Because we used a mixture of H_2 and argon in the actual experiments (as opposed to the control experiments), we carried out the same procedure for He/Ar mixtures. Furthermore, in the old experiments, we realized that one of the reference "good working points" was at an electric input power of 48 W. As a consequence, the results made public are mainly at the input power of 48 W (plus a few tests at 50 W). All the work performed at different input powers were not shown or discussed in public, and was only used for internal discussions in the working group.

The main disadvantage of this approach is the large indetermination, usually quantified in ± 1 W and in some specific conditions up to ± 2 W. The typical excess power when the system showed "useful" effects was in the range of 5–10 W during the test performed at Frascati Laboratories using the Agilent DAQ. In conclusion, the detected excess power, even in the worst situation, is real beyond any doubt.

As regards to the experiments of type B, they were actually a comparison of two different experiments. The two were at the same power level of 50 W, but with varying pressure of H_2 inside the chamber, from 3.8 down to 0.5 bar at low glass content and from 2.7 down to 0.1 bar for a large amount of glass. This was accomplished by dismantling the internal set up in order to add additional glass fibers. In this case the comparison is more difficult. Anyway, the difference on both the external (about 16°C, from 149 to 165°C) and highest internal temperatures (over 50°C, from 270 to 320°C) at 0.5 bar of pressure, is so large that one can hardly infer that the results are meaningless or invalid. In short, adopting for the calculation of the true effective area of the reactor just a value of length = 10 cm, diameter = 4 cm and for the (convective) Newton component a value of 15, we conservatively assumed that the experiment with a low amount of glass is the blank, i.e. equal to the input power of 50 W. In such conditions, the excess power of the configuration with large amount of glass seems to reach a value as large as 11.7 W.

In conclusion, even supposing a margin of error of ± 3 W, the effect seems to be real. About the experiment of type C, the comparison is performed in a very conservative way. We have two Constantan wires, with the same length and diameter, both with several knots, where the only difference is the addition of at most a few milligrams of iron inside the knots of the "active" wire. Iron is added through a very dilute solution of $Fe(NO_3)_2$ in HNO_3 (at about 5% concentration), 18 times, for 14 knots. In such a way iron is deposited in sub-micrometric layers. After such

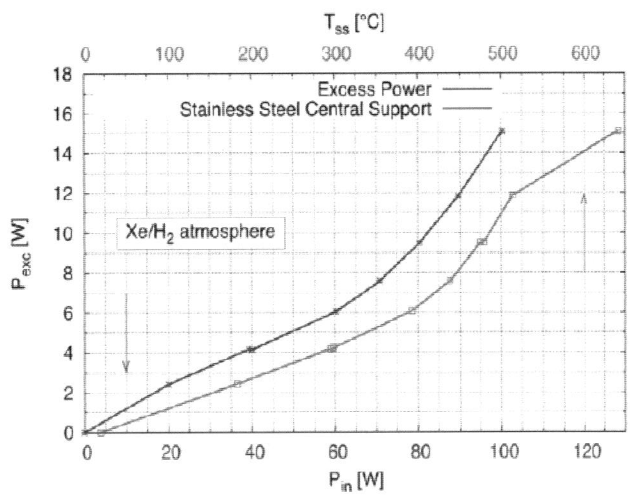

Figure 10. Tests performed in gas mixture atmosphere of xenon (1.4 bar) and H_2 (1.7 bar). The temperature of the SS internal support is showed in red. The difference in power between the configuration with iron and without iron is in blue.

treatments, the knots show strong ferromagnetic behaviors. Each wire is put inside the usual glass sheath and again inserted in an Al_2O_3 sheath ($T_{max} = 1200°C$). The sheaths are twisted around each other about every 2 cm. Moreover, they are inserted in another, more large diameter sheath. In such a geometrical configuration it is very difficult that temperature differences arise because of some geometric effect: only internally generated energy can be their source. For such kind of set-up power indetermination can be easily kept at less than 1 W.

9. Conclusions

The observation of single-wire spontaneous voltage generation in our experiment, in addition to achieving the aim of increasing AHE (using iron in the most recent specific case), represents one of the unexpected results of our recent research activity. The phenomenon shows itself as the generation of a voltage between the extremities of the non-powered wire in the reactor. The effect was first revealed serendipitously in the second-generation reactor with a large amount of glass. Furthermore, the successive addition of iron nanolayers and small knots in the wire (third generation reactor) increased and stabilized the magnitude of the generated voltage (up to 1900 μV) and current (up to 150 μA).

As concerns the anomalous heat production, iron has a beneficial effect because of its solubility of H_2 in the lattice largely increasing with temperature, while the knots are meant to create further non-equilibrium conditions inside the reactor. With such new wire features, we obtained larger excess power. From our observations and experiences we deduce that anomalous heat excess and spontaneous voltage generation are somehow related to the following conditions, or at least to some of them:

- Temperature as large as possible, obviously avoiding material sintering.
- Large hydrogen absorption/adsorption by means of catalytic materials, i.e. nanomaterials.
- Hydrogen flux as large as possible from regions at high concentration to regions at lower concentration.
- Presence of elements with hydrogen concentration increasing with temperature, such as iron.
- Presence of non-equilibrium conditions as large as possible: this is the main condition for getting any type of thermal or electrical anomaly.

We highlight that wires with good performances in terms of excess heat showed remarkable values of spontaneous voltage. One key aspect to be clarified is the role of strontium (Sr) in the generation of the spontaneous voltage. Supposing that the effect is due to some motion of electrons from the powered-wire toward the non-powered one, i.e. a thermionic emission or a process similar to it, an electron-emitting material like strontium (due to its low work function) might give a large contribution to it. Unfortunately, we performed no measurement of wire voltage for the old set-up with no strontium deposited on the glass sheaths, therefore, targeted tests are needed. Regarding iron, we found out that the fiberglass sheaths (SIGI Inc.) that we used may contain measurable amounts of iron (<2%) and iron oxide (<1%). Hence, a certain amount of iron was already present in the second-generation experiments. In this article we have also formulated some hypotheses about the mechanisms leading to thermal excesses in our experiments. We have focused on $H_2 \rightarrow 2H \rightarrow H_2$ reactions, individuating three combined processes:

- The catalytic properties of Constantan help molecular hydrogen dissociation.
- The hydrogen produced is partly adsorbed at fiberglass surface. Adsorption favors hydrogen recombination (an exothermic chemical reaction) but could also promote hydrogen absorption into the metallic lattice.
- If demonstrated, energy localization at nanoscales would enhance H_2 dissociation by converting thermal mechanical energy into a valuable chemical energy potential.

The next step of our activity is aimed at investigating in detail spontaneous voltage generation and possibly identify its nature and the variables that are related to this anomalous effect as well as to get its maximization. As concerns the better performance of the reactor after the accident with vapors of silicon oil, we think it is somehow related to the recent "Breakdown of Richardson's law ..." found in the electron emission from heated carbon nanotubes [14]. In this study, the authors measured an emission density at large temperature (1800–2000 K) more than one order of magnitude higher than that expected with the normal thermionic emission at macroscopic levels. We hypothesize that the heating of the wires in presence of silicon oil vapor may have led to the formation of meta-stable carbon nanostructures at the surface, acting as enhancers of the anomalous effects produced in our reactor. Further inspired by the experimental design of such an important paper, we observe that the pointed shape of the nanotube is also found in some of the nanostructures present at the surface of our Constantan wires after the thermal/electric treatments. As a consequence of the supplied power or even of the energy localization mechanism if real, hot pointed-shaped nanostructures could be also responsible for an electron emission whose current manifests even at temperatures (averaged on macroscopic extensions of the wire, as measured by the thermocouples) lower than 600°C, the threshold of the known thermionic emission (i.e. Richardson law). We note that also in Iwamura's experiments, the anomalous effects, specifically in the case of "transmutations", disappear if the size of the nanoparticles is too large or the Low Work Function material used (CaO) is changed to another with higher values of Work Function (MgO) [15].

Addendum

After the ICCF19 Conference, in May 2015, we became aware that Prof. Leif Holmlid (Göteborg University, Sweden) suggested the use of nanosized iron, as a cofactor in its deuterium catalyzer (the so-called "hydrogen transfer catalyst"), to increase the amount of the so-called, ultra-dense Rydberg matter in its specific reactor by Inertial Confinement Fusion (ICF) processes. He aimed, and demonstrated experimentally possible by high quality and sophisticated charge particle measurements, to produce very large excess power (energy gain over 1000) just using Deuterium gas (made ultra-dense by an appropriate catalyst, based on Fe–K) as initial fuel and a table-top laser as stimulator for the fusion processes. His main ideas are summarized in the recent patent application EP 2680271A1 [16].

Moreover, Prof. Friedwardt Winterberg (Germany–USA, Nevada Univ., USA), the well-known expert in ICF at the international level, suggested the use of a magnetic field to increase the amount of Rydberg matter in Holmlid's experiments [17]. We observe that our knots (carrying a sufficiently large current), filled with iron and/or iron oxides at nanosize, can represent a simple/first step, experimental set-up, to match some of these requirements. We guess that the ultra-short time of recombination reaction with a range of hundreds of femtoseconds could help the stimulation of LENR effects (extremely high peak power), especially in our set-up where we have at the same time:

- Sub-micrometric structured Constantan (large production of hydrogen);
- Large amount of fiber glass where to "store" and recombine hydrogen;
- Large electron emission as a consequence of the use of low work function materials and of the increased field emission due to the low dimensionality (the trigger of the system).

In the final analysis, we conclude that in some tests where we used pure D_2 and/or Xe/D_2 mixtures (ratio 10/1) at mild pressures, we observed a certain increase of gamma radiation by a NaI(Tl) gamma detector (model LB125 by Berthold), in the energy range of 25–2000 keV. This increase happened mainly during the fluctuations of operating temperatures of the system, i.e. increasing or decreasing input power. The effect lasted for several hundreds of seconds. Again, this effect appears to be due to non-equilibrium conditions. The first detection of this phenomenon occurred on June 2012; it was reported in a meeting on LENR held at the Italian Parliament (July 2, 2012) and shortly after that discussed at both the NI Meeting and at the ICCF17 Conference (Daejeon, South Korea) in August 2012.

Considering the fact that such effects happened mainly using D_2, we suppose that even in our experimental set-up there are places where Rydberg Matter, as described by Leif Holmlid, might produce a low intensity signal in our experiment during non-equilibrium conditions. Further systematic work has been planned in the next experiments to explore this possibility.

Finally, some of the authors of this report, considering the experimental evidence that in several LENR experiments (performed worldwide since 1989) there is no consistency between the AHE detected and the usual nuclear particles emitted, have the feeling that some LENR are related, in some aspects, to the so-called Dark Matter (which is theoretically or experimentally necessary to explain the universe, although its existence has not yet been clearly demonstrated) and emission of new type/exotic particles like the WIMP (Weakly Interacting Massive Particles). Even studies devoted to explain the large discrepancies in observations involving heat and helium released from the earth – such as the finding that heat measured when compared to helium too large by a factor of 20 – indicate the existence of new types of particles and related reactions [18].

Acknowledgements

This work was partially supported, since 2011, by a high-tech metallurgical company in north-east Italy. We are indebted to Prof. Giorgio Vassallo (Univ. Palermo, Italy) for very stimulating suggestions/discussions/critics and Dr. Cesare Lorenzetti for in-deep analysis about catalysis problematics. Additional editing for English clarity undertaken by Robert W. Greenyer B.Eng. Volunteer, MFMP.

References

[1] F. Celani et al., Cu–Ni–Mn alloy wires, with improved submicrometric surfaces, used as LENR device by new transparent, dissipation-type, calorimeter, in *ICCF17*, Daejeon, South Korea, published in *J. Condensed Matter Nucl. Sci.* **13** (2014) 56–67, ISBN: 978-1-63439-142-9, ISSN: 2227–3123.

[2] W. Bruckner et al., Oxidation behaviour of Cu–Ni(Mn) (Constantan) films, *Thin Solid Films* **258** (1995) 252–259.

[3] Y. Arata and Y.C. Zhang, Development of "DS-Reactor" as the practical reactor of "Cold Fusion" based on the "DS-cell" with "DS-Cathode", in *ICCF12*, Yokohama, Japan, published in *Condensed Matter Nucl. Sci.*, 2006, pp. 44–54, World Scientific, Singapore, ISBN: 978-981-256-901-1. http://www.worldscientific.com/doi/abs/10.1142/9789812772985_0004.

[4] F. Celani et al., Are specific glass surfaces co-factors for generation of anomalous effects, by catalytic materials, under H_2 gas at high temperatures? in 2014 MIT Colloquium on Cold Fusion effects, March 21–23, 2014, http://www.coldfusionnow.org/wp-content/uploads/2014/03/Celani-2014-CF-LANR-MIT.pdf.

[5] I. Langmuir, Flames of atomic hydrogen, *Ind. Eng. Chem.* **19**(6) (1927) 667–674.

[6] F. Celani et al., Further progress/developments, on surface/bulk treated Constantan wires, for anomalous heat generation by H_2/D_2 interaction, in *ICCF18*, Columbia, MO, USA, 2013, http://hdl.handle.net/10355/37367.

[7] The Nobel Prize in Chemistry 1999, http://www.nobelprize.org/nobel_prizes/chemistry/laureates/1999/press.html.

[8] E. Fermi, J. Pasta and S. Ulam, Studies of Nonlinear Problems, Document LA-1940, 1955.

[9] B. Ahern, Energy Localization: The key to Understanding Energy in Nanotechnology and Nature, http://lenr-canr.org/acrobat/AhernBSenergyloca.pdf.

[10] P.B. Umbahnhowar, F. Melo and H.L. Swinney, Localized excitations in a vertically vibrated granular layer, *Nature* **382** (1996) 793–796.

[11] Ed Gerstner, Second law broken, *Nature* (2002, July), http://www.nature.com/news/2002/020722/full/news020722-2.html.

[12] G.M. Wang, E.M. Sevick, E. Mittag, D.J. Searles and D.J. Evans, Experimental demonstration of violations of the second law of thermodynamics for small systems and short time scales, *Phys. Rev. Lett.* **89**(5) (2002) 050601. doi:10.1103/PhysRevLett.89.050601.

[13] J. Gieseler, R. Quidant, C. Dellago and L. Novotny, Dynamic relaxation of a levitated nanoparticle from a non-equilibrium steady state, *Nature Nanotechnology AOP* **9** (2014) 358–364.

[14] X. Wei, S. Wang, Q. Chen and L. Peng, Breakdown of Richardson's law in electron emission from individual self-Joule-heated carbon nanotubes, *Scientific Reports* **4** (2014) 5102. http://www.nature.com/srep/2014/140529/srep05102/full/srep05102.html.

[15] Y. Iwamura, S. Tsuruga and T. Itoh, Deuterium permeation induced transmutation experiments using nano-

structured Pd/CaO/Pd multilayer thin film, in *2014* MIT Colloquium on Cold Fusion effects, March 21–23, 2014, http://newenergytimes.com/v2/conferences/2014/MIT/Iwamura-MIT-2014.pdf.

[16] L. Holmlid, Method and apparatus for generating energy through inertial confinement fusion, EP2680271A1, Jan. 01, 2014, https://data.epo.org/publication-server/pdf-document?pn=2680271(&)tabki=A1(&)cc=EP.

[17] F. Winterberg, Ultra-dense deuterium and cold fusion claims, *Phys. Lett.* A **374** (2010) 2766–2771.

[18] F.J. Mayer and J.R. Reitz, Thermal energy generation in the earth, *Nonlin. Processes Geophys.* **21** (2014) 367–78.

[19] F. Celani et al., Experimental results on sub-micro structured Cu–Ni alloys under high temperatures hydrogen/deuterium interaction, in *Tenth Int. Workshop on Anomalies in Hydrogen–Metal Systems*, Pontignano, Italy, 2012, published in *Chem. Materials Res.* **3**(3) (2013) 27–56, ISSN 2224–3224 (Print), ISSN 2225-0956 (Online).

[20] S. Romanowski et al., Density functional calculations of the hydrogen adsorption on transition metals and their alloys. An application to catalysis, *Langmuir* **15**(18) (1999) 5773–5780.

J. Condensed Matter Nucl. Sci. 19 (2016) 46–55

Research Article

Off-mass-shell Particles and LENR

Mark Davidson*
Spectel Research Corp., Palo Alto, CA, USA

Abstract

A recent and somewhat radical theoretical explanation for LENR is reviewed. It is based on variable mass theories of relativistic quantum mechanics that date back to the 1930s in works by Fock and Stueckelberg, and up to the present by many others. It explains a large number of observed anomalous effects in LENR by positing that nuclear rest-masses can vary in "nuclear active environments" in condensed matter settings. The varying masses modify the kinematic constraints of the nuclear reactions. It also offers a mechanism for enhancing electron screening and-or quantum tunneling rates, for allowing for resonant tunneling, and for modified radioactive decay rates by mass changes in the decaying isotopes.

Keywords: Horwitz-Piron, LENR, Nuclear anomalies, Stueckelberg, Variable mass

1. Introduction

There are a large number of experiments that cannot be explained easily or at all by conventional nuclear physics theory. These include most experiments from the LENR community along with a number of others concerning radioactive decay rates and other anomalous phenomena ([7,8] and references therein). Physicists have largely disengaged from this subject, although many study it secretly late at night, the problem being frustration over how these experimental results can possibly be true. If one reads enough of the experimental literature though, one begins to ask the mirror question - How can these experiments all be wrong? I found myself on this path and tried about a dozen ideas to explain and reconcile this dilemma over a number of years. Slowly, I felt cornered into the theory presented here, and in previous papers [7,8]. My conclusion was, and still is, that the only way the experiments can be explained is if the nuclear rest masses of particles are somehow varying in those "nuclear active environments" where the reactions are occurring. But how could that be? Aren't the rest masses fixed? Having been trained in theoretical particle physics, I knew that in relativistic quantum field theory, like the standard model of particle physics, the masses inside of Feynman diagrams – the "virtual particles" – can and do vary from their rest masses [50,49]. These particles are also called "off the mass shell" in the vernacular of particle physics for this reason. Actually, the standard model makes a very interesting prediction in this regard. It predicts that all interacting particles have some probability of being off the mass shell all of the time, even if the inter-particle interactions are very weak (just not exactly zero for all time).

*E-mail: mdavid@spectelresearch.com

This conclusion follows from the fact that Feynman integrals, the building blocks of perturbative solutions to quantum fields, involve integrations over the masses of particles in loops and in the interior portions of the diagrams, and since all particles in our universe have been interacting with other particles since the beginning of time and will continue to interact, they are always at least slightly virtual. In order to explain LENR, quite large off-shell deviations are required, often in the MeV range. These have never been directly observed, or expected. Thus this must be viewed as a highly speculative idea. I've proposed this theory because I see no other way to explain the experiments of LENR. I want the reader to understand this, and that I am not claiming that mass variation is in any way proven. I shall make arguments in favor of it, but proving that nature works this way will depend on future experiments.

Off-mass-shell behavior is an inherently relativistic effect. Usually it's associated with hgih velocity motion. In ordinary matter at near to room temperature we expect that the velocities of the particles to be non-relativistic, since otherwise the matter would not be stable. The usual relativistic correction factor of $\gamma = 1/\sqrt{1-\mathbf{v^2}/c^2}$ can be safely set to 1. Nevertheless, we consider the possibility that the rest masses can deviate from their usual values in some special systems. Inside of a Feynman diagram, consider a virtual particle with a 4-momentum P^μ. The velocity of the virtual particle (if it is timelike) can be taken to be $|\mathbf{v}| = c\left|\overrightarrow{P}\right|/\sqrt{P^\mu P_\mu}$ (I use the time-like metric signature here $(+,-,-,-)$). So it is possible for a virtual particle to have a slow velocity, but still go off the mass shell, since all values for P^μ are integrated over in the Feynman integral. Even though the motions of all the massive particles in a solid might be non-relativistic, we must still use a relativistic theory to examine this off-shell behavior. The laws of physics in a local Lorentz frame are Lorentz covariant to a high degree of accuracy (if we ignore curvature of space-time locally and gravity effects). The non-relativistic limit of the laws of physics, resulting in Galilean invariance, is only an approximation to this. So we should use a Lorentz-covariant theory, like the standard model of particle physics or quantum electrodynamics (QED), to study this sort of phenomenon. The problem is that this is very hard to do. We would prefer to use a relativistic wave mechanical theory, like Schrödinger's non-relativistic equation, to describe it instead, But the standard relativistic versions of wave mechanics, like the Dirac equation, Klein–Gordon equation, Proca equation, Rarita–Schwinger equation, etc. are all unacceptable as many body quantum theories because either their energy is not bounded from below, or they have negative probability states, or both. The modern view among particle physicists is that only a quantum field theory description is possible, and a wave mechanical particle description does not exist. Due to the complexity of the perturbation theory for the standard model in a solid, there exists no rigorous bound on how far the virtual masses can wander off-shell. Thus we are venturing into terra incognita, but I feel that the experiments require us to pursue this theoretical possibility.

The difficulties with developing a relativistic wave mechanics can be understood very easily. Relativity theory says that space and time are coordinates in a 4-dimensional manifold, and that they mix linearly under Lorentz transformations. So space and time are the same in some sense. But in quantum mechanics, time is a c-number parameter that precisely orders the evolution of causal history and events and therefore commutes with all the operators of the Hilbert space of quantum states, whereas the positions of quantum particles become non-commuting operators. So in quantum mechanics. time is very different from position. These two different requirements on the time variable produce lots of problems for relativistic quantum mechanics. The time variable is overloaded. This problem was recognized early on in the development of quantum theory, and a solution was proposed in the 1930s independently by Fock and Stueckelberg [13,47,46]. We discuss these theories more below.

2. The Time–energy Uncertainty Relation and Variable Mass

Some physicists might ask the following question.

In a solid, where the velocity of the particles are non-relativistic, the energy of a particle is simply given by $E \approx mc^2$. *But we expect an energy-time uncertainty relation of the form* $\Delta E \Delta t \approx \hbar$, *and therefore (taking* $c = 1$)

$\Delta m \Delta t \approx \hbar$. *If the particle is stable, so it is lifetime is infinite, then we can take* Δt *to be* ∞ *and consequently why is not* $\Delta m \approx 0$, *i.e. on the mass shell?*

Therefore, they expect weakly interacting particles to be always on the mass shell. This argument is reinforced by the fact that most of the time when we measure a particle's mass in condensed matter it is on the mass shell. I believe that the experimental evidence in LENR suggests that in very rare and special nuclear active environments, the masses are varying unexpectedly, and so let's examine the theoretical basis for this Heisenberg uncertainty argument a bit deeper.

First of all, the Heisenberg uncertainty relation deals with the uncertainty for E, it says nothing about the expectation value $\langle E \rangle$

$$\Delta E = \sqrt{\left\langle (E - \langle E \rangle)^2 \right\rangle}. \tag{1}$$

Second of all, the Heisenberg uncertainty relation is an inequality, and so for energy and time it would be

$$\Delta E \Delta t \geq \hbar/2 \tag{2}$$

and therefore, if $\Delta t = \infty$ all we can say is that $\Delta E \geq 0$, which is not a very significant result.

Third, the origin of the Heisenberg uncertainty is the canonical quantization rule $[p, q] = -i\hbar$, but time t in conventional quantum mechanics is a c-number which commutes with all the operators in the Hilbert space of states. Therefore the commutator$[E, t] = 0$. So how could there be a time-energy uncertainty relation? This unresolved question has occupied many articles over the years, a partial list being [5,3,6,9,26,43,36].

Finally, there is no acceptable relativistic wave mechanical theory, and so we can't even define a Hamiltonian or position operators covariantly for a many-body system.

Consequently, I conclude that the Heisenberg uncertainty principle argument against variable masses in a condensed matter setting are not conclusive or even relevant. We expect the nuclear active environment to be a non-equilibrium mixed-state quantum open system with complex morphology, unknown catalytic processes, and possibly with external electromagnetic fields, and proton, deuteron, and electron currents flowing in the material. In other words, it is basically a perfect storm of complexity about which it is very difficult to say anything with rigor and confidence. One thing we can say for sure though, and that is that the energy–momentum tensor density is conserved, since this is a property of the standard model, or any other relativistic model we might consider. Consequently, even though the mass of a particular particle may change, this does not mean that energy is not conserved. Just as in the Feynman diagrams of relativistic perturbation theory where the particles are off shell, the total energy and momentum are always exactly conserved.

3. Fock–Stueckelberg Covariant Wave Mechanics

The idea of Fock and Stueckelberg was to add a second time variable and make the space five dimensional (for a single particle) [13,47,46]. Sometimes these are called historical-time or proper-time models. The theory was utilized by Feynman, Schwinger, and Nambu [11,41,34]. For example, Feynman used it to carry out his path integration approach to quantum mechanics for the Klein–Gordon equation [11]. This is the simplest example of a historical time theory.

The usual Klein–Gordon equation is

$$(i\partial - A)^{\mu} (i\partial - A)_{\mu} \Psi = -M^2 \Psi, \tag{3}$$

where M is the particle's rest mass, and where A is the vector potential for an external classical electromagnetic field. The conserved current is $j_{\mu} = i\left[\phi^{*}\partial_{\mu}\phi - \phi\partial_{\mu}\phi^{*}\right]$, but j_0 takes on both positive and negative values, and so it cannot be taken to be a probability current. Moreover, a localizable Schrödinger type position operator cannot be defined for this equation in the sense of Newton and Wigner [35]. The idea is to add a second invariant time τ, and consider the modified Klein–Gordon equation

$$i\frac{\partial\varphi(x,\tau)}{\partial\tau} = \frac{1}{2}(i\partial - A)_\mu (i\partial - A)^\mu \varphi(x,\tau), \quad \text{where } x = \{x^0, x^1, x^2, x^3\}. \tag{4}$$

It is very similar to the time dependent Schrödinger equation, but with the "historical time" τ replacing the usual time variable t, and with the four coordinates of space-time x^μ replacing the usual three coordinates of space. Because of this, it is quite easy to formulate the path integral method for this system by following the same steps as used for the Schrödinger equation. Feynman points out that if A^μ does not depend on τ, then separable solutions exist so that $\varphi(x,\tau) = \Psi(x)\exp(i\frac{1}{2}M^2\tau)$ and ψ is the solution to the usual Klein-Gordon equation. Equations of this type were first studied by Fock and Stueckelberg. The path integral solution includes all paths in space-time connecting two space-time points and parametrized by τ, with no restrictions on the path, so that off-mass-shell paths are included in the path integral solutions. This reflects the fact that Feynman diagrams contain virtual particles which are not on the mass shell. Horwitz and Piron [19–23] extended the theory to multiple particles. This theory is no longer simply a reformulation of the standard quantum mechanics. It is now different, and in particular the off-shell behavior is much more prevalent. In the quantum version of this theory, the on-mass-shell state vectors are not complete in the whole quantum Hilbert space. Rather, off shell states are also required for completeness.

Another approach that has been taken is to develop a 5-dimensional generalization of QED, called pre-Maxwell theory, for which the Feynman diagram rules have been worked out and applied to various scattering processes [29,28].

So the assumption is that in the "nuclear active environments" of LENR, a variable mass theory such as one of these has become an effective theory for the solid state behavior. Perhaps this assumption can be derived from just the standard model of particle physics, exploiting its off-shell virtual particle qualities, but in the meantime we can take a phenomenological approach and try and deduce what experimental conditions seem to lead to off-shell behavior in LENR. More discussions along these lines were presented in [7,8].

4. Could the Standard Model of Particles Allow Large Virtual Mass Variations in Condensed Matter?

The standard model is believed by most physicists to describe all phenomenon observed in nature except for gravity [37,50]. Thus all of nuclear and condensed matter physics are believed to be derivable from the standard model. For example, the neutron–proton mass difference has recently been accurately calculated, and agrees with the measured value, by using lattice QCD plus QED [2].

The first question then is, are there any theorems that would rule out the possibility of large mass variations? The usual arguments based on the Heisenberg uncertainty relations, as we have seen above, are not conclusive on this point. There is the equivalence theorem in field theory which states that changing field variables will change off-shell Green's functions while leaving the S-matrix invariant [10,48]. But this does not prove that in condensed matter which is undergoing some arbitrary complex non-equilibrium process, that large excursions off the mass shell cannot occur and contribute significantly to the resulting probability amplitudes for different reactions. I have not found any theorem that rigorously limits off-shell mass excursions of this type in condensed matter that would be relevant.

In quantum electrodynamics, charged particles do not have a simple fixed mass due to the interaction with infrared photons as in [16,4,33] and references therein. Charged particles continually interact with massless infrared photons, and therefore cannot be assigned fixed masses. The charged particle surrounded by its cloud of infrared photons is called an infraparticle. This infrared problem poses an interesting and as yet unsolved challenge to QED and the standard model. When you consider these results in the framework of an interacting, non-equilibrium condensed matter system, it seems quite plausible that in some special circumstances a large mass variation for charged particles may be possible. Perhaps, in some approximation, the Fock–Stueckelberg models might describe such an interacting infraparticle.

We can say for certain that the Fock–Stueckelberg type of models are compatible with the basic principles of relativity and quantum mechanics, and that they do allow for large mass variation. It is conceivable that in some special circumstances in a solid, the matter there might be better described by them rather than by the usual Schrödinger theory. What environmental conditions are required to bring about such a state of matter? I do not know, but the experiments of LENR shed some light on it. The nuclear active environments tend to occur near the metallic surface, in non-equilibrium situations, possibly in cracks or crevices, and in the presence of fluxes of deuterium, hydrogen, and electrons through the metal. External stimulation with electrical, optical, or other stimuli of the metal often helps, as do catalytic elements.

5. Enhanced Tunneling Rates Caused by Increased Electron Masses

In the Born–Oppenheimer approximation for molecules and solids, the size scale is determined by the inverse of the electron mass [7]. Therefore, an electron mass increase will reduce the inter-nuclear distance, and greatly enhance quantum tunneling and fusion rates. This is a form of electron screening. The enhancement of deuterium fusion in a D_2 molecule was rigorously quantified by Koonin and Nauenberger [27], where it was shown that a mass of ten electron masses was necessary to explain the reaction rates measured by Fleischmann and Pons, and a mass of five electron masses was needed to explain Jones' results. With simply a large electron mass, one cannot explain the discrepancy in the branching ratio, but with a simultaneous reduction in the deuterium mass, one can explain that as well, and in this case the electron mass increase factor can be much smaller than 10 since the resonant tunneling effect greatly enhances the fusion rate [7]. An electron mass increase factor of 5 is sufficient to explain the results of Jones, and this could be adequate, when combined with resonant tunneling, to explain the excess heat experiments that have been observed [7].

Another form of tunneling which might have been observed in some experiments is neutron transfer. In a test of the commercial E-Cat reactor [30] it was reported that the fuel consisting of powder containing nickel and lithium in their natural isotopic ratios, was modified by the reaction and that the ash consisted mainly of lithium-6, most of the lithium-7 having apparently lost a neutron, and the only isotope observed in the ash of nickel was nickel-62, the heaviest stable isotope. So apparently the neutrons from lithium-7 had been transferred to nickel isotopes. Neutron tunneling has been proposed as an explanation for this [15]. If electron masses were to increase in a molecule binding together lithium and nickel atoms, then the tunneling rate could increase by many orders of magnitude. Therefore, this neutron transfer mechanism seems quite plausible as a candidate for explaining these results provided that masses can vary in this way. I take this experimental result as quite tentative though.

6. Some Anomalous Nuclear Reactions that can be Interpreted as Allowed by Mass Variation

Most of the nuclear anomalies that have been observed in LENR experiments and elsewhere cannot be explained by conventional nuclear physics. I believe that they can all be interpreted as being enabled by variable mass behavior. So, here is a partial list of various effects and how they can be explained. I do not claim that these experimental claims are all correct. I do think that there is some non-zero probability in the Bayesian sense that any of these experiments might be true though, and since they are persistently reported in the literature I have tried to explain them.

(1) Deuterium fusion $\mathrm{d+d} \to \alpha$. This reaction was studied in detail in [7]. It takes place between two neighboring deuterons in a palladium lattice. The electron mass must increase for at least one electron which is binding the two deuterium particles together, and the deuteron masses must subsequently decrease. When the two deuterons, imagined as continuously going off-shell in [7], reach a combined mass which is very close to the

mass of an alpha particle, a resonant tunneling effect can be expected to occur, and this gives the observed behavior of Pons and Fleischmann. Fusion is enabled by mass-tuned quantum tunneling in this picture, but only excess heat is produced.

(2) In the Widom–Larsen theory [51] it is argued that protons and "heavy" electrons can react to form a neutron and a neutrino $e + p \to n + \nu_e$ at near room temperatures, because the electron's rest mass has increased. The neutron activation caused by the presence of these neutrons then can produce nuclear reactions. The general variable mass theory contains this Widom–Larsen effect as a special case. We allow masses for all charged particles to decrease as well as increase, unlike Widom–Larsen, and this leads to many more reactions being possible.

(3) Oppenheimer–Phillips processes whereby a deuteron gives up a neutron to a metal nucleus through quantum tunneling. This effect has been advocated by Passell [38]. Enhanced electron screening is required for the cross sections of these reactions to be large enough to account for observed effects. Increased electron mass would provide such a mechanism.

(4) Neutron transfer involving two deuterons $d + d \to p + t$. This reaction would be made possible by an increased electron mass which would increase the electron's screening ability, along with possibly a variation in the deuteron mass to allow for resonant tunneling of the neutron when the summed mass of the two deuterons were to equal the sum of the on-shell masses of the proton and tritium. This reaction could help explain why there is much more tritium observed in Fleischmann–Pons type experiments than neutrons [45].

(5) The 'Reifenschweiler effect' [40] is the observation that the beta-decay rate of tritium (half-life 12.5 years) is reduced reversibly by about 25–30% when the isotope is adsorbed into 15 nm titanium-clusters in a temperature window between 160–275°C. Remarkably at 360°C the original radioactivity reappears. First discovered circa 1960/1962 at Philips Research, Eindhoven. The reported decay rate reduction can be explained if the tritium mass and consequently the kinematic phase space for the decay were reduced [8]. Since the Q for tritium decay is about 18.6 keV, if the mass of the tritium drops by this amount, its decay rate would go to zero.

(6) Other radioactive decay effects that might indicate mass variation [8]. Radioactive isotopes implanted into metals at low temperature $\sim$12 K show variations of decay rates in some cases, which could be due to enhanced electron screening, which in turn could be due to mass variation [31,32,14]. These results have not been completely reproducible so far, and the theory proposed based on conventional electron screening has been criticized. It is another controversial LENR phenomena .

(7) Time varying radioactive decay rates may indicate mass variation [8]. Experiments show decay rates varying with time for a number of isotopes. Frequency analysis has shown annual, diurnal, and approximate monthly variations [12]. Once again, these results are controversial. It has been suggested that solar neutrinos might be the cause of this decay-rate variation.

(8) With varying particle masses, transmutations can in theory at least occur in a number of ways in nuclear active environments. Enhanced electron screening caused by electron mass increases can modify alpha decays, beta decays, and electron capture rates. Mass changes of nuclei can change reaction rates or make reactions possible which would normally be forbidden. Resonant fusion of hydrogen or deuterium with other nuclei, resonant fusion of alpha particles and other nuclei, and even fission of heavier nuclei might occur after a mass change. Also, there is the possibility of neutron creation and subsequent capture as in the Widom–Larsen theory, or neutron stripping or hopping reactions, leading to many possible transmutations. In short a world of possibilities exist, and a managerie of transmutations have already been observed experimentally in LENR [44,45]. The number of transmutations observed in LENR are so large, that many scientists have concluded

that it is proof that the experiments must be wrong. But maybe not, given the multitude of processes made possible by variable masses.

More details may be found in [7,8].

7. Suggestions for Experiments to Test the Variable Mass Idea

If radioactive nuclei are placed into the nuclear active environments, their decay rates should change if their masses change, and this might be observable by monitoring the real-time radiation during an experiment. Probably gas loading experiments would be the easiest ones to monitor in this way, and some imaging capability would be desirable because the nuclear active environments tend to be very localized on the surface of the metal. Tritium is an excellent candidate to serve as a tracer element to reveal mass variations of hydrogen isotopes. Tritium can be added to deuterium or hydrogen in LENR experiments. The tritium in the nuclear active volumes should experience a mass change like deuterium if the variable mass theory is correct. This should be detectable as a change in the tritium decay rate in those zones while they remain in a nuclear active state. Other radioactive isotopes might be tried as well. A replication of the Reifenschweiler experiments would be very desirable too.

8. Implications for other Theories of LENR

The possibility of variable mass does not rule out any of the other prevalent theories of LENR, but actually can provide a mechanism for them to be valid. For example, the Widom-Larsen effect is enabled by the type of variable mass behavior we are contemplating here. Also, the collective interaction of the lattice with the nucleus as proposed by Hagelstein [17,18], Preparata [39], and Schwinger [42] could be understood as mediated by mass variation over an extended time leading up to the nuclear reaction. The lattice could slowly take up or give up energy to the nuclear environment as the masses of some of its particles changed prior to a nuclear reaction. There are a number of theories which rely on the existence of small atoms which can be understood if electrons can increase their mass. A theory due to Kim proposes a Bose–Einstein mechanism [25], and the existence of such a state of matter might be enabled by resonant tunneling as in [7] where deuterium resonant tunneling was enabled by mass reduction of the deuterium particles.

9. Choices for Physics

As I see it, physics should re-engage in a serious way with this subject. Here are the choices facing mainstream physics: (1) Dismiss the experimental claims and the entire field of LENR (more than 1500 international research papers); (2) Do nothing, but wait and see what happens next; (3) Continue to try and explain the claims with existing theories after 24 years of failed effort; or (4) Develop a new theory, based on relativistic quantum mechanics, that modifies existing theory and explains the experiments. Obviously I prefer the fourth choice.

Much of modern physics has separated from close laboratory experimental feedback. Fields like quantum gravity, string theory, interpretations of quantum mechanics, emergent theories of gravity, etc. are examples. Here, with LENR, we have a large discrepancy between theory and experiment which may require a fundamental modification of our theories for nuclear physics and condensed matter, and we also have lots of laboratory feedback to aid us.

If off-mass-shell quantum mechanics is needed in order to understand LENR results, then our understanding of quantum mechanics will be affected at a fundamental level. The standard on-shell wave equations would have to be considered then as approximations to more general off-shell theories. The standard relativistic wave equations (Klein–Gordon, Dirac, Proca, etc.) all have problems with a localized position operator, or negative energies, or negative probabilities. Thus, modern physics regards quantum fields as preeminent over wave mechanics. But a resurgence of the Fock–Stueckelberg theories could change this, and make wave mechanics prominent once again. The practical implications if LENR is real are obviously potentially enormous. Only the re-engagement of the physics community will allow the full promise of this technology to be realized. The downside monetary risk of coming up dry seems insignificant compared to the potential benefits both to mankind and to fundamental physics. What gambler would not take this bet? Even if all the experiments in LENR turn out to be wrong, physics could still benefit if by examining this issue we can learn to engineer the control of rest masses. This would then open doors to finding new ways to make LENR devices and provide a powerful tool in chemistry and condensed matter physics.

10. Conclusion

We have listed here some of the anomalous LENR reactions which have not been explained by conventional nuclear physics. If nuclear rest masses can vary in special nuclear active environments, then all of these reactions can be allowed with the right mass variation.

After surveying the literature on mass variation in relativistic physics, the historical time models pioneered by Fock and Stueckelberg are prominent. The modern refinements of this theory especially by Horwitz and co-workers stand out as the most likely candidate theories around which to construct a model for LENR. The task before us is to try and quantify what physical effects cause a nuclear active environment to come into existence.

The theories for mass variation are not yet complete. In particular, we do not have a theory which interpolates between theories in which the rest mass is quite freely variable, like for instance pre-Maxwell theory, and ones that have fixed mass, like conventional QED or the standard model applied to few particle systems. I have in mind a new phenomenological parameter that would depend on position inside a material and control how easy or hard it is for particles to drift off the mass shell. In ordinary matter or in vacuum, this parameter would make deviation from standard rest masses exceedingly small, but in the nuclear active environments it would allow significant deviation of mass from the expected value.

Many of the objections to LENR experimental claims in the past were based on the fact that neutrons and tritium were not observed in experiments involving deuterated palladium in the proper amounts, energies, and ratios as deuterium fusion would require [1,24]. If masses of electrons and deuterons can vary, then this not a valid justification for rejecting these results out of hand. So, if one is to rule out a theory like this one, the change in the ratios of tritium to neutrons or their energies cannot be used to dismiss experiments. Rather, they might be viewed as experimental evidence in support of variable masses.

Much of modern physics has separated from close laboratory experimental feedback. Fields like quantum gravity, string theory, interpretations of quantum mechanics, emergent theories of gravity, etc. are examples. Here, with LENR and related anomalies, we have a large discrepancy which may be resolved by a fundamental modification of our theories for nuclear physics and condensed matter, and we also have lots of laboratory feedback to aid us. It seems that physicists ought to re-examine this subject in the light of these new developments, and determine with certainty whether or not the nuclear anomalies are real.

Acknowledgments

I would like to thank Larry Horwitz, Martin Land, Peter Hagelstein, David Nagel, David Katinsky, David Kidwell, Michael McKubre, and Tom Grimshaw for valuable conversations and correspondence, and Antonio LaGatta and TSEM for their hospitality as hosts for ICCF19.

References

[1] Cold Fusion Research, November 1989 - A Report of the Energy Research Adivsory Board to the United States Department of Energy, DOE Report DOE/S-0071 (August, 1989) and DOE/S-0073, DE90, 005611, Department of Energy, 1989.

[2] S. Borsanyi, S. Durr, Z. Fodor, C. Hoelbling, S.D. Katz, S. Krieg, L. Lellouch, T. Lippert, A. Portelli, K.K. Szabo and B.C. Toth, Ab initio calculation of the neutron–proton mass difference, *Science* **347**(6229) (2015) 1452–1455, arXiv: 1406.4088.

[3] J.S. Briggs, A derivation of the time-energy uncertainty relation, *J. Phys.: Conf. Ser.* **99**(1) (2008) 012002.

[4] D. Buchholz, Gauss' law and the infraparticle problem, *Phys. Lett. B* **174**(3) (1986) 331–334.

[5] M. Bunge, Virtual processes and virtual particles: real or fictitious? *Int. J. Theoret. Phys.* **3**(6) (1970) 507–508.

[6] P. Busch, The Time–Energy Uncertainty Relation, In J.G. Muga, R.S. Mayato and Í.L. Egusquiza (Eds.), *Time in Quantum Mechanics*, No. 734, in Lecture Notes in Physics, Springer, Berlin, Heidelberg, 2008, pp. 73–105.

[7] M. Davidson, Theories of variable mass particles and low energy nuclear phenomena, *Foundations Phys.* **44**(2) (2014) 144–174.

[8] M. Davidson, Variable mass theories in relativistic quantum mechanics as an explanation for anomalous low energy nuclear phenomena, *J. Phys.: Conf. Ser.* **615**(1) (2015) 012016.

[9] V.V. Dodonov and A.V. Dodonov, Energy–time and frequency–time uncertainty relations: exact inequalities, *arXiv:1504.00862 [quant-ph]*, Apr. 2015.

[10] H.W. Fearing and S. Scherer, Field transformations and simple models illustrating the impossibility of measuring off-shell effects, *Phy. Rev. C* **62**(3) (2000) 034003.

[11] R.P. Feynman, Mathematical formulation of the quantum theory of electromagnetic interaction, *Phy. Rev.* **80**(3) (1950) 440–457.

[12] E. Fischbach, J.B. Buncher, J.T. Gruenwald, J.H. Jenkins, D.E. Krause, J.J. Mattes and J.R. Newport, Time-dependent nuclear decay parameters: new evidence for new forces? *Space Sci. Rev.* **145**(3–4) (2009) 285–335.

[13] V. Fock, Die Eigenzeit in der klassischen und in der Quantenmechanik, *Phys. Z. Sowjetunion* **12** (1937) 404–425.

[14] L. Gang, S. Jun, W. Bao-Xiang, J. Chao, B. Xi-Xiang, Z. Sheng, Z. Yong-Nan, Z. Sheng-Yun, Z. Li-Hua, L. Wei-Ping, L. Zhi-Hong, W. You-Bao, G. Bing, L. Yun-Ju and Q. Xing, Enhancement of β^+ -Decay Rate of ^{22} Na in Metal Pd at Low Temperature, *Chinese Phys. Lett.* **25**(1) (2008) 70–72.

[15] C.-O. Gullstrom, Low radiation fusion through bound neutron tunneling, Oct. 2014.

[16] S. Gulzari, J. Swain and A. Widom, Asymptotic infrared fractal structure of the propagator for a charged fermion, *Modern Phys. Lett. A* **21**(38) (2006) 2861–2871.

[17] P.L. Hagelstein and I.U. Chaudhary, Energy exchange in the lossy Spin–Boson model, *J. Condensed Matter Nucl. Sci.* **5** (2011) 52–71.

[18] P.L. Hagelstein and I.U. Chaudhary, Including nuclear degrees of freedom in a lattice hamiltonian, *J. Condensed Matter Nucl. Sci.* 7 (2012) 35–50.

[19] L. Horwitz, Spin, angular momentum and spin-statistics for a relativistic quantum many-body system, *J. Phys. A: Math. and Theoret.* **46**(3)(2013) 035305.

[20] L.P. Horwitz and R. Arshansky, On relativistic quantum theory for particles with spin 1/2, *J. Phys. A: Math. General* **15**(12)(1982) L659–L662.

[21] L.P. Horwitz and Y. Lavie, Scattering theory in relativistic quantum mechanics, *Phy. Rev. D* **26**(4) (1982) 819–838.

[22] L.P. Horwitz and C. Piron, Relativistic dynamics, *Helv. Phys. Acta* **46**(3) (1973) 316–326.

[23] L.P. Horwitz, C. Piron and F. Reuse, Relativistic dynamics for the spin 1/2 particle, *Helv. Phys. Acta* **48**(4) (1975) 546–548.

[24] J.R. Huizenga, *Cold Fusion: The Scientific Fiasco of the Century*, Oxford University Press, Oxford, New York, Jan. 1994.

[25] Y.E. Kim, Theory of Bose–Einstein condensation mechanism for deuteron-induced nuclear reactions in micro/nano-scale metal grains and particles, *Die Naturwissenschaften* **96**(7) (2009) 803–811.

[26] D.H. Kobe and V.C. Aguilera-Navarro, Derivation of the energy–time uncertainty relation, *Phy. Rev. A* **50**(2) (1994) 933–938.

[27] S.E. Koonin and M. Nauenberg, Calculated fusion rates in isotopic hydrogen molecules, *Nature* **339**(6227) (1989) 690–691.

[28] M.C. Land, Pre-Maxwell electrodynamics, *Foundations Phys.* **28**(9) (1998) 1479–1487.

[29] M.C. Land and L.P. Horwitz, Off-Shell Quantum Electrodynamics, *arXiv:hep-th/9601021*, Jan. 1996.

[30] G. Levi, E. Foschi, B. Hoistad, R. Pettersson, L. Tegner and H. Essen, Observation of abundant heat production from a reactor

device and of isotopic changes in the fuel, Technical report, Elforsk, Oct. 2014.

[31] B. Limata, F. Raiola, B. Wang, S. Yan, H.W. Becker, A. D'Onofrio, L. Gialanella, V. Roca, C. Rolfs, M. Romano, D. Schürmann, F. Strieder, and F. Terrasi, First hints on a change of the 22na βdecay half-life in the metal Pd, *Eur. Phy. J. A – Hadrons and Nuclei* **28**(2) (2006) 251–252.

[32] B.N. Limata, L. Gialanella, A.D. Leva, N.D. Cesare, A.D'Onofrio, G. Gyurky, C. Rolfs, M. Romano, D. Rogalla, C. Rossi, M. Russo, E. Somorjai and F. Terrasi, ^{7}Be radioactive beam production at CIRCE and its utilization in basic and applied physics, *Nucl. Instr. Methods Phys. Res. Sect. B* **266**(10) (2008) 2117–2121.

[33] G. Morchio and F. Strocchi, Infrared Problem, Higgs Phenomenon and Long Range Interactions, In G. Velo and A. S. Wightman (Eds.), *Fundamental Problems of Gauge Field Theory*, No. 141 in NATO ASI Series, Springer, New York, US, 1986, pp.301–344.

[34] Y. Nambu, The use of the proper time in quantum electrodynamics–I, *Prog. Theoret. Phy.* **5**(1) (1950) 82–94.

[35] T. D. Newton and E.P. Wigner, Localized states for elementary systems, *Rev. Modern Phy.* **21**(3) (1949) 400–406.

[36] H. Nikolic, Quantum mechanics: Myths and facts, *Foundations Phy.* **37**(11) (2007) 1563–1611, arXiv: quant-ph/0609163.

[37] R. Oerter, *The Theory of Almost Everything: The Standard Model, the Unsung Triumph of Modern Physics*, Plume, New York, London, reprint edition, Sept. 2006.

[38] T. O. Passell, The case for deuteron stripping with metal nuclei as the source of the Fleischmann–Pons excess heat effect, *J. Condensed Matter Nucl. Sci.* **15** (2014) 1–7.

[39] G. Preparata, *QED Coherence in Matter*, World Scientific, Singapore, Jan. 1995.

[40] O. Reifenschweiler, Reduced radioactivity of tritium in small titanium particles, *Phys. Lett. A* 184 (1994) 149–153.

[41] J. Schwinger, On gauge invariance and vacuum polarization, *Phy. Rev.* **82**(5) (1951) 664–679.

[42] J. Schwinger, *Infinite Energy Magazine*, Issue 1, 1995, p. 10.

[43] W. D. Smith, *The Energy-Time Uncertainty Principle*, CiteseerX, 1998.

[44] M. Srinivasan, G. Miley and E. Storms, Low-energy nuclear reactions: transmutations, In S.B. Krivit, J.H. Lehr and T.B. Kingery (Eds.), *Nuclear Energy Encyclopedia*, Wiley, New York, 2011, pp. 503–539.

[45] E. Storms, *Science of Low Energy Nuclear Reaction: A Comprehensive Compilation of Evidence and Explanations about Cold Fusion*, World Scientific, Singapore, July 2007.

[46] E. Stueckelberg, La signification du temps propre en mécanique ondulatoire, *Helv. Phys. Acta* **14** (1941) 322–323.

[47] E. Stueckelberg, Remarque à propos de la création de paires de particules en théorie de la relativité, *Helv. Phys. Acta* **14** (1941) 588–594.

[48] I. V. Tyutin, Once again on the equivalence theorem, *Phy. Atomic Nuclei* **65**(1) (2002) 194–202.

[49] M. Veltman, *Diagrammatica: The Path to Feynman Diagrams*, Cambridge University Press, Cambridge, First edition, July 1994.

[50] S. Weinberg, *The Quantum Theory of Fields, Vol. 3 Volume, Hardback Set*, Cambridge University Press, Cambridge, First edition, Feb. 2000.

[51] A. Widom and L. Larsen, Ultra low momentum neutron catalyzed nuclear reactions on metallic hydride surfaces, *Eur. Phys. J. C – Particles and Fields* **46**(1) (2006) 107–111.

J. Condensed Matter Nucl. Sci. 19 (2016) 56–67

Research Article

Quantum Tunneling in Breather 'Nano-colliders'

V.I. Dubinko*

NSC Kharkov Institute of Physics and Technology, 1 Akademicheskaya St., Kharkov 61108, Ukraine

Abstract

In many crystals with sufficient anharmonicity, special kinds of lattice vibrations, namely, discrete breathers (DBs) can be excited either thermally or by external triggering, in which the amplitude of atomic oscillations greatly exceeds that of harmonic oscillations (phonons). Coherency and persistence of large atomic oscillations in DBs may have drastic effect on quantum tunneling due to *correlation effects* discovered by Schrödinger and Robertson in 1930. These effects have been applied recently to the tunneling problem by Vysotskii et al., who demonstrated a giant increase of sub-barrier transparency during the increase of the correlation coefficient at special high-frequency periodic action on the quantum system. In the present paper, it is argued that DBs present the most natural and efficient way to produce correlation effects due to time-periodic modulation of the potential well width (or the Coulomb barrier width) and hence to act as breather 'nano-colliders' (BNC) triggering low energy nuclear reactions (LENR) in solids. The tunneling probability for deuterium (D–D) fusion in 'gap DBs' formed in metal deuterides is shown to increase with increasing numbers of oscillations by ~190 orders of magnitude, resulting in the observed LENR rate at extremely low concentrations of DBs. Possible ways of engineering the nuclear active environment based on the present concept are discussed.

Keywords: Correlation effects, Discrete breathers, Low energy nuclear reactions, Nuclear active sites

1. Introduction

The problem of tunneling through the Coulomb potential barrier during the interaction of charged particles is the key to modern nuclear physics, especially in connection with low energy nuclear reactions (LENR) observed in solids [1–4].

The tunneling (a.k.a. transmission) coefficient (TC) first derived by Gamow (1928) for a pure Coulomb barrier is the Gamow factor, given by

$$G \approx \exp\left\{-\frac{2}{\hbar}\int_{r_1}^{r_2} \mathrm{d}r\sqrt{2\mu\left(V\left(r\right)-E\right)}\right\}, \tag{1}$$

where $2\pi\hbar$ is the Planck constant, E is the nucleus CM energy, μ is the reduced mass, r_1, r_2 are the two classical turning points for the potential barrier, which for the D–D reaction are given simply by $\mu = m_{\mathrm{D}}/2$, $V(r) = e^2/r$.

*E-mail: vdubinko@hotmail.com

For two D's at room temperature with thermal energies of $E \sim 0.025$ eV, one has $G \sim 10^{-2760}$, which explains the pessimism about LENR and shows the need for some special conditions arising in solids under typical LENR conditions (D_2O electrolysis [1–3], E-cat [4], etc.), which help to overcome the Coulomb potential.

Corrections to the cross section of fusion due to the screening effect of atomic electrons result in the so-called "screening potential", which acts as additional energy of collision at the center of mass [5]. The screening potential was measured by the yields of protons or neutrons emitted in the D(d, p)T or D(d, n)^{3}He reactions induced by bombardment of D-implanted solid targets with deuterons accelerated to kinetic energies of several keV, equivalent to heating up to $\sim 10^7$ K [6]. However, even the maximum screening potentials found in Pt (675 eV), PdO (600 eV) and Pd (310 eV) are far too weak to explain LENR observed at the temperatures of these experiments, which are below the melting point of solids (in the E-cat) or boiling point of liquids (with electrolysis). Besides, the absence of significant radiation under typical LENR conditions indicates that other reactions should take place, based on interactions between 'slow' particles, which may be qualitatively different from the interactions between accelerated ones.

The most promising and universal mechanism of the stimulation of nuclear reactions running at low energy is connected with the formation of *coherent correlated states* of interacting particle, which ensures the high probability of the nuclear reactions under conditions, where the ordinary tunnel effect is negligible. These states minimize a more general uncertainty relation (UR) than Heisenberg UR usually considered in quantum mechanics, namely, Schrödinger–Robertson UR [7,8], which takes into account correlations between coordinate and momentum operators. Correlation effects have been applied to the tunneling problem by Dodonov et al. [9] and more recently by Vysotskii et al. [10–13] who demonstrated a giant increase of sub-barrier transparency (up to hundreds of orders of magnitude!) during the increase of correlation coefficient at special high-frequency periodic action on quantum system.

In this paper, we argue that such an action can be most naturally and effectively realized due to time-periodic modulation of the width of potential wells for atoms oscillating in the vicinity of discrete breathers (DBs). DBs are spatially localized large-amplitude vibrational modes in lattices that exhibit strong anharmonicity [14–23]. Due to the crystal anharmonicity, the frequency of atomic oscillations increase or decrease with increasing amplitude so that the DB frequency lies outside the phonon frequency band, which explains the weak DB coupling with phonons and, consequently, their robustness at elevated temperatures. DBs can be excited either thermally or by external driving, as was observed experimentally [17,18] and modelled in various physical systems [19–24].

Studies of DBs in three-dimensional crystals by means of molecular dynamics simulations using realistic interatomic potentials include ionic crystals with NaCl structure [16,19,20], diatomic A_3B crystals [21], graphene [22], semiconductors [23] and metals [24,25]. DBs in biopolymers such as protein clusters have been studied using the coarse-grained nonlinear network model [26].

Presently the interest of researchers has shifted to the study of the *catalytic impact* of DBs on the reaction rates in solids and on the biological functions of biopolymers [26,27]. The excitation of DBs in solids have been shown to result in a drastic amplification of the reaction rates in their vicinity. Two cases considered up to now include chemical reactions [27–30] and LENR [31]. In the former case, the amplification mechanism is based on modification of the classical Kramers escape rate from a potential well due to a periodic modulation of the well depth (or the reaction barrier height), which is an archetype model for chemical reactions since 1940 [27].

In the latter case, the so-called *gap DBs*, which can arise in diatomic crystals such as metal hydrides/deuterides (e.g. palladium deuteride, PdD), have been argued to be the LENR catalyzers [31]. The large mass difference between H or D and the metal atoms provides a gap in the phonon spectrum, in which DBs can be excited in the H/D sub-lattice resulting in time-periodic closing of adjacent H/D atoms, which should enhance their fusion probability. The main problem with this mechanism was that unrealistically small separation between atoms (~ 0.01 Å) must be attained in order to increase TC (Eq. (1)) by ~ 100 orders of magnitude required for a noticeable LENR rate at the best choice of parameters. Such distances are considerably smaller than the range of conventional chemical forces. However, this estimate did not take into account correlations between coordinate and momentum operators arising in a DB due to the cooperative nature of its dynamics. This problem is addressed in the present paper.

This paper is organized as follows. In Section 2, formation of coherent correlated states (CCS) under time-periodic action on a particle in the parabolic potential is reviewed [11–13]. In Section 3, this model is applied to the evaluation of the TC for the atoms oscillating in the vicinity of DBs and of the corresponding increase of their fusion rate. In Section 4, based on the rate theory of DB excitation under D_2O electrolysis and on the modified TC for D–D fusion in the PdD lattice, the fusion energy production rate is evaluated as a function of temperature, electric current and material parameters and compared with experimental data. The results are summarized in Section 5.

2. Formation of Correlated States in Non-stationary Potential Well

2.1. Schrödinger–Robertson UR and TC

The tunneling effect for nuclear particles is closely related to the uncertainty relation (UR), which determines the limits of the applicability of the classical and quantum descriptions of the same object. It appears that the well-known and widely used Heisenberg UR is a special case of a more general inequality, discovered independently by Schrödinger [7] and Robertson [8], which can be written in the following form [9]

$$\sigma_x \sigma_p \geq \frac{\hbar^2}{4\left(1-r^2\right)}, \qquad \sigma_x = \left\langle \left(x - \langle x \rangle^2\right) \right\rangle, \qquad \sigma_x = \left\langle \left(p - \langle p \rangle^2\right) \right\rangle, \tag{2}$$

where

$$r = \frac{\sigma_{xp}}{\sqrt{\sigma_x \sigma_p}}, \qquad \sigma_{xp} = \langle \hat{x}\hat{p} + \hat{p}\hat{x} \rangle / 2 - \langle x \rangle \langle p \rangle \tag{3}$$

is the *correlation coefficient* between the coordinate, x, and momentum, p. At $r = 0$ (non-correlated state) Eq. (2) is reduced to the Heisenberg UR, while in a general case, a non-zero r in the UR can be taken into account by the formal substitution

$$\hbar \to \hbar_{\mathrm{ef}} = \frac{\hbar}{\sqrt{1-r^2}}, \tag{4}$$

which leads to the formal shift of the border between the classical and quantum descriptions of the same object in the transition from non-correlated to correlated state [13].

Given this, a natural question arises: can nonzero correlations lead to real physical effects? If yes, then the most impressive consequence would be a dramatic increase of the tunneling probability, if a true Planck constant in Eq. (1) could be replaced by the effective parameter (4). This substitution was justified by Vysotskii et al. [10–13], who solved the non-stationary Schrödinger equation for the particle wave function $\psi(x,t)$ with account of correlation effects for a very low barrier transparency (tunneling probability) in the initially uncorrelated state $G_{r=0} \ll 1$ that corresponds to the condition $E \ll V_{\max}$:

$$G_{r\neq 0} \approx \exp\left\{ -\frac{2}{\hbar_{\mathrm{ef}}} \int_{r_1}^{r_2} \mathrm{d}r \sqrt{2\mu \left(V(r) - E\right)} \right\} = \left(G_{r=0}\right)^{\sqrt{1-r^2}}. \tag{5}$$

This approximation is within an order of magnitude of the result of the exact calculation of the potential barrier transparency using rigorous quantum-mechanical methods [10–13]. From Eq. (5), it follows that when a strongly correlated state with $|r| \to 1$ is formed, the product of the variances of the particle coordinate and momentum increases indefinitely, and the barrier becomes 'transparent': $G_{|r|\to 1} \to 1$ even if $E \ll V_{\max}$.

Although the substitution $\hbar \to \hbar_{ef}$ is not quite correct, it clearly demonstrates the high efficiency of using coherent correlated states in solving applied tunneling-related problems in the case of a high potential barrier and a low particle energy $E \ll V_{max}$, which is typical for LENR.

CCS can be formed in various quantum systems. The most natural way to form such state is to place a particle in a *non-stationary potential well.*

2.2. Formation of CCS under time-periodic oscillation of harmonic potential well

A model system considered by Vysotskii et al. [11–13], for the evaluation of the correlation coefficient, is a particle with the mass M, coordinate $x(t)$ and momentum $p(t)$ in a non-stationary parabolic potential well (i.e. non-stationary harmonic oscillator),

$$V(x,t) = M(x(t))^2 (\omega(t))^2 / 2 \tag{6}$$

for which a change of the eigen frequency $\omega(t)$was shown to result in an increase of $|r(t)|$. Several scenarios of time evolution $\omega(t)$ have been investigated [11–13], including its monotonic decrease or periodic modulation. The latter regime can be provided, e.g. at a constant well depth V_{max} and the potential well width $L(t)$ that changes periodically resulting in a time-periodic modulation of the eigen frequency as follows:

$$L(t) = L_0 (1 + g_\Omega \cos\Omega t), \qquad L_0 = \sqrt{8V_{max}/M\omega_0^2}, \tag{7}$$

where L_0 and ω_0 are the initial parameters of the well before the action of correlated forces, g_Ω and Ω are the modulation amplitude and frequency, respectively.

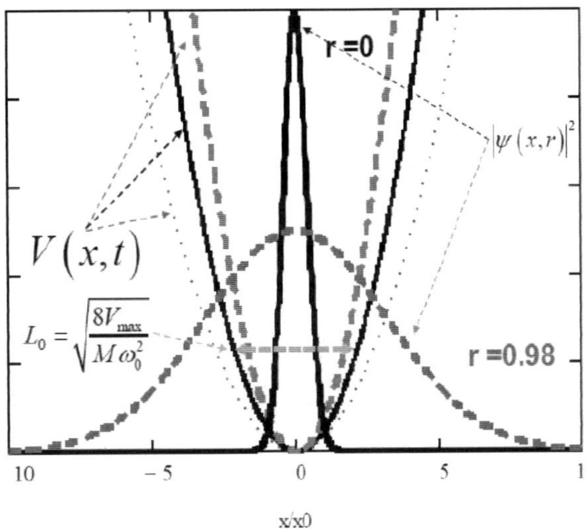

Figure 1. Sketch of the time-periodic parabolic potential $V(x,t)$ at different moments of time corresponding to the initial (solid black), minimal (dash red) and maximal (dot blue) well width. Probability density $|\psi(x,r)|^2$ for the particle localization in the well and in the sub-barrier region is shown schematically for uncorrelated state $r = 0$ (solid black) and for strongly correlated state $r = 0.98$ at the times of the maximal coordinate dispersion (dash green) [10]. $x_0 = \sqrt{\hbar/M\omega_0}$ is the half-width of the particle localization in the unperturbed well.

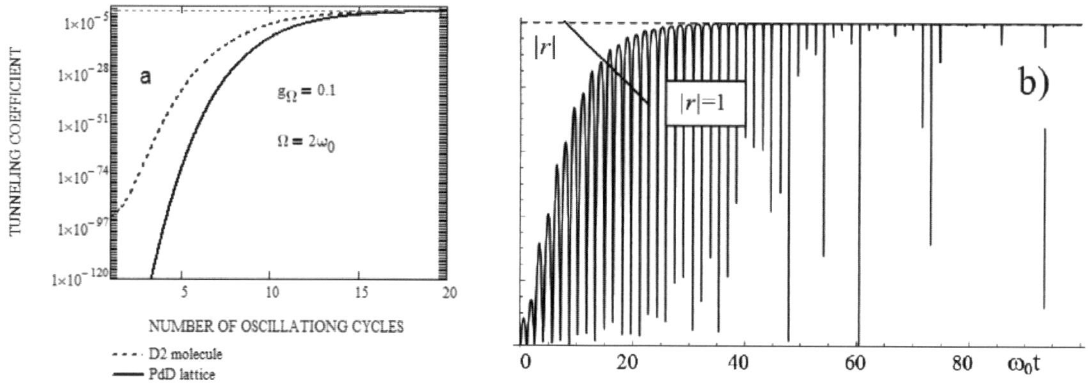

Figure 2. (a) Tunneling coefficient increase with increasing number of the well modulation cycles, $n = \omega_0 t/2\pi$, evaluated by Eq. (8) for the correlation coefficient amplitude increasing as shown in **(b)** at $\Omega \approx 2\omega_0$, $g_\Omega = 0.1$ [13]. D–D equilibrium spacing in a D_2 molecule ($L_0 = 0.74$ Å) and in the PdD crystal ($L_0 = 2.9$ Å).

Figure 1 shows that the probability density $|\psi(x,r)|^2$ for the particle localization in the time-periodic well is very narrow for uncorrelated state $r = 0$ (solid black), while it spreads significantly into the sub-barrier region for the strongly correlated state $r = 0.98$ at the times of the maximal coordinate dispersion (dash green) [10].

From a detailed analysis [11–13] it follows that the process of formation of strongly correlated coherent state with $|r|_{max} \to 1$ in response to the action of limited periodic modulation (Eq. (10)) is possible only at any of two conditions: (i) $\Omega = \omega_0$ (resonant formation) or (ii) Ω is close to $2\omega_0$ (parametric formation): $|\Omega - 2\omega_0| \leq g_\Omega \omega_0$.

Figure 2 shows the evolution of the correlation coefficient in time under the action of the harmonic perturbation with a parametric frequency $\Omega = 2\omega_0$ at $g_\Omega = 0.1$. The correlation coefficient oscillates with time but its amplitude $|r|_{max}$ increases monotonically with the number of modulation cycles, $n = \omega_0 t/2\pi$, resulting in a giant increase of the tunneling coefficient, as demonstrated in Fig. 3, which shows the TC evaluated by Eq. (8) that takes into account both the electron screening [31] and the correlation effects [13]:

$$G^*(L,r) = \exp\left\{-\frac{2\pi e^2}{\hbar_{ef}(r)}\sqrt{\frac{\mu}{2\left(E + e^2/L\right)}}\right\}, \tag{8}$$

where L is the minimum equilibrium spacing between D atoms determined by electron screening, E is their kinetic energy (~eV/40 at room temperature) $\ll$ screening energy $\sim e^2/L$. One can see that the difference in electron screening and the corresponding initial D–D distances in a D_2 molecule ($L_0 = 0.74$ Å) and in the PdD crystal ($L_0 = 2.9$ Å) leads to a huge tunneling difference in the initial (uncorrelated) state, in which TC is negligible in both cases. However, with increasing number of modulation cycles, the correlation coefficient amplitude increases according to Fig. 2 resulting in a giant increase of TC up to ~ 1 in dozens of cycles for parametric formation $\Omega \approx 2\omega_0$, which does not require an exact coincidence of the frequencies [13].

The most important and non-trivial practical question now is: how to realize such a periodic action at the atomic scale? Modulation of the frequency of the optical phonon modes via excitation of the surface electron plasmons by a terahertz laser suggested in [13] as a driving force for the CCS formation is very questionable [31], and it does not explain LENR observed in the absence of laser driving. In Section 3, we will consider a new mechanism based on the large-amplitude time-periodic oscillations of atoms naturally occurring in discrete breathers.

3. Breather-induced Time-periodic Action on the Potential Barrier

In order to develop a mechanism for DB-based LENR in metal hydrides/deuterides (e.g. PdD or NiH), let us consider their crystal structure in more detail. At ambient conditions, Ni/Pd hydrides/deuterides crystallize in FCC structure with the space group of the Rock-salt structure, which is called Fm3m in Hermann–Mauguin notation [32,33].

Molecular dynamic (MD) simulations have revealed that diatomic crystals with Morse interatomic interactions typically demonstrate *soft type* of anharmonicity [19], which means that DB's frequency decreases with increasing amplitude, and one can expect to find the so-called gap DBs with a frequency within the phonon gap of the crystal. The large mass difference between H or D and the metal atoms is expected to provide a wide gap in phonon spectrum (Fig. 3), in which DBs can be excited, e.g. by thermal fluctuations at elevated temperatures, as demonstrated by Kistanov and Dmitriev [20] for the different weight ratios and temperatures. The density of phonon states (DOS) of the NaCl-type crystal for the weight ratio m/M = 0.1 at temperatures ranging from 0 to 620 K is shown in Fig. 3 (a)–(d).

Figure 3 (e) shows that DOS for PdD and PdH measured experimentally are qualitatively similar to DOS calculated for the NaCl-type crystal. First-principle calculations [33] point out that phonon spectra in PdD and PdH are strongly renormalized by anharmonicity. The appearance of two additional broad peaks in the DOS of NaCl at elevated temperatures (starting from T = 310 K) can be seen in Fig. 3 (c), (d). One of them is in the gap of the phonon spectrum, while another one lies above the phonon spectrum. The appearance of the peak *in the gap* of the phonon spectrum can be associated with the spontaneous excitation of gap DBs at sufficiently high temperatures, when nonlinear terms in the expansion of interatomic forces near the equilibrium atomic sites acquire a noticeable role. As the temperature increases, the lifetime and concentration of gap DBs in the *light atom sub-lattice* increase [21]. The dynamic structure of gap DBs has been revealed in [19] where they have been excited simply by shifting one *light atom* or two neighboring light atoms from their equilibrium positions while all other atoms were initially at their lattice positions and had zero initial velocities.

There are two main peculiarities of DBs related to the formation of coherent correlated states, namely, oscillations of atoms comprising a DB are (i) *time-periodic* and (ii) *coherent*, i.e. they have different amplitudes but the same or

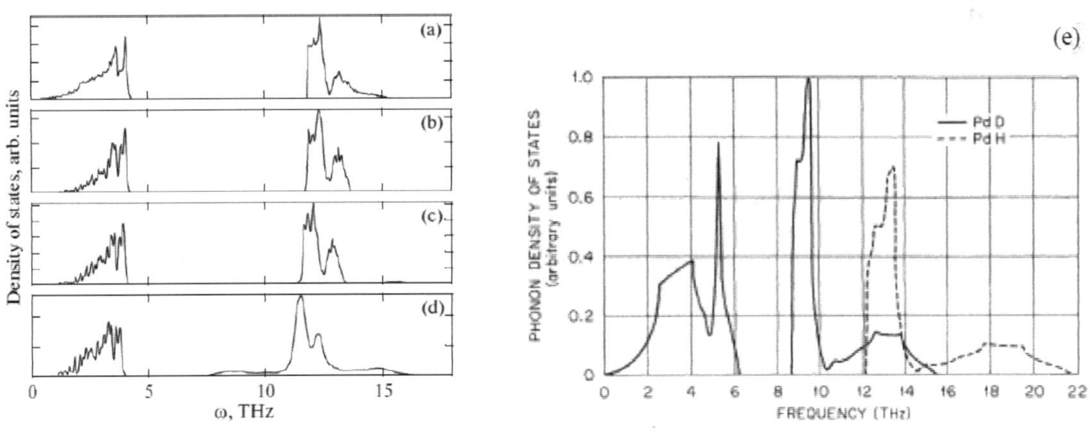

Figure 3. DOS of the NaCl-type crystal for the weight ratio $m/M = 0.1$ at temperatures T =(a) 0, (b) 155, (c) 310, and (d) 620 K. (e) DOS for PdD and PdH crystals based on the force constants [33] fitted to the experimental results for $PdD_{0.63}$ assuming that the forces in PdD and $PdD_{0.63}$ were identical. Reproduced from [20,33] Copyright by APS.

commensurate frequency. In the extreme case of DB localized at one light atom, it oscillates with a large amplitude, A in the anharmonic potential well, which determines its frequency Ω as follows [19]

$$\Omega(A) = \sqrt{\alpha + \frac{3}{4}\beta A^2}, \quad \alpha = \frac{2\gamma_1}{m}, \quad \beta = \frac{2\gamma_2}{m}, \quad \gamma_1(A) = R_1 A + S_1, \quad \gamma_2(A) = R_2 A + S_2, \tag{9}$$

where α determines the quasi-harmonic eigenfrequency of the potential well, and β describes its anharmonicity. Both of them depend on the DB amplitude, since it changes the force constants $\gamma_{1,2}$ of the potential. β is positive, which corresponds to hard type of non-linearity with frequency increasing with A. However, the central atom oscillating with large amplitude shifts positions of neighboring atoms so that α decreases with A resulting in the observed decrease of $\Omega(A)$[19].

Let us apply this model to a DB in the PdD lattice, dispersion curves of which [33] are shown in Fig. 3(e). Figure 4(a) shows DB$\langle 110 \rangle$ frequency Ω and eigen frequencies of (quasi-harmonic) potential wells for neighboring D atoms, ω_0 as functions of the DB amplitude, evaluated with the force constants assumed to fit dispersion curves of $PdD_{0.63}$ (Fig. 3 (e)):

$$\omega_0(A) = \sqrt{\frac{2\gamma_1(A)}{m}}, \quad \gamma_1(A) = R_1 A + S_1, \tag{10}$$

$$R_1 = -0.026\ \text{eV}/\text{Å}^3, \quad S_1 = 0.008\ \text{eV}/\text{Å}^2, \quad R_2 = -0.017\ \text{eV}/\text{Å}^5, \quad S_2 = 0.035\ \text{eV}/\text{Å}^4. \tag{11}$$

The frequency of the optic modes at the zone center ($\sim$9 THz), which determines the maximum DB frequency, is low in comparison with other hydrides, thus implying a weak nearest-neighbor Pd–D force constant in $PdD_{0.63}$ [33].

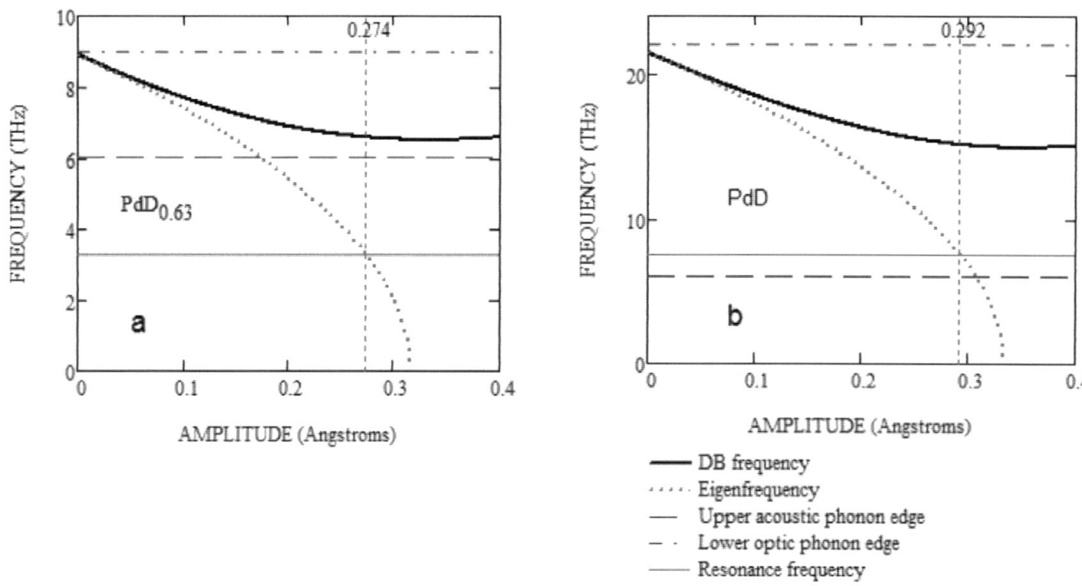

Figure 4. DOS of the NaCl-type crystal for the weight ratio $m/M = 0.1$ at temperatures T =(a) 0, (b) 155, (c) 310, and (d) 620 K. (e) DOS for PdD and PdH crystals based on the force constants [33] fitted to the experimental results for $PdD_{0.63}$ assuming that the forces in PdD and $PdD_{0.63}$ were identical. Reproduced from [20,33] Copyright by APS.

Figure 4(b) shows $\Omega(A)$ and $\omega_0(A)$ evaluated with the force constants fitted to raise the lower optic phonon edge for stoichiometric PdD up to 22 THz and to broaden the phonon gap accordingly:

$$R_1 = -0.143\ \mathrm{eV}/\mathrm{\AA}^3, \qquad S_1 = 0.048\ \mathrm{eV}/\mathrm{\AA}^2, \qquad R_2 = -0.013\ \mathrm{eV}/\mathrm{\AA}^5, \qquad S_2 = 0.143\ \mathrm{eV}/\mathrm{\AA}^4. \tag{12}$$

In the $PdD_{0.63}$ case (Fig. 4 (a)), increase of the DB amplitude up to 0.274 Å leads to the excitation in neighboring wells of the harmonic with the frequency $\sim$ 3.3 THz equal to half of the main DB frequency ($\sim$6.6 THz), which interacts with acoustic phonons below the gap and makes the DB unstable, similar to the NaCl case for $m/M > 0.2$ [19].

In the PdD case (Fig. 4(b)), increase of the DB amplitude up to the critical value $A_{\mathrm{cr}} \approx 0.292$ Å leads to the excitation in neighboring wells of the harmonic with the frequency $\sim$7.5 THz equal to half of the main DB frequency ($\sim$15 THz), which lies above the upper acoustic phonon edge and does not interact with phonons. Such DBs are stable, and they lead to the parametric formation of CCS of deuterons in the neighboring quasi-harmonic potential wells subjected to time-periodic modulation of their eigenfrequencies $\omega_0(A_{\mathrm{cr}}) \approx 7.5$ THz by the DB frequency $\Omega(A_{\mathrm{cr}}) \approx 15$ THz. As a result of such modulation, D–D fusion is expected to occur in several dozens of DB cycles (Fig. 2(a)) since the modulation amplitude

$$g_\Omega \approx A_{\mathrm{cr}} \Big/ \left(a_{\mathrm{PdD}} \sqrt{2} \big/ 2 \right) \approx 0.1.$$

Three main resonances of excess energy released under joint action of two low-power laser beams with variable beat frequency ranging from 3 to 24 THz at $\sim 8 \pm 1$ THz, 15 ± 1 THz and 21 ± 1 THz [34] correlate in our model with the DB-induced harmonic frequency, $\omega_0(A_{\mathrm{cr}})$ $\approx$7.5 THz, DB parametric frequency $\Omega(A_{\mathrm{cr}})$ $\approx$15 THz and DB initial frequency, 21 THz, respectively.

4. LENR Rate under Heavy Water Electrolysis

The DB excitation occurs by thermal fluctuations and by external driving displacing atoms from equilibrium positions. The rate of thermal excitation of DBs having energy E is given by the Arrhenius law [27,31]

$$K_{\mathrm{DB}}^{\mathrm{th}}(E) = \omega_{\mathrm{DB}} k_{\mathrm{DB}}^{\mathrm{ef}} \exp\left(-\frac{E}{k_{\mathrm{B}} T} \right), \tag{13}$$

where k_{B} is the Boltzmann constant, T is the temperature, and $\omega_{\mathrm{DB}} \approx \Omega(0)$ is the attempt frequency that should be close to the edge of the phonon band, from which DBs are excited. In the case of gap DBs under consideration, it is about 21 THz (Fig. 4(b)). $k_{\mathrm{DB}}^{\mathrm{ef}}$ is the efficiency coefficient for the excitation of DBs with parameter fitting parametric resonance conditions.

External driving of the DB excitation can be provided by knocking of surface atoms out of equilibrium position by energetic ions or molecules under non-equilibrium deposition of deuterium under electrolysis. This produces focusing collisions and moving DBs (a.k.a. quodons) that can transfer vibration energy in the crystal bulk [25]. The amplitude of the quasi-periodic energy deviation of atoms along the quodon pathway, V_{ex}, can reach almost 1 eV with the excitation time, τ_{ex}, of about 10 oscillation periods, which results in the amplification of the DB generation rate proportional to the electric current density J [31]:

$$K_{\mathrm{DB}}^{J}(E) = K_{\mathrm{DB}}^{\mathrm{th}}(E)\left(1 + \left\langle I_0\left(\frac{V_{\mathrm{ex}}}{k_{\mathrm{B}} T} \right) \right\rangle \omega_{\mathrm{ex}} \tau_{\mathrm{ex}} \right), \qquad \omega_{\mathrm{ex}}(F_{\mathrm{q}}) = F_{\mathrm{q}} b^2 \frac{3 l_{\mathrm{q}}}{R_{\mathrm{P}}}, \qquad F_{\mathrm{q}} = \frac{J}{2e}, \tag{14}$$

where ω_{ex} is the mean number of excitations per atom per second caused by the flux of quodons, e is the electron charge, b is the atomic spacing. The product $F_{\mathrm{q}} b^2$ is the frequency of the excitations per atom within the layer of a thickness l_q equal to the quodon propagation range, while the ratio $3l_{\mathrm{q}}/R_{\mathrm{P}}$ is the geometrical factor that corresponds to the relative number of atoms within the quodon range in a PdD particle of a radius R_{P}. The coefficient of proportionality between F_q and the electron flux J/e assumes that each electrolytic reaction that involves a pair of electrons releases a vibrational energy of ~1 eV, which is sufficient for generation of one quodon with energy $V_{ex} < 1$ eV.

Multiplying the DB generation rate (14) by the tunneling probability in a DB, $G^*(L,r)$(8) and integrating over DB energies one obtains the D–D fusion rate per PdD unit cell:

$$FuR = \frac{1}{\Delta E} \int_{E^*_{DB}-\Delta E}^{E^*_{DB}+\Delta E} K^J_{DB}(E)\, G^*(r)\, dE \tag{15}$$

that dramatically depends on the correlation coefficient, r, which, in its turn, strongly depends on the DB amplitude ~ DB energy and the number of DB cycles before decay, n_{DB}.

Only a small fraction of DBs can form CCS in their vicinity and act as effective *breather nano-colliders* (BNC). They must have some particular energies, $E^*_{DB} \pm \Delta E$, in order to cause the parametric resonance producing CCS. If $\Delta E \ll k_B T$ Eq. (15) is reduced to

$$FuR \approx 2K^J_{DB}(E^*_{DB})\, G^*(r(E^*_{DB}, n_{DB})) \underset{n_{DB}>n^*_{DB}}{\longrightarrow} K^J_{DB}(E^*_{DB}), \qquad G^*(r(E^*_{DB}, n^*_{DB})) \approx \frac{1}{2}, \tag{16}$$

where the number of DB cycles required to make the Coulomb barrier 'transparent', $n^*_{DB} \approx 100$ at $g_\Omega = 0.1$.

We consider the following reaction [2]: D + D $\rightarrow$ ^{4}He + 23.8 MeV$_{lattice}$, which is based on the experimentally observed production of excess heat correlated with production of "nuclear ash", i.e. ^{4}He [2,3]. Multiplying the DB-induced fusion rate (16) by the energy $E_{D\text{-}D} = 23.8$ MeV, produced in D–D fusion one obtains the excess energy production rate per atom, $P_{D\text{-}D}$ as a function of temperature and electric current: $P_{D\text{-}D}(T,J) = K^J_{DB}(E^*_{DB}, T, J)\, E_{D\text{-}D}$. Usually, the output power density is measured per unite surface of a macroscopic cell, $P^S_{D\text{-}D}$, as a function of the electric current density at a fixed temperature and at temperature increasing with J, as illustrated in Fig. 5. This is given by the product of $P_{D\text{-}D}$, the number of atoms per unit volume, $1/\upsilon_{PdD}$(υ_{PdD}) being the atomic volume of PdD) and the ratio of the cell volume to the cell surface:

$$P^S_{D\text{-}D}(T,J) = P_{D\text{-}D}(T,J)\, \frac{L_S}{\upsilon_{PdD}}, \tag{17}$$

where L_S is the cell size, if cubic, or thickness, in case of a plate .

Figure 5 shows the LENR output power density DBs as a function of electric current density and temperature evaluated by Eq. (17) assuming material parameters listed in Table 1. Comparison with experimental data shows that the present model describes quantitatively the observed linear dependence of $P^S_{D\text{-}D}$ on the current density at a constant temperature as well as the deviation from the linear dependence, if temperature increases with increasing electric current density. Thermally activated nature of the reactions leading to LENR has been noted for quite a long time [3], and the activation energy was estimated in some cases to be ~0.65 eV. The present model not only explains these observations, but also reveals that the underlying physics is a consequence of the synergy between thermally activated and externally driven mechanisms of the DB excitation in deuterated palladium.

5. Conclusions and Outlook

The main message of this paper is that DBs present the most efficient way to produce CCS due to time-periodic modulation of the potential well width (or the Coulomb barrier width) and hence to act as BNC triggering LENR in solids. The BNC concept proposed in a previous work [31] did not take into account correlation effects, and hence, unrealistically small separation between atoms (~ 0.01 Å) would have to be attained in order to enhance the LENR rate up to a noticeable level. Figure 2 demonstrates effect of CCS in the BNC model manifested by a number of DB

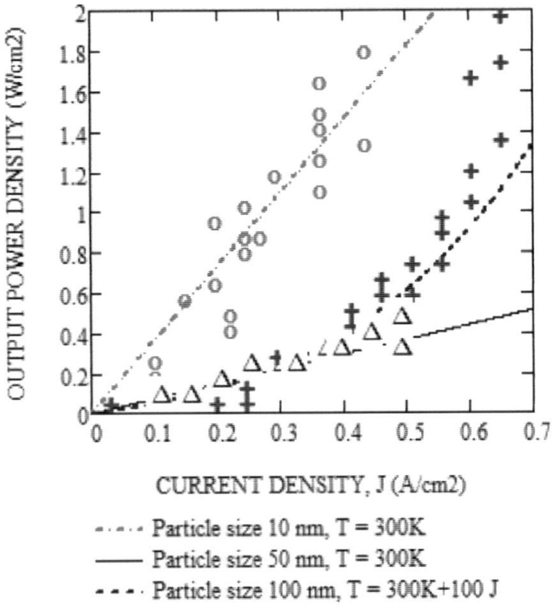

Figure 5. LENR output power density according to the model Eq. (17) and material parameters listed in the table as a function of electric current density at constant T and T increasing with J as $T = 300\text{K} + 100J$. Experimental data [3] Fig. 42: **o** - Bush (constant T), Δ – Storms (constant T), + Aoki et al. (T increasing with J).

cycles required to produce experimentally observed LENR rate ~1 W/cm^2. It can be seen that in the modified model, the DB lifetime plays a much more important role than the tunneling D–D spacing, and that DB amplitude of several fractions of an angstrom is sufficient to produce the required effect, if CCS parametric conditions are met.

The present model describes the observed linear dependence of the excess power output on the current density under heavy water electrolysis at a constant temperature, as well as its exponential increase with increasing temperature, which can be the basic LENR mechanism in the hot CAT-type installations.

Table 1. Material parameters used for the plot in Fig. 5.

Parameter	Value
D–D spacing in PdD, b (Å)	0.29
DB parametric amplitude, A_{cr} (Å)	0.292
DB parametric energy, E^*_{DB} (eV)	1
DB initial frequency, $\Omega(0)$ (THz)	21
DB parametric freq, $\Omega(A_{cr})$ (THz)	15
Harmonic frequency $\omega_0(A_{cr})$ (THz)	7.5
DB lifetime, $\tau^*_{DB} = n^*_{DB}/\omega_{DB}$(s)	6×10^{-12}
DB excitation efficiency, k^{ef}_{DB}	4×10^{-11}
Quodon excitation energy V_{ex} (eV)	0.8
Quodon excitation time, τ_{ex}(s)	6×10^{-13}
Quodon range, $l_q = 10b$ (nm)	2.9
Cathode size/thickness (mm)	5

The proposed mechanism of CCS formation near the gap DBs requires a sufficiently broad phonon gap that is not observed below the critical D loading ~ 0.83 examined so far. Our hypothesis is that mechanical stresses arising in a fine powder of PdD_x above the critical loading $x > 0.83$ can make the phonon band similar to that shown in Fig. 4(b), thus switching on the DB-induced formation of CCS. Further investigations of DOS and DBs in the extreme conditions of LENR are required.

An alternative mechanism of the DB-induced CCS formation may involve high frequency (hard type) DBs, manifested by the peak *above* the phonon spectrum in NaCl type crystals. Atomic modeling of DBs of various types in metal hydrides/deuterides is an important outstanding problem since it may offer ways of *engineering* the nuclear active environment.

Acknowledgements

The author is grateful to Sergey Dmitriev and Vladimir Vysotskii for helpful discussions and valuable criticism.

References

[1] M. Fleischmann, S. Pons and M. Hawkins, *J. Electroanal. Chem.* **261** (1989) 301.

[2] M. McKubre, F. Tanzella, P. Hagelstein, K. Mullican and M. Trevithick, The need for triggering in cold fusion reactions, in *10th Int. Conf. on Cold Fusion*, MA, Cambridge, 2003, LENR-CANR.org.

[3] E.K Storms, *The Science of Low Energy Nuclear Reaction*, World Scientific, Singapore, 2007.

[4] A.G. Parkhomov, *Int. J. Unconventional Sci.* **7**(3) (2015) 68.

[5] H.J. Assenbaum, K. Langanke amd C. Rolfs, *Z. Phys.* A **327** (1987) 461.

[6] J. Kasagi, Screening potential for nuclear reactions in condensed matter, in *ICCF-14 Int. Conf. on Condensed Matter Nucl. Sci.*, 2008.

[7] E. Schrödinger, *Ber. Kgl. Akad. Wiss.*, Berlin, S. (1930) 296.

[8] H.P. Robertson, *Phys. Rev.* **34** (1930) 163.

[9] V.V. Dodonov and V.I. Man'ko, *Phys. Lett.* A **79**(2/3) (1980) 150.

[10] V.I. Vysotskii and S.V. Adamenko, *J. Tech. Phys.* **55** (2010) 613.

[11] V.I. Vysotskii, M.V. Vysotskyy and S.V. Adamenko, *J. Exp. Theoret. Phys.* **141** (2012) 276.

[12] V.I. Vysotskii, S.V. Adamenko and M.V. Vysotskyy, *J. Exp. Theoret. Phys.* **142** (2012) 627.

[13] V.I. Vysotskii and M.V. Vysotskyy, *Euro. Phys. J.* A **49**(8) (2013) 1–12, DOI 10.1140/epja/i2013-13099-2.

[14] A.J. Sievers and S. Takeno, *Phys. Rev. Lett.* **61** (1988) 970.

[15] S. Flach and A.V. Gorbach, *Phys. Rep.* **467** (2008) 1.

[16] V. Hizhnyakov, D. Nevedrov and A.J. Sievers, *Physica* B, **316–317** (2002) 132.

[17] M.E. Manley, A.J. Sievers, J.W. Lynn, S.A. Kiselev, N.I. Agladze, Y. Chen, A. Llobet and A. Alatas, *Phys. Rev.* **B** 79 (2009) 134304.

[18] M.E. Manley, *Acta Materialia* **58** (2010) 2926.

[19] L.Z. Khadeeva and S.V. Dmitriev, *Phys. Rev.* B **81** (2010) 214306.

[20] AA. Kistanov and S.V. Dmitriev, *Phys. Solid State* **54** (2012) 1648.

[21] L.Z. Khadeeva and S.V. Dmitriev, *Phys. Rev.* B, **84** (2011) 144304.

[22] S.V. Dmitriev, A.P. Chetverikov and M.G. Velarde, *Physica Status Solidi* (b) **54** (2015) 1648.

[23] N.K. Voulgarakis, G. Hadjisavvas, P.C. Kelires and G.P. Tsironis, *Phys. Rev.* B **69** (2004) 113201.

[24] M. Haas, V. Hizhnyakov, A. Shelkan, M. Klopov and A.J. Sievers, *Phys. Rev.* B **84** (2011) 14430.

[25] D. Terentyev, A. Dubinko, V. Dubinko, S. Dmitriev, E. Zhurkin and M. Sorokin, Interaction of discrete breathers with pr imary lattice defects in bcc Fe, *Modelling Simul. Mater. Sci. Eng.* **23** (2015) 085007-085030.

[26] F. Piazza and Y.H. Sanejouand, *Phys. Biol.* **5** (2008) 026001.

[27] V.I. Dubinko, P.A. Selyshchev and J.F.R. Archilla, *Phys. Rev.* E **83**(4) (2011) 041124.

[28] V.I. Dubinko, F.M. Russell, *J. Nucl. Materials* **419** (2011) 378.

[29] V.I. Dubinko and A.V. Dubinko, *Nucl. Inst. and Methods in Phys. Res.* B **303** (2013) 133.
[30] V.I. Dubinko and F. Piazza, *Lett. Materials* **4** (4) (2014) 273.
[31] V.I. Dubinko, *J. Condensed Matter Nucl. Sci.* **14** (2014) 87.
[32] J.M. Rowe, J.J. Rush, H.G. Smith, M. Mostoller and H.E. Flotow, *Phys. Rev. Lett.* **33** (1974) 1297.
[33] I. Errea, M. Calandra1 and F. Mauri, *Phys. Rev. Lett.* **111** (2013) 177002.
[34] P.L. Hagelstein, D. Letts and D. Cravens, *J. Condensed Matter Nucl. Sci.* **3** (2010) 59–76.

J. Condensed Matter Nucl. Sci. 19 (2016) 68–87

Research Article

Final Report on Calorimetry-based Excess Heat Trials using Celani Treated NiCuMn (Constantan) Wires

Arik El-Boher*, William Isaacson, Orchideh Azizi, Jinghao He, Dennis Pease and Graham Hubler

The Sidney Kimmel Institute for Nuclear Renaissance (SKINR), Department of Physics and Astronomy, University of Missouri, Columbia, MO, USA

Abstract

Sensitive mass flow calorimetry was used in a series of tests to evaluate eight treated NiCuMn (Constantan) wires in a gas loading cell. The goal of this testing was to reproduce and confirm the production of excess heat when using an experimental setup similar to that demonstrated in 2013 by Dr. Francesco Celani at National Instrument's NI week and ICCF 17. Six of the eight Constantan wires used in the SKINR tests were provided by Dr. Celani and the remaining two were obtained from Mathieu Valat. Initial tests in the series closely followed the set-up, operation, and heating protocols used by Dr. Celani with a significantly modified stainless test cell to enable the use of mass flow calorimetry. Later tests used the same modified test cell but added additional heating protocols using pulsed or highly modulated electric currents (SuperWaves) to enhance hydrogen loading and create additional thermal gradients within the Constantan wires. No excess heat was observed in a calorimeter of a sensitivity <10 mW when running either the initial or later test protocols during $\sim$ 200 days of testing.

Keywords: Calorimetry, Gas loading, Hydrogen, Micro-nano surface, Nickel, Nickel hydride

1. Introduction

In 2005 Baranowski and Filipek summarized 45 years of Nickel Hydride history and prospective [1]. Although palladium hydride was discovered by Graham in 1866 [2], it took almost 100 years to prove experimentally that nickel is also a hydride forming metal [3–4]. The limited penetration depth at room temperature and the fast kinetics of hydrogen desorption at normal conditions defines the Ni–H interactions [3–5]. Experiments of charging nickel of different layer thicknesses (from few micro-meters to more than 100 μm) deposited on copper wires, demonstrated that hydrogen is mainly located near the wire. These experiments showed clearly that hydrogen penetrates only about 30 μm into the bulk, below the surface [6].

*Corresponding author. E-mail: elbohera@missouri.edu

© 2016 ISCMNS. All rights reserved. ISSN 2227-3123

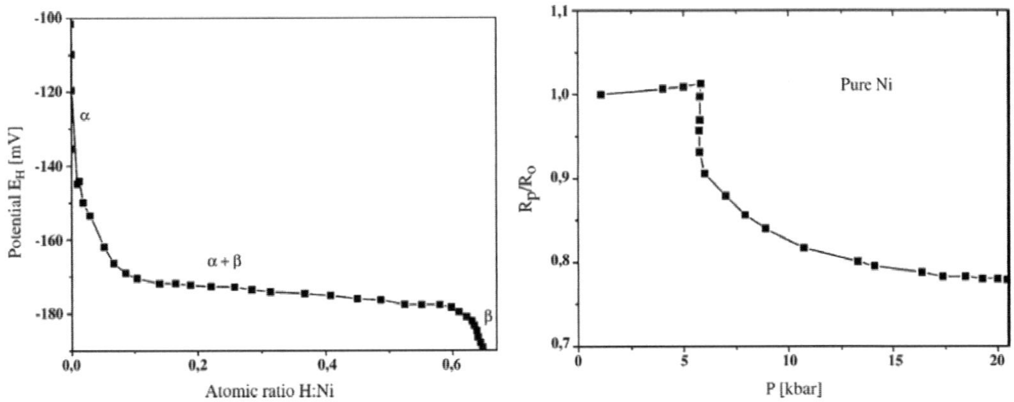

Figure 1. Data from Ref. [1].

Baranowski and Filipek [1] showed that the d-band of nickel is at least partially filled by the electrons from hydrogen during hydrogenation. Similarly, as in the Pd–H system, this filling of the d-band vacancy has radical consequences for electronic and magnetic properties of the hydride formed. For example, this results in a strong decrease of the electrical resistance, due to the reduction of the electron scattering at the d-band vacancy or/and to a decrease of the electron–phonon coupling.

On the right-hand side of Fig. 1, as shown by Baranowski [1], the relative electrical resistance ratio of a nickel sample as a function of gaseous hydrogen pressure at 25°C is presented. Before the hydride phase starts to form – that is below 6 kbar of gaseous hydrogen – the electrical resistance ratio rises a few percent above the initial value, due to the uptake of some hydrogen in the range of the α-phase; the dissolved hydrogen forms new scattering centers for the conduction electrons. As soon as the hydride phase is formed, the resistivity is reduced by more than 20%, becoming more metallic than in the hydrogen-free nickel. Even more radical are the changes of the magnetic properties of nickel. The ferromagnetism is lost, when going over to the hydride phase. In further electrochemical studies the electrode potential of the nickel cathode was measured as a function of the atomic ratio H/Ni of the nickel layer involved. The figure on the left presents such an example. The electrode potential measured was that of the cathode, taking the normal hydrogen electrode as reference. Clearly, three-phase regions are evident, α-phase for H/Ni atomic ratio ranging from 0 to below 0.1, the mixed $\alpha + \beta$ phase (plateau region) for H/Ni ratios up to about 0.6 and the β-phase region, in which a small change in H/Ni ratio is accompanied by steep change of electrode potential.

Starting in February 2011, Celani [7] studied the feasibility of new Nickel based alloys that are able to absorb Hydrogen (H_2) and/or Deuterium (D_2) and that have, in principle a possibility to generate anomalous heat effects at temperatures > 100°C.

Reports by Focardi and Piantelli [8], Miley 1996 [9] and, claims by A. Rossi and (later on) by Defkalion Company led Celani to investigate the family of NiCu alloys under high temperature H_2 gas loading. Due to theoretical considerations in a paper on catalysis (not related to LENR studies) by Romanowski et al. [10], Celani decided to explore the "large family" of CONSTANTAN alloys.

In the selection of materials, the figure of merit was the ability to decompose H_2 into H. A Constantan alloy ($Ni_{37}Cu_{63}$), among the materials studied, has the highest decomposition value (i.e. 3.2 eV); in comparison, pure Ni and Pd have values of 1.74 and 0.42 eV, respectively. Even if the alloy composition is changed to $Ni_{62}Cu_{38}$, the decomposition value remains almost constant (2.86 eV).

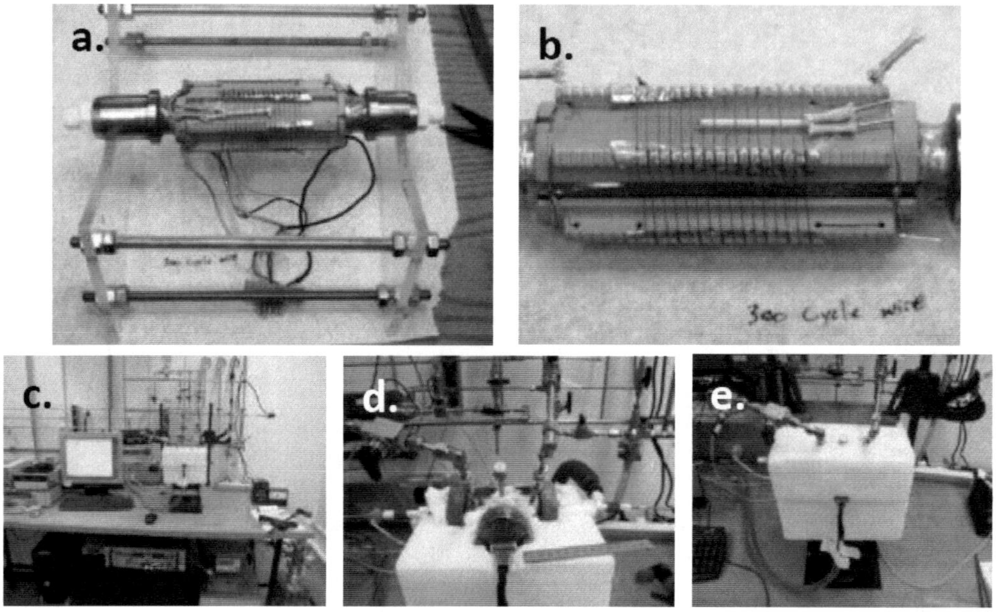

Figure 2. SKINR Test Cell.

Therefore Celani focused on a commercial (low cost) material, called ISOTAN44, with an alloy composition $Cu_{55}Ni_{44}Mn_1$, developed many years ago by Isabellenhutte Heusler, GmbH, KG-Germany. The ISOTAN 44 was selected according to the following, overall, considerations: high diffusion coefficient, low cost, good mechanical properties, high decomposition values-catalytic power (ΔE, in eV) and the ability to produce nano-micro structures on the wire's surface.

In NI week and ICCF17 Celani presented a working borosilicate Schott Duran glass cell with ISOTAN 44 wire wound on a MICA core which according to Celani's reports shows anomalous heat with a high reproducibility rate for the whole duration of the conferences. That experiment attracted the interest of SKINR which was committed to conduct an accurate calorimetric replication experiments with Celani's treated wires.

2. SKINR Test Cell and Experimental Facility

The experimental cell, as seen in Figs. 2(a)–(e) and 3, is made of a T-shaped stainless steel pressure chamber having inner diameter of 70 mm and a total length of 126 mm. Two horizontal opening of 40 mm in diameter are sealed with stainless steel flanges pressed by brass threaded nuts, against Viton O-rings. The two stainless steel flanges are supporting the inner centered ceramic (LAVA) wire holder which is riding on a stainless steel water cooled inner core. A vertical stainless steel opening is sealed with stainless steel CF multipin connector flange (Figs. 2(d) and 3). Two pins are used for measuring the ISOTAN 44 resistance when the ISOTAN wire is heated indirectly by the Nickel Chrome wire, and two pins are for delivering input electric power. In case of direct heating of the ISOTAN wire, its resistance is calculated by dividing the measured voltage by the current. Two remaining pins are for measuring the ISOTAN 44 wire temperature. All wires are connected to the pins using gold plated cylindrical 1 mm sliding

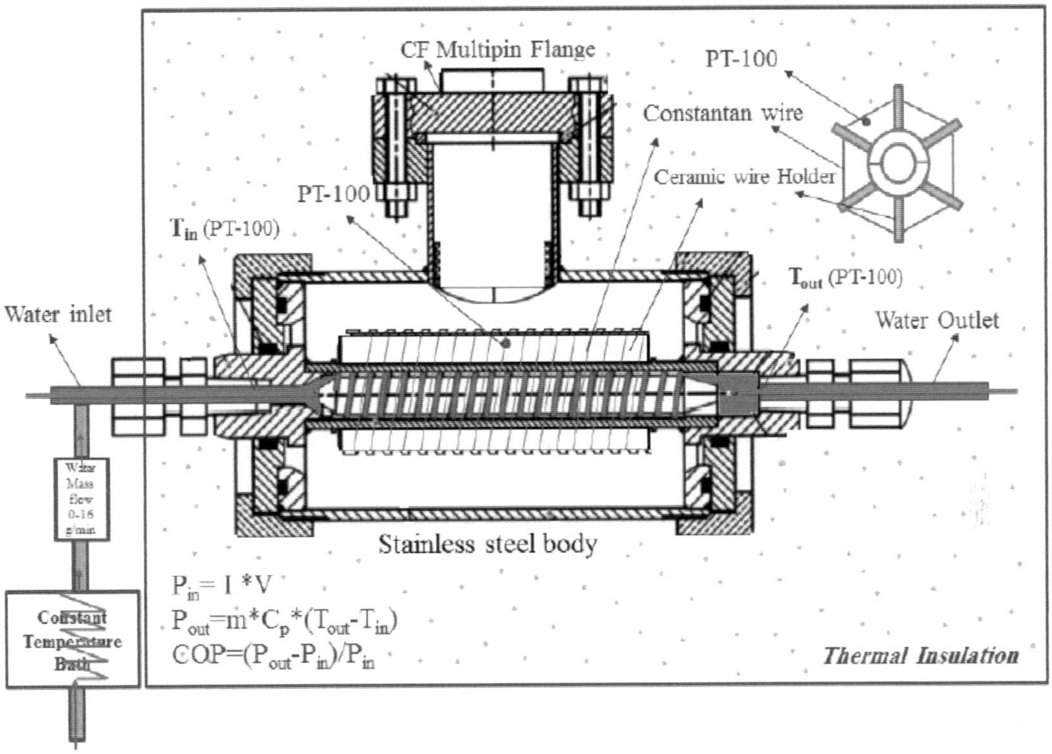

Figure 3. Experimental facility.

connectors. Cooling water flows along a spiral rectangular duct specially formed in the stainless steel core (Figs. 2(a) and 3). Heat transfers radially by convection and by radiation from the hot wires to the LAVA holder and by conduction to the stainless steel water-cooled core, which allows for accurate mass flow calorimetry. The LAVA ceramic wire's holder as seen in Fig. 2(b) has six longitudinal legs to support wires hanged freely in Hydrogen atmosphere. The legs are grooved with two nested spirals to separate the active wire from the heater wire. MICA coated glass tape are put on the face of each grooved leg to allow for further thermal barrier. The wire holder wound with treated ISOTAN 44 wire and with a 32 AWG Nickel Chrome heater wire. A PT100 RTD ceramic sheath is placed within the wires. The core assembly prior to insertion into the stainless steel chamber is shown in Fig. 2(a) and (b).

The assembled cell and the experimental facility are presented in Figs. 2(c)–(e) and 3. Figure 2(c) shows the experimental facility connected to a gas manifold containing Hydrogen, Deuterium, Argon, and Helium. Each gas can be used either separately or as a component of mixtures. Some of the experiments have been conducted with pure Hydrogen and some with Hydrogen–Argon mixture at different mixture rates to allow higher wire temperatures. The gas supply system has a mechanical pressure regulator to control the gas pressure accurately. The facility is equipped with turbine vacuum pump, so that the cell as well as the gas tubing can be evacuated down to 10^{-6} Torr before switching to a different gas or mixture, or for wire de-loading under vacuum. The cooling treated water flow rate is

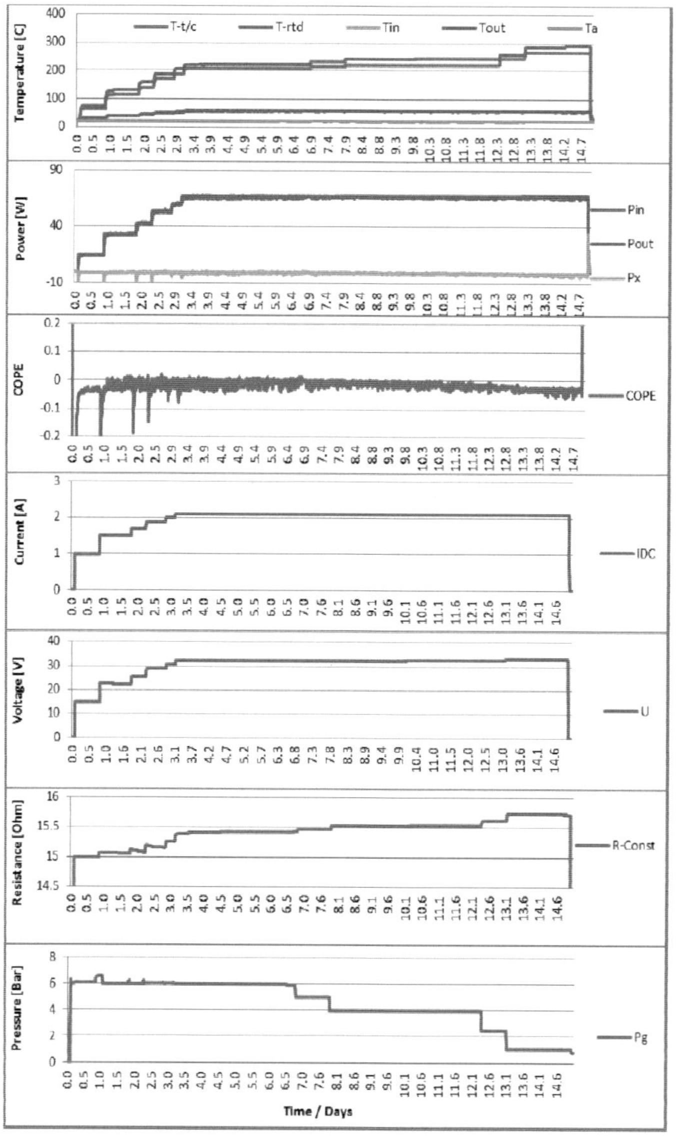

Figure 4. Wire number 270912B: 2-layer wire experiment.

controlled by a Porter Mass Flow Controller as shown in Fig. 3. Cooling water before entering the cell passes through stainless steel heat exchanger immersed in water control bath, in order to stabilize the water inlet temperature to the set point within $\pm 0.01^{\circ}$C, as shown in Fig. 3. National Instruments, in-house developed, LabVIEW data acquisition

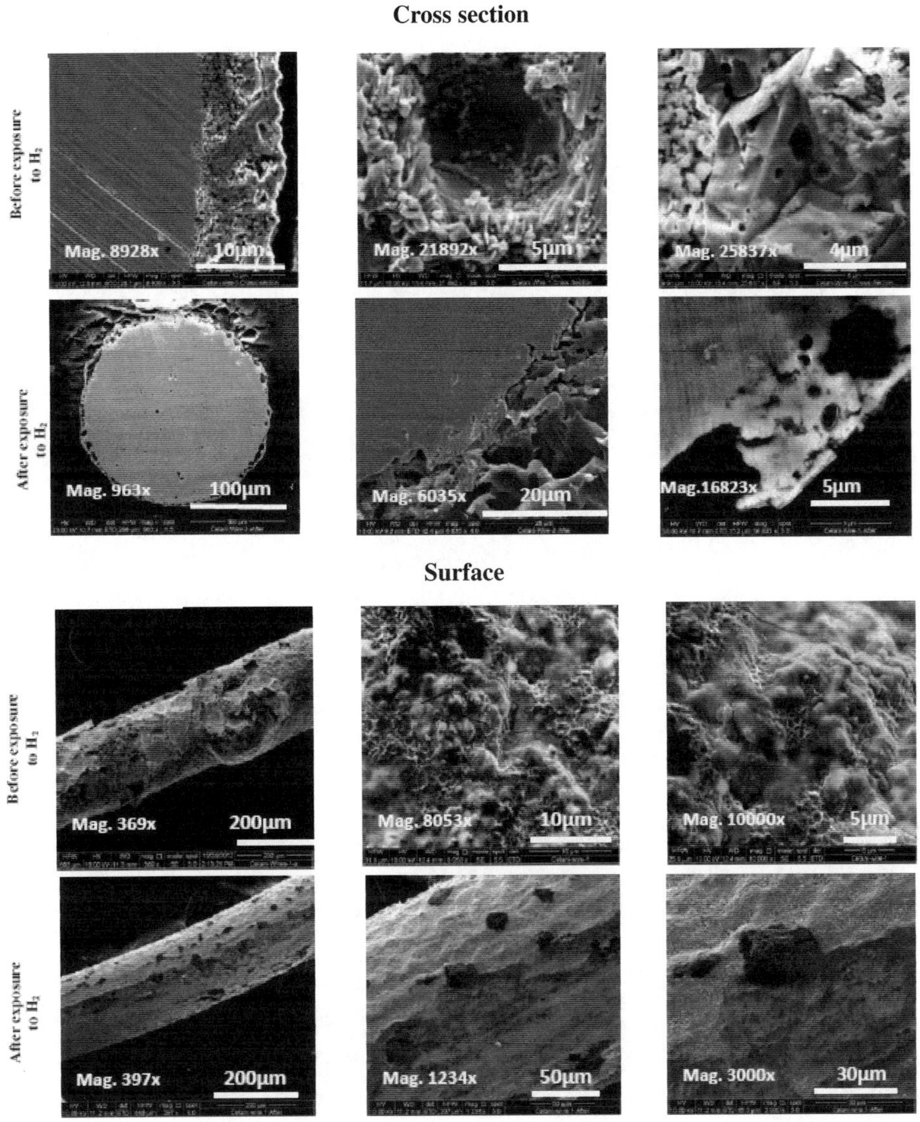

Figure 5. Cross section and surface images before and after exposure to H_2 of two layer wire.

program is used. The program allows to measure and store the cell body temperature using TC, wire, and ambient temperatures, the water inlet and outlet temperatures are measured by RTD PT-100 temperature elements. RTD PT-100 temperature elements used in experiments have an accuracy better than $\pm 0.1\%$ and with sensitivity of $0.001^\circ C$ as well as the wire resistance, the water mass flow rate, the cell pressure and the electrical input parameters. The

Before exposure

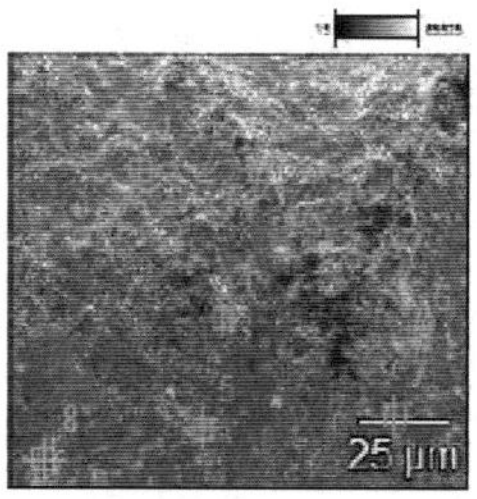

Before (wire, as received)	C-K	O-K	Si-K	S-K	Ca-K	Mn-K	Ni-L	Cu-L	W-M
Celani-1_pt1	0.59	5.35				5.23	17.96	40.87	
Celani-1_pt2	1.09	33.87	0.16			2.35	8.04	4.49	
Celani-1_pt3	4.49	33.44				1.98	8.63	1.47	
Celani-1_pt4	2.65	32.53				7.75	18.65	8.42	
Celani-1_pt5	0.54	32.63				2.45	12.89	51.49	
Celani-1_pt6	27.60	30.69	0.09	7.47	6.72		8.02	19.42	
Celani-1_pt7	0.86	31.11				2.74	10.62	54.22	0.45
Celani-1_pt8	0.41	34.71				3.72	21.10	40.06	

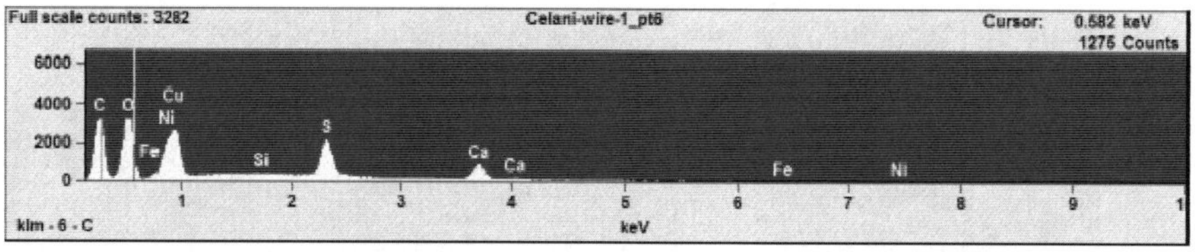

After exposure

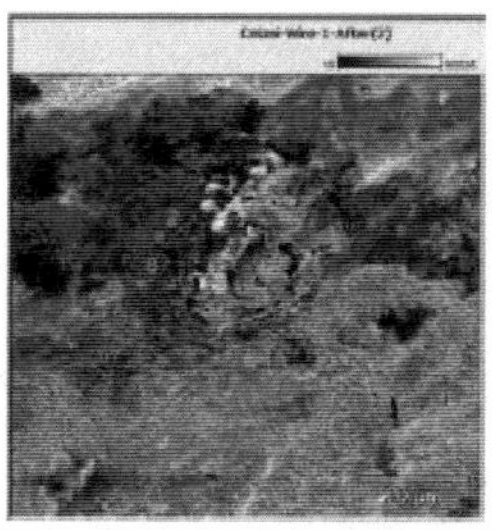

Element	Wt%	Wt% error	Atom%	Atom% error
C K	1.914	±0.21	6.91	±0.73
O K	9.46	±0.33	25.62	±0.90
Mg K	0.15	±0.06	0.26	±0.11
Si K	4.54	±0.12	7.0	±0.19
Mn K	19.42	±0.74	15.31	±0.58
Ni L	16	±2.38	11.81	±1.76
Cu L	48.52	±2.97	33.09	±2.03

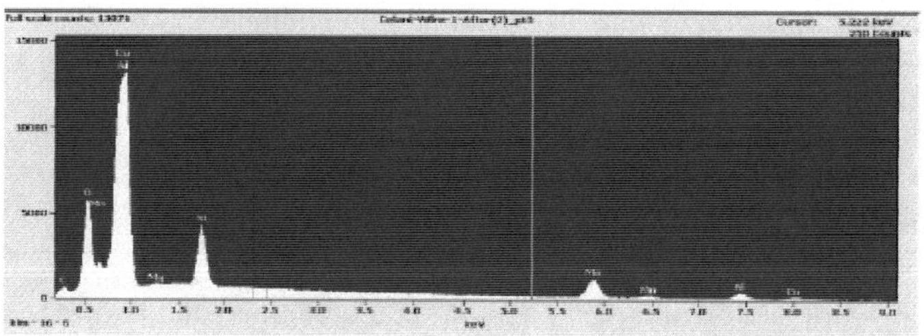

Figure 6. EDX analysis before and after exposure to H_2 of two-layer wire.

cell is energized by Kepco amplifier BOP 50-8D, which can be operated in a constant current or voltage mode. The amplifier is remotely controlled by the LabVIEW wave generator which is capable of generating DC, AC, positive and negative pulses as well as highly modulated waves (known as SuperWaves). The data acquisition system samples the temperatures, pressure and mass flow at a rate of 0.5 Hz (slow data). Simultaneously pairs of current (I) and voltage

(V) and $I \times V$ at a rate of 50 kHz are being sampled (fast data). I and V and $I \times V$ are stored on a buffer memory and then they are averaged and stored in a hard drive. Averaged input power calculated as $P_{\text{in}} = I \times V$ is compared with the output power being calculated as $P_{\text{out}} = mC_{\text{p}}(T_{\text{out}} - T_{\text{in}})$. If the difference between P_{out} and P_{in} is positive, then there is excess heat (P_x). The power gain is also calculated as COP $= (P_{\text{out}} - P_{\text{in}})/P_{\text{in}}$. Energies are calculated by integrating P_{in}, P_{out} and P_x over time and are presented as E_{in}, E_{out} and E_x. Figure 2(d) shows the assembled

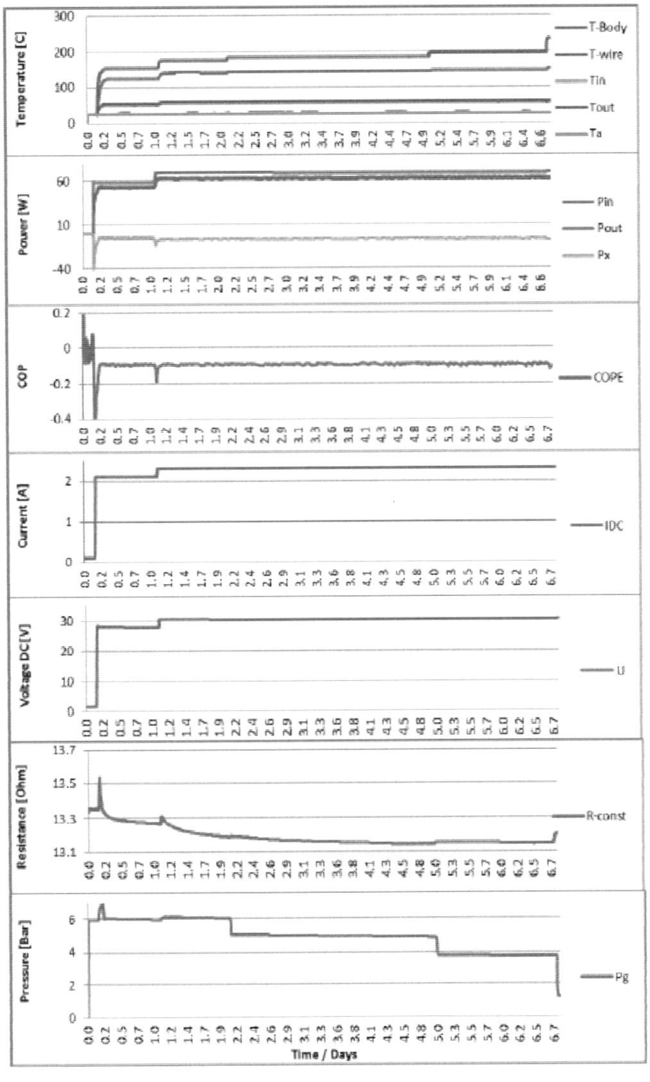

Figure 7. Wire number 280912A: 480-layer wire – initial loading.

insulated cell exposed on its upper side, in which one can identify the brass tightening nuts as well as the conflate flange with the multi-pin connector's plug. Also the gas intake/vacuum tubing and the gas vent through a needle valve can be seen. The thermally insulated cell, tubing and wiring are shown in Fig. 2(e). Figure 3 describes the water mass flow calorimeter having an accuracy of ±1% with sensitivity better than 10 mW. The mass flow calorimeter's

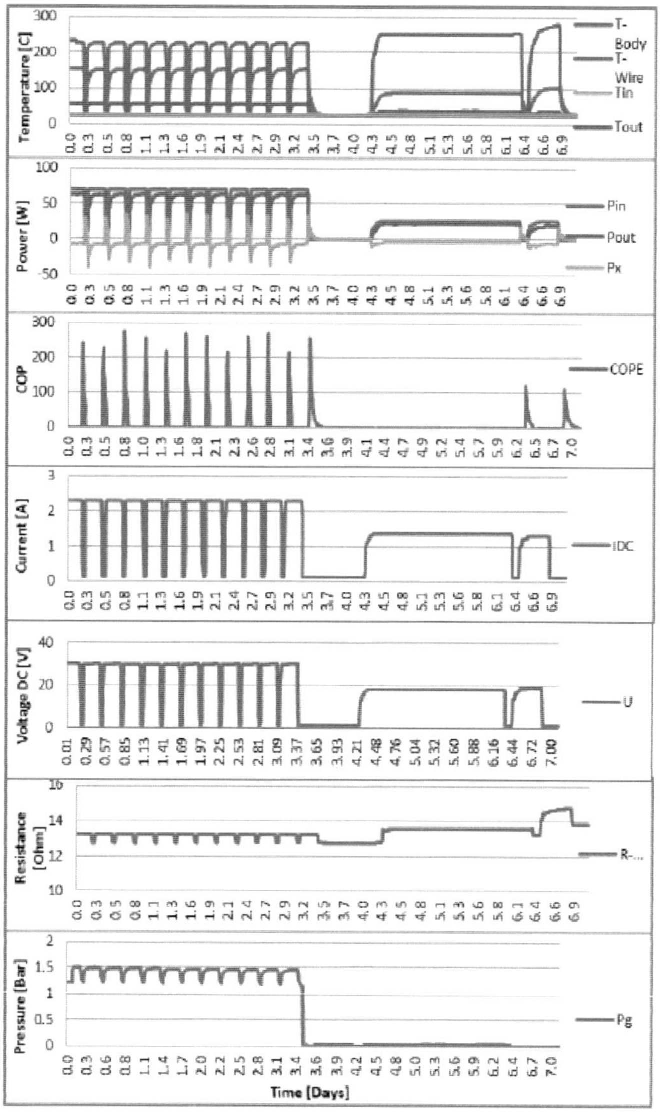

Figure 8. Pulsed power cycling – wire number 280912A: 480-layer wire 480-layer wire – initial loading.

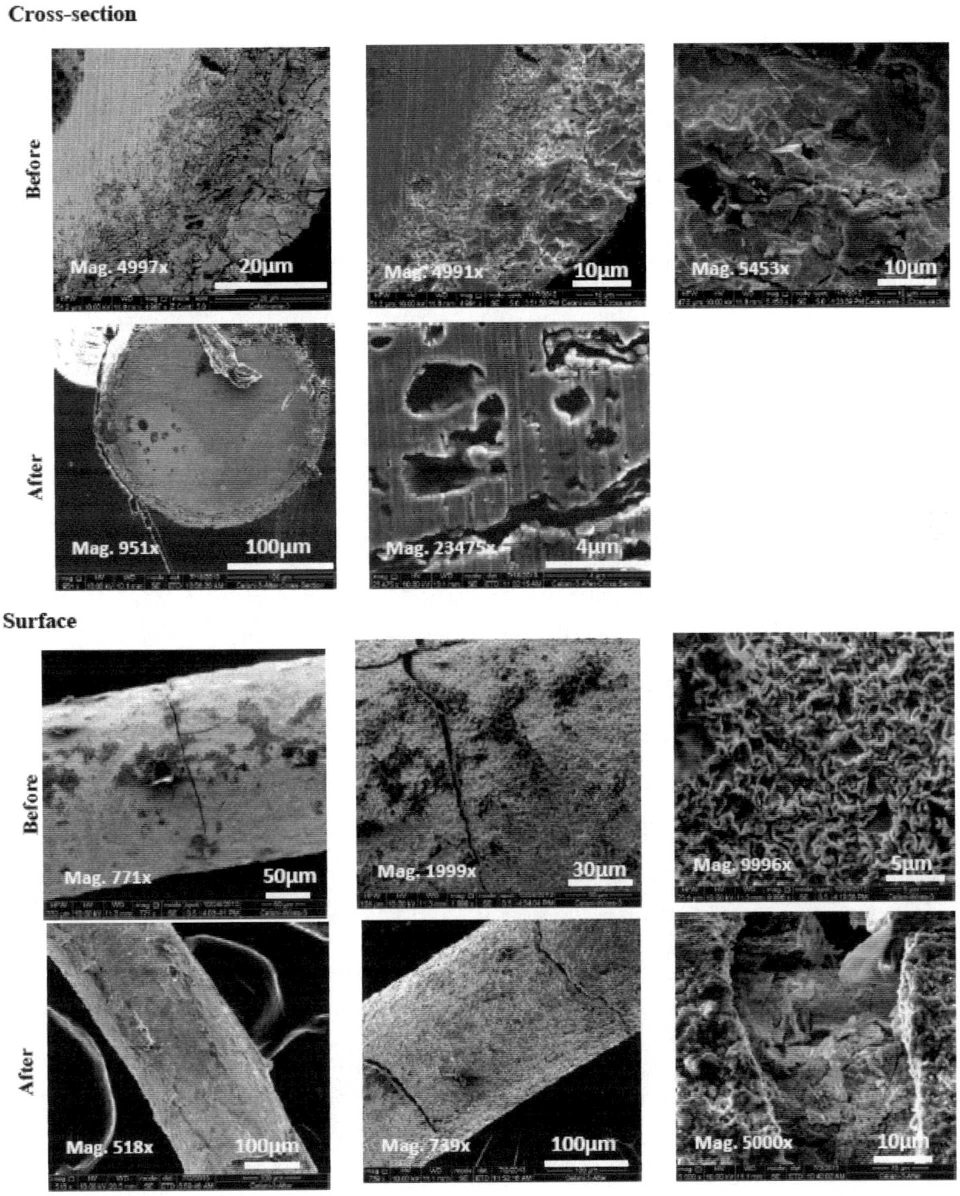

Figure 9. Cross section and surface images of 480-layer wire before and after exposure to H_2.

effectiveness varies from 95% to 90%, due to imperfect thermal insulation.

3. Results

3.1. Wire number 270912B: 2-layer wire experiment (Fig. 4)

The wire treating procedure was not disclosed to SKINR, but it is believed that the number of layers indicates the number of cycles of wire oxidation at elevated temperatures under air atmosphere.

Experiment was conducted under the following conditions: 25% Hydrogen /75% Argon atmosphere, DC current stepped to 2.1 A, electrical input power stepped up to 67 W and kept constant while gas pressure lowered in steps from 6 to 1.1 bar in order to increase the wire temperature. The initial wire resistance was 15.01 Ω. Constantan (ISOTAN 44) wire was heated indirectly by the 32 AWG Nickel Chrome wire up to a temperature of about 300°C. It can be seen from the R/R_0 vs. pressure graph in the introduction that the electrical resistance increases in α-phase and decreases after transition from α-phase to $\alpha + \beta$ and to β-phase. In this experiment a resistance increase of up to 15.7 Ω ($R/R_0 = 104\%$) was observed probably because the wire remained in α-phase region at a low hydrogen

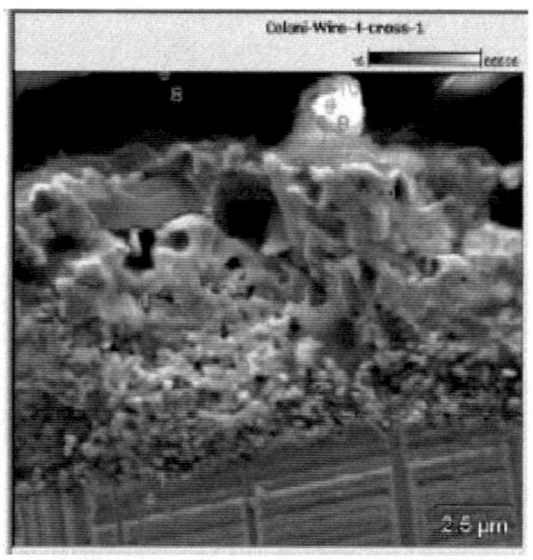

Element	Wt%	Wt% error	Atom%	Atom% error
C K	78.95	±1.46	89.47	±1.65
O K	7.61	±1.06	6.48	±0.90
Si K	0.71	±0.23	0.34	±0.11
S K	1.04	±0.21	0.44	±0.09
Cl K	2.33	±0.27	0.90	±0.10
K K	1.66	±0.29	0.58	±0.10
Ca K	1.12	±0.33	0.38	±0.11
Cu L	6.58	±0.69	1.41	±0.15

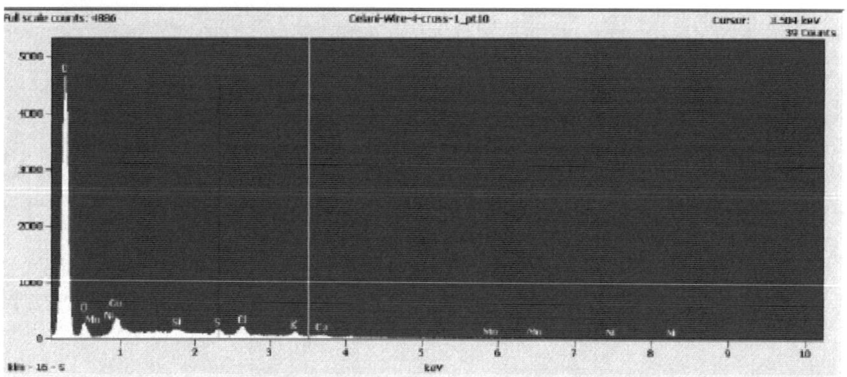

Figure 10. Cross section of 480-layer wire (number 280912A) before experiment.

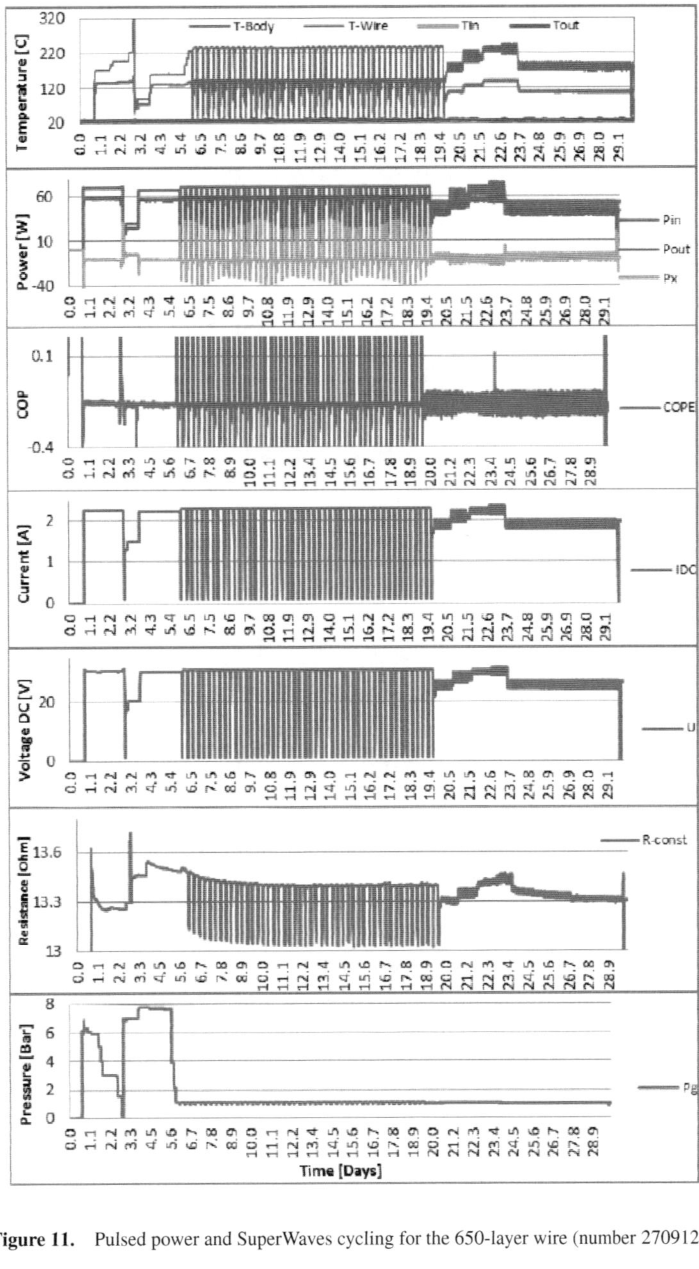

Figure 11. Pulsed power and SuperWaves cycling for the 650-layer wire (number 270912C).

loading of less than 10% H/Ni atomic ratio on average. It is believed that H/Ni ratio is higher near the surface and even higher in the nano-structures developed on the surface. No excess heat was observed ($P_{\text{out}} < P_{\text{in}}$). Experimental results with 300 L wire are similar to those with 10 L wire and therefore were not presented. The wire resistance is slightly increased, which is again an indication of relatively low loading, presumably in α-phase, which refers to atomic loading ratio (H/Ni) of less than 10%. No excess heat was observed.

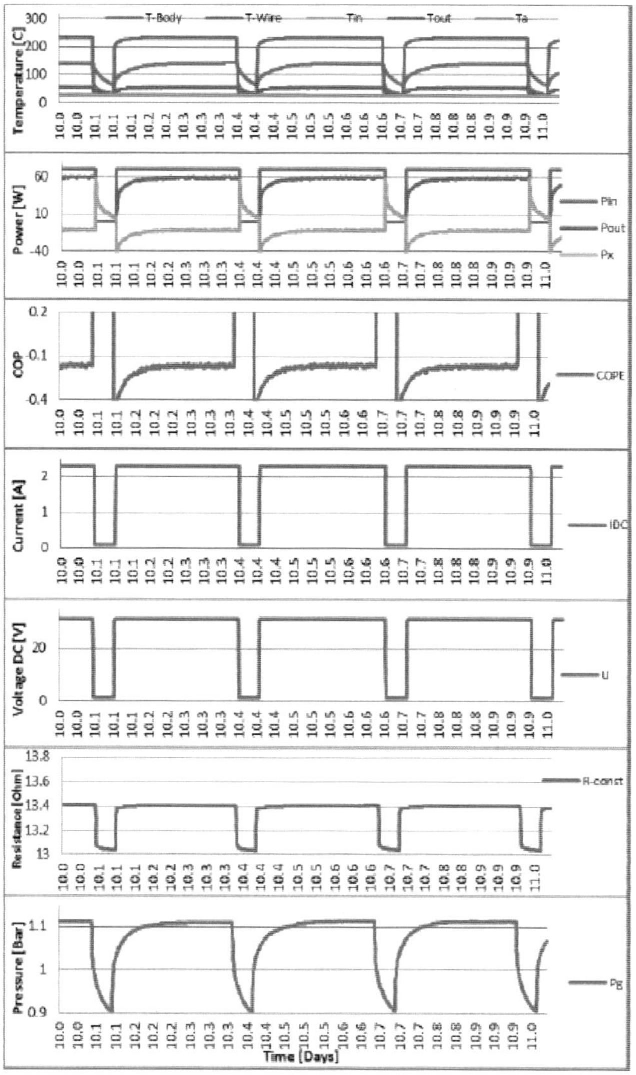

Figure 12. Pulsed power cycling for the 650-layer wire (number 270912C).

Figure 5 shows clearly that the treated wire (number $270912B$) with two layers has a higher modified surface before exposure than that after exposure to H_2. It seems that the exposed wire oxides undergo reduction in the H_2 atmosphere.

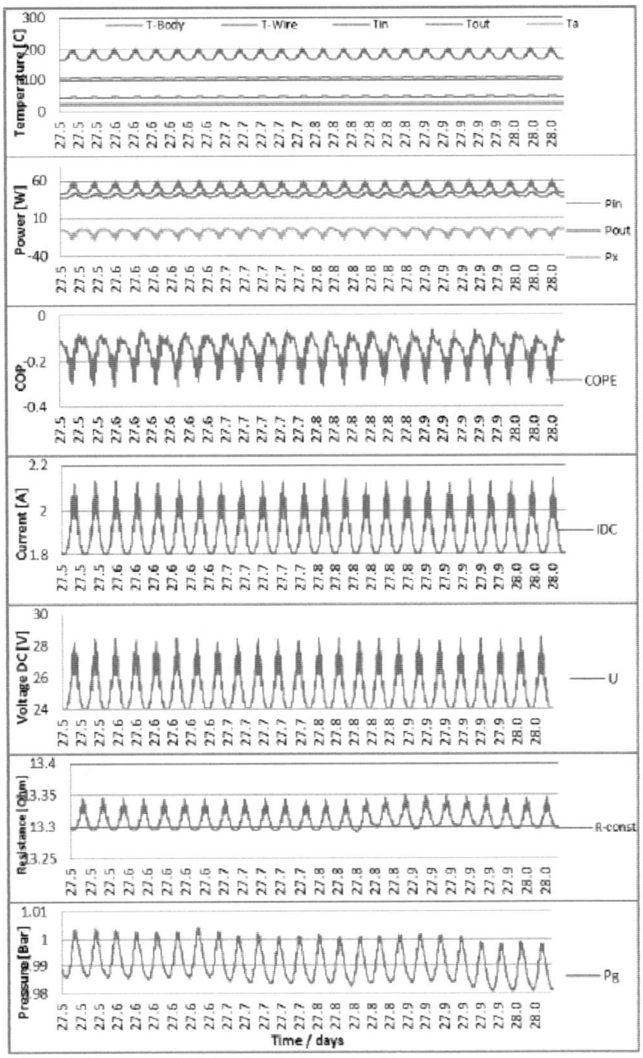

Figure 13. SuperWaves cycling for the 650-layer wire (number 270912C).

3.1.1. *The sulfur question*

Celani claimed that no excess heat achieved due to the fact that stainless steel contains 0.03% of Sulfur which is poisoning the treated wire. The stainless steel chamber was heated up to 600°C under vacuum many times prior to this set of experiments. EDX shows that the as-received ISOTAN 44 wire contains locally 7.47 wt.% Sulfur (in point 6) before exposure to H_2 as seen in the Fig.6 table; while in the other seven points analyzed, sulfur was not detected. However, no Sulfur was found after exposure to H_2 in the stainless steel chamber, suggesting that the lack of excess heat events cannot be simply explained by the use of stainless steel as confinement material although Celani used a borosilicate Schott Duran glass pipe.

3.2. Wire number 280912A: 480-layer wire – initial loading experiment (Fig. 7)

Wire running conditions are listed as follows: 100% Hydrogen atmosphere, initial wire resistance 13.37 Ω, DC current stepped up to 2.1 A and then to 2.3 A, input power stepped up to 69 W, gas pressure lowered in steps from 6 to 1.1 bar, wire temperature at 4 bar was about 200°C and at 1.1 bar was about 250°C. First run with the 480 layers wire showed that hydrogen loading was not high and electrical resistance ratio decreased to a value of about 98% ($\alpha + \beta$ phases). No excess heat was observed.

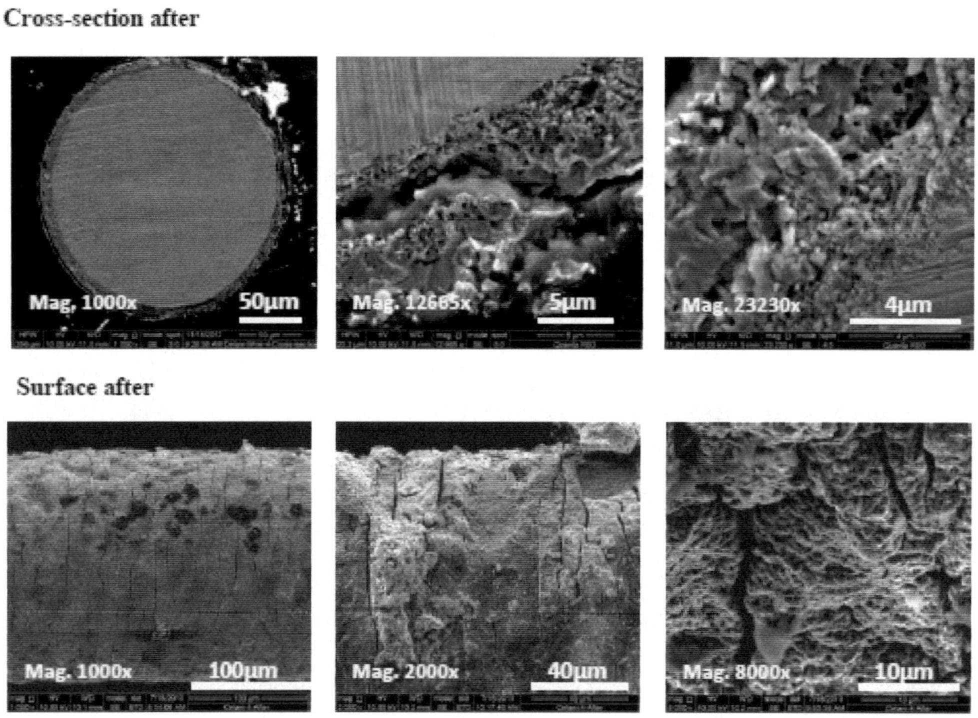

Figure 14. 650-layer wire, wire number 270912C.

3.3. Wire number 280912A: 480-layer wire – pulsed power cycling (Fig. 8)

An error in programming of input power led to an approximately 10 min disconnection between the first run (Fig. 7) and second run (Fig. 8) with the same 480 layers wire. Wire running conditions are: 100% Hydrogen atmosphere, hot cell pressure raised to 1.5 bar, pressure at room temperature was about 1.2 bar. In order to apply dynamics (excitation) to the Ni–H system, the DC current stepped up to 2.3 A for 6 h (maximum pulse input power of 69 W for 6 h) and then current reduce down to 0.1 A for 1 h. (minimum input power of 0.1 W for 1 h) Wire was de-loaded twice after the pulsed input power terminated. First de-loading was conducted under static vacuum and second under active vacuum.

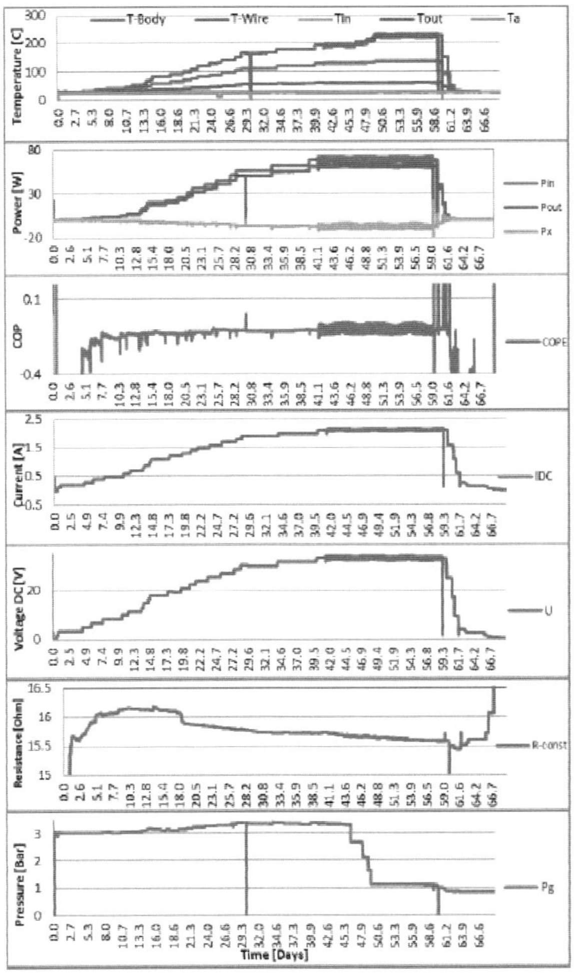

Figure 15. Cross section and surface images before and after exposure to H_2 of two layer wire.

Resistance ratio decreased to about 95%, which corresponds to low loading in $\alpha + \beta$ phase region. No excess heat was observed.

SEM images in Fig. 9 show clearly that the treated wire (number 280912A) with 480 layers has a cracked brittle surface before exposure to H_2. It appears that after exposed to H_2 atmosphere the wire surfaces partially flaked off.

Figure 10 shows the cross section of the treated 480-layer wire before experiment where micro- and nano-structures are clearly seen. Sulfur was detected again (1.04 wt.%), on the as-received wire. The oxidized treated zone penetrates up to about 9 μm deep.

3.4. Wire number 270912C: 650-layer wire (Figs. 11–13)

Figure 11, shows the results obtained at 100% hydrogen atmosphere, DC current stepped to 2.1 A, gas pressure lowered in steps from 6 to 1.1 bar, current pulses applied with 6 h at 2.3 A (with maximum pulse input power of 71 W), followed

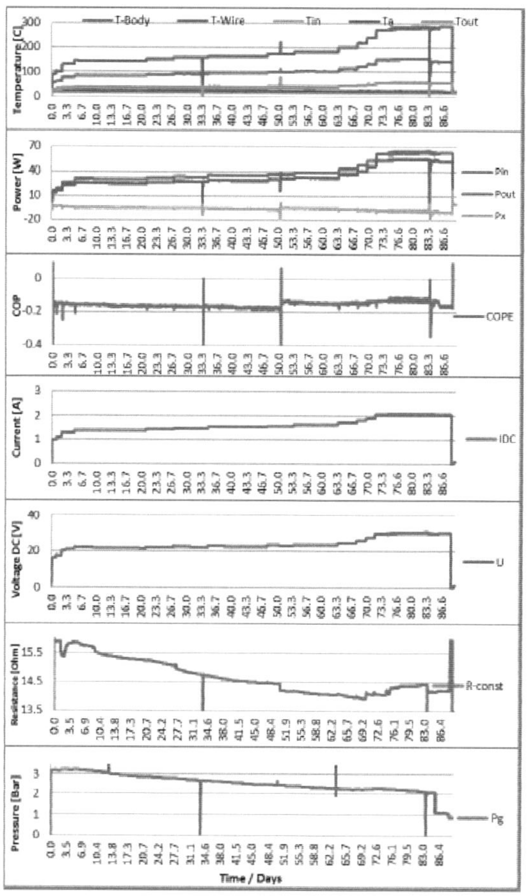

Figure 16. Cross section and surface images before and after exposure to H_2 of two layer wire.

by 1 h at 0.1 A (with minimum input power of 0.1 W). Pulses were applied for a total duration of ~327 h, then switched to SuperWaves. After trial and error in first few waves, final wave parameters of 30 min period, 1.8 A offset, and 2.1 A average current were chosen. SuperWaves was composed of five wave modulations, with total duration of ~243 h. Initial wire resistance was 13.67 Ω. Resistance ratio decreased to about 98%, which corresponds to low loading in $\alpha + \beta$ phase region. No excess heat was observed in DC, pulses and SuperWaves current modes. Figures 12 and 13 present a zoom in of Fig. 11 in time ranges of 10–11 days and 27.5–28 days, respectively.

In Fig. 14, the micro/nano structures due to the surface treatment can be seen with penetration depth of about 10 μm. Flakes and cracks on the surface can be seen on the surface images.

Table 1. Summary of experimental results.

Wire test no.	Treated wire	Max. R/Ro (%)	Signals used	Test duration (days)	Excess heat?	Notes:
1	2 layers #270912B	103.97	DC	15	No	
2	300 layers #021012A	103.33	DC	8	No	Wire burned through at defect
3	300 layers #021012A	101.5	DC	17	No	
4	480 layers #208912A	98.35	DC	7	No	
5	480 layers #208912A	95.32	DC and Pulsed DC	7	No	
6	650 layers #270912C	97.69	DC, Pulsed DC and SW	20	No	Performed with Dr. Celani in attendance
7	10 layers #020813	95.70	DC and SW	69	No	Lot-2 wire received from Dr. Matthew Valat
8	200 layers #050813	87.00	DC and SW	90	No	Lot-2 wire received from Dr. Matthew Valat

In Fig. 14, the micro/nano structures due to the surface treatment can be seen with penetration depth of about 10 μm. Flakes and cracks on the surface can be seen on the surface images.

3.5. Wire number 020813: 10-layer wire (Fig. 15)

This wire is from a second lot provided by Dr. Matthew Valat. The experiment was conducted at the conditions of 100% hydrogen atmosphere, DC current stepped up to 2.1 A, followed by current SuperWaves with 2.1 A average, the input power at maximum was 68 W and gas pressure lowered in steps from 3 to 1.1 bar. The initial wire resistance was 15.01 Ω. Resistance ratio reduced to 95.7%. No excess heat was observed in this experiment.

3.6. Wire number 050813: 200-layer wire (Fig. 16)

This wire is again from a second lot provided by Dr. Matthew Valat. The experiment of hydrogen loading in this wire was conducted under the same conditions as the 10-layer wire (number 020813). The initial wire resistance was 16.10 Ω and it dropped by about 13% (R/R_0= 0.87), due to hydrogen loading. The loading ratio was higher than in previous experiments and is clearly in the mixed $\alpha + \beta$ phase with H/Ni ratios less than 0.6. In spite the fact that the resistance ratio is in the same range as achieved in Dr. Celani's experiments (0.9–0.8), no excess heat was observed. Under similar R/R_0, Dr. Celani reported the observation of 10–20 W excess heat with an input power of 48 W.

4. Conclusions

In short, we list all experimental results in Table 1 showing wire numbers, maximum measured hydrogen resistance ratios, applied power modes, and test durations. We conclude our tests as follows.

- Dr. Celani's wire preparation process appears to be a selective oxidation/reduction of the Nickel/Copper/Manganese wire.
- The number of layers equates to the number of pulsed oxidation cycles that the wire has been put through and correlates with the thickness of the treated surface and with the micro-nano surface morphology.
- Obtaining an accurate temperature reading of such a small wire is problematic, actual temperature is higher.
- Loading studies in pure nickel have been done by Baranowski and Filipek. Unfortunately calibration curves of R/R_0 vs. H/Ni atomic ratio, for pure nickel and definitely for Constantan are not available, as are available for H/Pd and for D/Pd. This tool is essential for understanding the mechanism of excess heat. Further studies on gas loading in Ni systems are most important, especially due to researchers (Rossi, Parkhomov, and others) which are reporting on positive results in Ni/H systems. Furthermore, when the wire reaches certain temperatures, there is a reduction of the Nickel oxides which affects the wire resistance.
- Loading as indicated by resistance and by pressure reduction was much better for the second lot of wires.
- In seven out of eight runs conducted in SKINR's replication study, using wires treated by Celani and applying Celani's running protocol, R/R_0 values achieved were higher than 92%, which corresponds to low loading ratios of less than 10% atomic ratio. Celani is reporting in his experiments, on R/R_0 values of 92–80%, followed by excess heat. It is most probable that in the first seven experiments, no excess heat was observed, maybe due to low hydrogen loading. High loadings (atomic ratios) are accepted as a necessary condition, in Pd/D systems, for increasing the probability to achieve excess heat.
- Only in the last Experiment #050813 with 200 layers wire, the resistance ratio achieved was in within the range Dr. Celani obtained, but no excess heat was observed. At this level of loading Dr. Celani showed excess heat. It is most recommended that Dr. Celani will conduct future experiments with mass flow calorimetry.

Acknowledgment

This work is fully supported by Mr. Sidney Kimmel.

References

[1] B. Baranowski and S.M. Filipek, *J. Alloys Compounds* **404–406** (2005) 2–6.
[2] T. Graham, Philos. *Trans. Roy. Soc.* **156** (1866) 415.
[3] B. Baranowski and M. Smiaşowski, *Bull. Polon. Acad. Sci.* 7 (1959) 663.
[4] B. Baranowski and M. Smiaşowski, *J. Phys. Chem. Solids* **12** (1959) 206.
[5] A. Stroka and B. Baranowski, *Pol. J. Chem.* **76** (2002) 1019.

[6] B. Baranowski, Z. Szklarska-Smiaşowska and M. Smiaşowski, *Bull. Polon. Acad. Sci.* **6** (1958) 179.
[7] F. Cellani, E.F Marano, A. Spallone, A. Nuvoli, E. Purchi, M. Nakamura, B. Ortenzi, S. Pella, E. Righi, G. Trenta, S. Bartalucci, G.L. Zangari, F. Micciulla and S. Bellucci, Cu—Ni–Mn alloy wires, with improved sub-micrometric surfaces, used as LENR device by new transparent, dissipation-type, calorimeter, in *Proc 17th Int. Conf. on Condensed Matter Nuclear Science* (*ICCF 17*), South Korea, Daegeon, 2012.
[8] S. Focardi, V. Gabbani, V. Montalbano, F. Piantelli and S. Veronesi, Asti Workshop on Hydrogen/Deuterium loaded metals, *Conf. Proc. 64*, W.J.M.F. Collis editor, 1999, p. 35.
[9] G.H. Miley and J.A. Patterson, *J. New Energy* **1** (1996) 5.
[10] S. Romanowski, W. M. Bartczak and R. Wesoşkowski, *Langmuir* **15** (1999) 5773.

J. Condensed Matter Nucl. Sci. 19 (2016) 88–97

Research Article

Integrated Policymaking for Realizing Benefits and Mitigating Impacts of LENR

Thomas W. Grimshaw*

Energy Institute, The University of Texas at Austin, Austin, TX, USA

Abstract

The potential benefits of LENR as a source of energy have been well understood since its announcement in 1989. The improved prospects of LENR in the past few years are indicated by the significant numbers and varied locations of researchers in several countries, a large body of accumulated evidence, major advances in explanation and theory development, and recent favourable occurrences, including a plethora of proposed energy-producing devices. The changing landscape creates policymaking opportunities for supporting LENR to realize its potential benefits, planning proactively to deal with anticipated adverse secondary impacts, and integrating the updates in a comprehensive policy program. The first policymaking opportunity, updates of current policies for LENR support, may best be accomplished in a framework of evidence-based policymaking. The level of evidence for the existence of LENR is at least a preponderance of evidence (greater than 50% probability). There may even be clear and convincing evidence (greater than 70%). Correspondingly, the policy response should be at least reinstatement and research comparable to other emerging energy technologies. Enhanced support, perhaps on a par with hot fusion, is indicated if there is clear and convincing evidence. The second policymaking opportunity related to LENR's changing landscape is to address potential adverse secondary impacts with proactive planning. Broad deployment of LENR for energy supply may be expected to have major secondary impacts as a disruptive technology. Direct impacts will take place on all phases of the energy chain – supply, transport, storage, and consumption. Indirect impacts will be felt most by the components of society that are closely tied to the energy cycle, such as sectors of the workforce and energy-based communities. Technology assessment is a mature and readily available methodology for identifying secondary impacts and developing mitigative measures. A third policymaking opportunity is to integrate LENR policy updates that are needed as demonstrated by its changing landscape. Policies for mitigating adverse secondary impacts, for example, can be formulated based on policies for LENR support and the resulting pace of its development and deployment. New and updated policies may also be integrated at the national and international level and between the public and private sectors. The public interest will be served by update and integration of LENR policies for its development and mitigating its impacts. However, many challenges must be overcome to achieve the update and integration objectives. The world desperately needs new sources of clean and inexpensive energy. The case of cold fusion would perhaps be a curiosity in the history of science if the stakes were not so high.

Keywords: Improved LENR prospects, Integrated LENR policies, LENR impact mitigation, LENR research policies, Potential LENR benefits

*E-mail: thomaswgrimshaw@gmail.com

© 2016 ISCMNS. All rights reserved. ISSN 2227-3123

1. Introduction

LENR presents major opportunities to enhance the public interest as a potential new source of cheap and clean energy. Its improved prospects in recent years have resulted in a need for updates in LENR policies. Policymaking opportunities are emerging in three main areas – supporting LENR to realize its potential benefits, planning proactively to deal with its anticipated adverse secondary impacts, and integrating the updates in a comprehensive policy program. The objectives of this paper are to:

- Review the changing landscape of LENR.
- Describe opportunities for updating policies for support of LENR development.
- Delineate potential policies for mitigating adverse secondary impacts.
- Analyze opportunities for integrating LENR policies both nationally and internationally.
- Summarize the benefits and challenges of achieving updated and integrated policies.

The case of cold fusion would perhaps be a curiosity in the history of science was the world not in such desperate need of new sources of clean and inexpensive energy.

2. The Changing Landscape of LENR

Improved LENR prospects are indicated by at least four lines of argument – the significant numbers and varied locations of researchers in several countries, the resulting large body of accumulated evidence, advances in theory development, and recent favorable events.

2.1. Continued research worldwide

Unlike most claims of new phenomena that are not accepted by mainstream science, LENR research was not discontinued after it was rejected. On the contrary, many investigators have continued to work in the field, resulting in a large body of evidence for LENR reality. For example, at least 50 investigators in nine countries (including the US, Italy, Japan, India, and China) have continued LENR research. An international LENR society, International Society of Condensed Matter Nuclear Science (ISCMNS), was formed a few years ago, and an affiliated journal dedicated to LENR research reporting (*J. Condensed Matter Nucl. Sci.*) is published quarterly. International conferences are held in countries around the world every 1–2 years, with a typical attendance of about 200. Nineteen conferences have been held since they were begun in 1990. Attendance at the latest conference in 2015, which took place in Italy, was nearly 600 [1]. A substantial community of LENR researchers and other interested parties has emerged. Its size is indicated by the CMNS Google Group, which was formed about 10 years ago and currently has over 300 participants.

Although the US Department of Energy has not provided leadership in LENR research, investigations continued at several other US agencies after the 1989 rejection. For example, the US National Aeronautics and Space Administration (NASA) has conducted research at both the Glenn and Langley research centers [2,3]. Elements of the US Department of Defense (DoD) have also continued research and related interests. Several components of the US Navy have also had active LENR research efforts. The US Naval Research Laboratory (NRL), for example, has been working on LENR since the field started in 1989. Other Navy organizations that have pursued LENR research and related activities are the US Space and Naval Warfare Systems Command (SPAWAR), US Naval Air Weapons Station (China Lake), and the US Naval Postgraduate School.

2.2. Large and growing body of evidence

The substantial research in LENR has resulted in a large accumulation of evidence for its reality. One indicator of this evidence is a website dedicated to collecting LENR publications (LENR-CANR.org), which has a bibliography of more than 3500 journal papers and related items. As of January 2015, an estimated 3.2 million papers had been downloaded from the website [4].

Storms [5] has documented 380 papers reporting LENR just up to about 2007 as indicated by four signatures – anomalous heat (184 reports), tritium (61), transmutation (80), and radiation (55). Many more reports have been prepared in the subsequent years. Storms and Grimshaw [6] examined the evidence for LENR in relation to published criteria for distinguishing science from pseudoscience by Langmuir [7], Sagan [8], and Shermer [9]. Twenty-seven criteria were compiled, and LENR was examined in relation to each criterion. It was found that the criteria were satisfied, and it was concluded that LENR research is science and not pseudoscience.

2.3. Advances in theory development

Progress has also been made in developing an explanation of LENR. Many hypotheses have been advanced, but much remains to be done to converge on a full explanation. Two well-known examples are the hypotheses advanced by Peter Hagelstein of MIT and Edmund Storms, who is retired from Los Alamos National Laboratory.

Hagelstein [10] notes that LENR is indicated by the large amount of energy produced, the absence of expected chemical products, and the presence of expected amounts of new helium-4 in palladium deuteride experiments where LENR is observed. He observes that there appears to be no other conclusion besides a nuclear origin for the observations, but that there is a lack in LENR of the usual radiation signals that are used to study nuclear reactions. Hagelstein's hypothesis includes both conventional and new physics. In palladium deuteride systems reactions occur in vacancies in the lattice. The reactions involve fractionation of a large nuclear quantum combined with a coupling mechanism involving vibration and nuclei. Hagelstein utilizes the fundamental relativistic Hamiltonian in the explanation. The approach thus uses new concepts on a foundation of established physics.

Storms' hypothesis [11] proposes small sites, termed "nuclear active environments" (NAEs), that are located at or close to the surface rather than in the lattice, as is postulated by Hagelstein. These NAEs form in microcracks that are typically caused by stress relief in the material. Hydrogen ions migrate into the NAEs and form linear structures called "hydrotons". Vibration of the ions in the hydroton results in nuclear reactions, with release of energy as photons that are absorbed in the lattice. The mechanism of the nuclear reactions in the hydroton has not yet been explained but would almost certainly involve new physics.

2.4. Recent developments

The case for LENR is strengthened by several recent occurrences in the field. One of the most significant of these was the emergence of research centers at several universities. The Sidney Kimmel Institute for Nuclear Renaissance (SKINR) was formed at the University of Missouri in 2012 to perform fundamental research aimed at discovery of the mechanisms of the anomalous heat effect, a term used for LENR. Experiments are performed in four areas – nuclear mechanism, general mechanism, solid state theory, and cathode development (for electrolytic cells) [12]. The Center for Emerging Energy Science (CEES) was founded at Texas Tech. University in 2015 to explore critical parameters in the observation of the anomalous heat effect [13]. The intent of its work is to gain fundamental understanding of the LENR mechanisms. A Condensed Matter Nuclear Reactions Division was also recently formed at Tohoku University in Sendai, Japan. Three purposes have been advanced for the Division – fundamental LENR research, development of a new energy generation method, and determination of a new approach for nuclear waste decontamination [14].

Further indication that cold fusion potential may be realized is the significant number of LENR-based devices that have been introduced in recent years. Although many of the claims have not been verified (and significant issues have been identified in some cases), the devices represent a high level of interest and activity that may signify a need for policy updates. One major example is Andrea Rossi's E-Cat (for "energy catalyzer"), which is apparently based on a nickel–hydrogen setup. Several demonstrations of this device were held in 2011, culminating in a multiple-unit test in October 2011. About 2350 kWh of energy was reported for this test [15]. A three-part test of a high-temperature version of Rossi's device (E-Cat HT or "Hot Cat") was subsequently performed [16]. The first part of the test was not considered successful because the reactor melted before meaningful data could be obtained. The second test reportedly produced 195 kWh of energy. The third part was indicated to produce 95 kWh.

Another set of experiments, consisting of two phases, was subsequently performed with a different E-Cat design [17]. These experiments are frequently referred to as the "Lugano test" for the location in Switzerland where they were performed. During the 32-day test, 1.4 MWh of net energy was reported. The experiments also included analyses of the isotopes of in the energy-producing contents of the E-Cat. Observed shifts in the isotope composition before and after the tests were inferred to be the result of nuclear reactions. The large amounts of energy produced, high ratios of output to input power, and changes in isotope content were interpreted as evidence of LENR. It was announced in 2014 that the firm Industrial Heat had acquired partial rights to Rossi's E-Cat technology [18].

Investigation of devices apparently similar in design to the E-Cat has continued, notably in Russia and China. Parkhomov [19,20], a retired researcher from Lomonosov Moscow State University, reported experiments performed with two different but related designs. Both devices were configured to approximate the Lugano test of Rossi's E-Cat, but with significant differences, including the method of heat measurement. A principal conclusion was that the devices, described as "similar to (the) high-temperature Rossi heat generator ... produce more energy than they consume" above temperatures above about 1100°C. It was also concluded that the second device produced more than 40 kWh of excess energy.

Jiang is a retired researcher affiliated with the Ni–H Research Group at the China Institute of Atomic Energy in Beijing. His reactor design and materials are somewhat similar to those of Parkhomov with a setup approximating the Lugano test [21]. The experiment was apparently performed for over 12 h, during which 600 W of excess heat was observed for a portion of that time. The reported ratio of the 600 W to the input power of 780 W was 0.77. Jiang concluded that "the origin of excess heat cannot be explained by chemical energy".

JET Energy, Inc. has conducted LENR research with two types of devices called the NANOR and PHUSOR [22], both of which utilize deuterium. The PHUSOR is an aqueous configuration that uses palladium or nickel with the deuterium. The NANOR is non-aqueous and uses nano-scale particles consisting mainly of palladium, zirconium, and nickel [23]. JET Energy maintains close collaboration with the Energy Production and Conversion Group at MIT [24].

Brillouin Energy [25] has developed LENR-based technology for energy production using hydrogen and nickel (or other metal with appropriate properties). The technology is referred to as "Controlled Electron Capture Reaction" (CECR). Hydrogen is brought into contact with nickel, and reactions are stimulated with electromagnetic pulses. The energy is reported to be in the form of heat that is absorbed by the metal and captured for beneficial use. An apparently updated version of the Brillouin approach and technology (HHT^{TM}) was recently reported [26].

None of these examples is yet a working device having practical applications or commercial production. However, particularly when considered in aggregate, they provide further evidence that cold fusion may yet fulfill its potential as a source of energy.

In summary, a changing landscape for LENR is indicated by the substantial number of researchers, the accumulated body of research, and progress in developing theories. The recent emergence of academic research entities and the proposed LENR energy devices also seem to strengthen the cold fusion case. Three goals must be achieved for LENR and its benefits to be realized - more consistent reproducibility, fuller explanation of the process, and demonstration of its ability to produce usable amounts of energy. These goals may be achieved with affirmative policies for increased R&D support.

3. Policy Updates for LENR Support

The first policymaking opportunity resulting from LENR's changing landscape is revision of current policies for LENR support. Updates in these policies may best be accomplished in a framework of evidence-based policymaking [27,28]. The policy options are:

(1) Discontinue research entirely (unlikely given the continuing interest).
(2) Business as usual – continued marginalization.
(3) Reinstatement and development with other emerging energy technologies.
(4) Enhanced support, perhaps on a par with hot fusion.
(5) Crash program, possibly like the Manhattan Project during World War II, to realize LENR's benefits.

Selecting the alternative that best serves the public interest may be challenging because of the history and continuing rejection of LENR. Policymaking is further complicated by a need for improved reproducibility and a better explanation of the LENR phenomenon. To deal with these complications, LENR policy may be analyzed and established in terms of level of evidence for its existence:

(1) Preponderance of evidence (>50% probability).
(2) Clear and convincing evidence (>70%).
(3) Beyond a reasonable doubt (>90%).

The level of evidence may then be further interpreted for decisions on appropriate policy responses. At least a preponderance of evidence may reasonably be inferred from the large number researchers, the major body of evidence that has been accumulated, and the progress in achieving LENR explanation. Clear and convincing evidence is indicated by the emergence of LENR-dedicated research centers at several universities and by the significant number of proposed devices that purport to produce energy from LENR. When an adequate explanation is achieved (or reliable empirical devices emerge), it may be asserted that the evidence is sufficient to demonstrate LENR beyond a reasonable doubt.

Policy responses (PR) to these proposed levels of evidence may also be suggested. If LENR is indicated with a preponderance of evidence, it should be fully reinstated and pursued with other emerging energy technologies. If there is clear and convincing evidence, a higher level of support is needed, perhaps comparable to hot fusion support over the past five decades. If LENR is indicated beyond a reasonable doubt, it may be appropriate to institute a crash program similar to the Manhattan Project, which resulted in the atomic bomb in World War II.

In summary, it appears based on the level of evidence that LENR should at a minimum be reinstated and researched fully. It may in fact warrant investigation and development at a level similar to hot fusion research. Figure 1 shows diagrammatically how the changing landscape of LENR leads to the need for policy updates for its support. The changing landscape began sometime after LENR's 1989 announcement and rejection (AR). The four lines of argument (4L) for its improved prospects led to a need for policy updates (NPU). The updates are founded on evidence-based policymaking (EBP). The five policy options (PO) are evaluated by the level of evidence (LOE) for LENR existence, leading to the appropriate policy responses (PR) – reinstate and research fully or provide more enhanced support.

4. Policies for Mitigating Adverse Secondary Impacts

The second policymaking opportunity resulting from LENR's changing landscape is to address potential adverse secondary impacts (ASI) with proactive planning. Broad deployment of LENR for energy supply may be expected to have major secondary impacts as a disruptive technology. Direct impacts are anticipated for all phases of the energy chain – supply, transport, storage, and consumption. Indirect impacts will be felt most by the components of society that are closely tied to the energy cycle, such as the affected sectors of the workforce and the communities that rely on energy activities (e.g., coal mining towns).

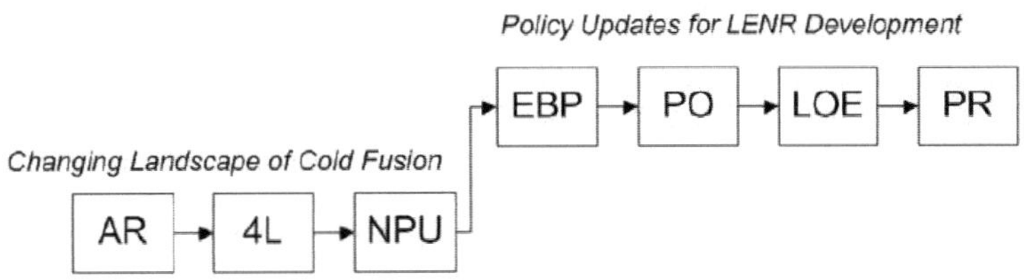

Figure 1. The Changing Landscape of LENR and Resulting Need for Policy Updates for Its Support. AR – Announcement and Rejection (1989); 4L – Four Lines of Argument; NPU – Need for Policy Updates; EBP – Evidence-Based Policymaking Framework; PO – Five Policy Options; LOE – Level of Evidence for LENR; PR – Policy Responses (Updates).

Technology Assessment (TA) is a mature and well-established method for addressing both direct and indirect secondary impacts and may readily be applied to cold fusion case [29,30]. The stages of a TA application are generally as follows:

(1) Identify impacts.
(2) Determined affected parties.
(3) Develop mitigation strategy.
(4) Define sources of assistance (e.g., agencies).
(5) Engage representatives (e.g., advisory group).
(6) Define mitigation measures (MM) for both direct and indirect impacts.
(7) Develop and implement mitigation plan (MP).

TA enables proactive planning to mitigate impacts and has ample precedent for application to energy-related issues [31,32]. Figure 2 summarizes how the changing LENR landscape leads to the need for policies for mitigating adverse secondary impacts in addition to required policy updates for supporting LENR development. Adverse secondary impacts (ASI) are addressed by technology assessment methodology (TA). Mitigating measures (MM) are defined, leading to an overall mitigation plan (MP).

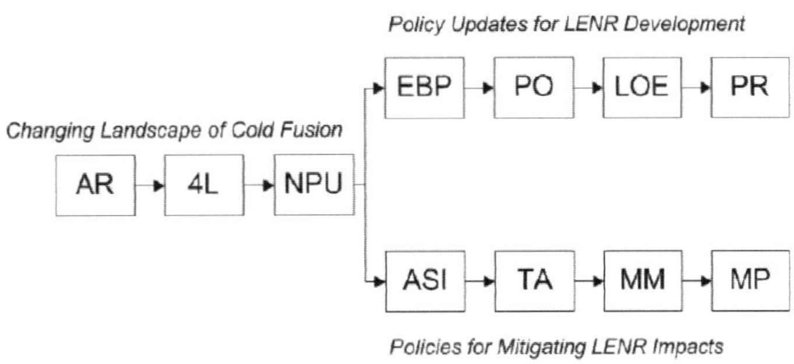

Figure 2. Illustration of Need for Policies to Mitigate Adverse Secondary Impacts Resulting from the Changing LENR Landscape. ASI – Adverse Secondary Impacts; TA – Technology Assessment Methodology; MM – Mitigating Measures; MP – Overall Mitigation Plan.

5. Opportunities for Integrating LENR Policies

A third policymaking opportunity for LENR is to integrate the policy actions and updates. For example, policies for mitigation planning for secondary impacts can be coordinated with the pace of LENR development and deployment. Policies can also be integrated among agencies at the national level, between the public and private sectors, and among nations.

5.1. Integration of mitigation planning with LENR development support

As LENR prospects improve as a result of increased support, mitigation planning can be adjusted for the changing imminence and rate of deployment. This adjustment would be necessary to achieve the objectives of proactive planning for mitigation. Figure 3 shows how the policy response for LENR development (PR) and resulting rate of deployment guides planning for mitigation (GP) as the overall plan (MP) is prepared.

5.2. Integration of LENR policies among agencies, nations, and the private sector

A focus on integrated LENR policymaking results in opportunities in several other policy areas. Public agency policy integration (PA) may take place at the local, state, and national levels and requires alignment and effective communication of the policymaking entities within the agencies. Formal arrangements, such as inter-agency agreements, may be used, or integration may be achieved by informal measures, such as regular inter-agency meetings. While these measures have been used to some extent by agencies for various issues in the past, they may become increasingly important as LENR deployment progresses.

LENR development – and dealing with its impacts – may be enhanced with stronger integration between the public and private components of society (PP). For example, LENR may benefit from government policies and measures to address "market failures", in a similar vein to current laws and regulations for environmental protection. Existing programs, such as small-business research support and provisions for technology transfer from government labs to privately held companies, could increase in importance if the government becomes more active in LENR research. Public-private partnerships (PPPs) may provide another vehicle for supporting LENR development and realization. An improved stance among patent and trademark entities would also substantially enhance efforts in the private sector to realize the benefits of LENR. Opportunities may be found for integrating these policy changes and updates in the public and private aspects of LENR development.

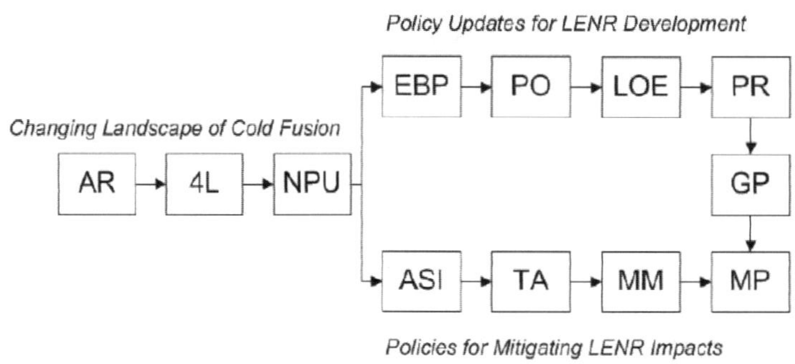

Figure 3. The Pace of Mitigation Planning is Guided by Policies for LENR Development and the Resulting Rate of Its Deployment. GP – Guidance for Mitigation Planning.

At the international level, programs may be established for supporting LENR research (IN). As LENR reaches the stage of worldwide deployment, bilateral and multi-lateral agreements may be made or updated to enhance its availability. For example, the United Nations may implement programs for making small LENR units available in a dispersed manner in Third World nations. World Bank loans may be made to nations needing support in acquiring LENR technology for the benefit of human health and the environment. The World Trade Organization may consider LENR and its humanitarian benefits for special rulemaking to enhance availability worldwide. Again, opportunities may be found for integration of policy changes or updates among these international entities.

5.3. Summary path to updated and integrated LENR policies

As policies are updated for LENR support and for mitigating adverse secondary impacts, and as they are integrated at various levels, the public interest will be served for the humanitarian benefits of LENR. Figure 4 illustrates the full path from the present situation of LENR's changing landscape to the prospective future of fully updated and integrated LENR policies. Integrated policymaking (IPM) is the foundation for public agency (PA), public-private (PP), and international (IN) policy integration. The desired result is fully updated and integrated policies (UIP) for LENR support and impact mitigation.

6. Benefits and Challenges of LENR Policy Integration

Achieving integration of LENR policies as updates are accomplished will have substantial public interest benefits. But many challenges must be overcome as well. LENR policy integration will help avoid conflicts and actions that are at cross-purposes among interested parties. Correspondingly, there will be increased

efficiency in achieving the policy objectives of the parties as well as increased cost effectiveness where entities have common interests. Policy integration may also achieve improved social equity, with more rational decisions and less influence of purely political considerations.

A principal challenge for policy update and integration is the historical barrier to LENR acceptance after its initial rejection. This barrier seems likely to be surmounted as LENR continues to be investigated and the evidence continues to show that it is real – and that its potential benefits are attainable. Another challenge is the sheer immensity of

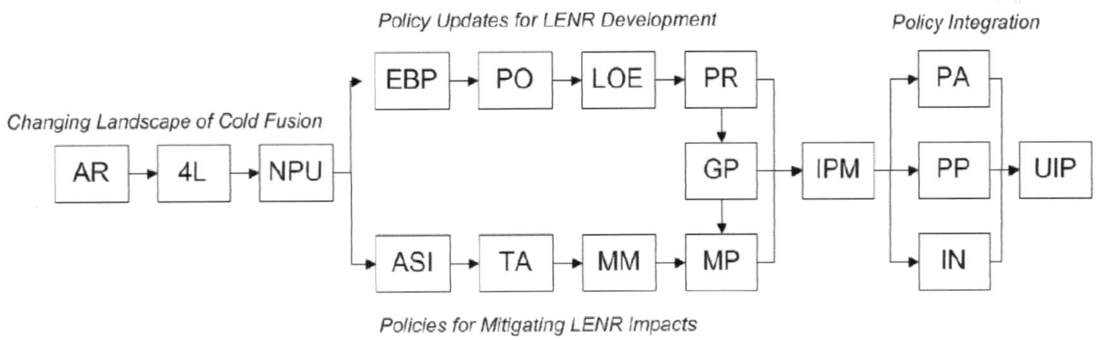

Figure 4. Path to Updated and Integrated LENR Policies. IPM – Integrated Policymaking Framework; PA – Integration among Public Agencies; PP – Policy Coordination between Public and Private Sectors; IN – Integration among Nations at the International Level; UIP – Updated and Integrated LENR Policies.

the expected direct and indirect secondary impacts. Proactive planning to address these impacts will be a major undertaking. The existence of the long-standing and well-established energy policy framework may also present a barrier to effective LENR policies and their integration. This framework includes many conflicting interests and agendas that will have to be considered as policies are updated and integrated.

7. Summary and Conclusions

Despite being rejected not long after it was announced, LENR has continued to be pursued in many venues, resulting in improved prospects and the need for policy updates. Updates are needed both for support of LENR development and preparation to mitigate its anticipated adverse secondary impacts. As these updates are accomplished, there are opportunities to integrate the policies to support and realize LENR with mitigation planning for anticipated impacts. There are also policy integration opportunities among public and private entities and at many levels within nations and internationally. The benefits of updating and integrating LENR policies are substantial, but the challenges for doing so are also very large. Opportunities for policy updates and integration may be set forth conceptually, but realization in the "real world" will be much more difficult. Nevertheless, it is clear that the public interest will be served by updating LENR policies and achieving their integration.

References

[1] International Society of Condensed Matter Nuclear Science. Date unknown. Online. Available: http://www. iscmns.org/

[2] S. Wrbanek et al., NASA Glenn Research Center Experience with "LENR Phenomenon", Presentation at Interagency Advanced Power Group (IAPG), Mechanical Working Group (MWG). May 2012.

[3] P. Douglas and D. Wells et al., Low Energy Nuclear Reaction Aircraft – 2013 ARMD Seedling Fund Phase I Project. Langley Research Center, Hampton, Virginia. NASA/TM – 2014-218283, 2013.

[4] LENR-CANR.org, a Library of Papers about Cold Fusion, Online, Available: http://lenr-canr.org/wordpress/ ?page_id=1213.

[5] E. Storms, Science of low energy nuclear reaction: a comprehensive compilation of evidence and explanations about cold fusion, World Scientific, Singapore, Tables 2, 6, 8, 11, 2007.

[6] E. Storms and T. Grimshaw, Judging the validity of the Fleischmann–Pons effect, *J. Condensed Matter Nucl. Sci.* **3** (2010) 9–30.

[7] I. Langmiur, Colloquium on pathological science. held at the Knolls research laboratory, December 181953, Reproduced as "Pathological Science", *Phys. Today* **42**(10) (1989) 36–48.

[8] C. Sagan, The fine art of baloney detection, Chapter 13 in *The Demon-Haunted World: Science as a Candle in the Dark*, Random House, New York, NY, 1995.

[9] M. Shermer, *The Borderlands of Science – Where Sense Meets Nonsense,* Oxford Univ. Press, Oxford, UK, 2001.

[10] P. Hagelstein and I. Chaudhary, Phonon models for anomalies in condensed matter nuclear science, *Current Sci.* **108**(4) (2015) 507–513.

[11] E. Storms, *The Evaluation of Low Energy Nuclear Reaction: an Explanation of the Relationship between Observation and Explanation*, Infinite Energy Press, Concord, NH, 2014.

[12] G. Hubler, Sidney Kimmel institute for nuclear renaissance, *Current Sci.* **108**(4) (2015) 562–564.

[13] T. Scarbrough et al., The center to study the anomalous het effects, Poster Presented at ICCF-19, Padua, Italy, April. 2015.

[14] Y. Iwamura et al., The launch of a new plan on condensed matter nuclear science at Tohoku University, Paper Presented at ICCF-19. Padua, Italy, April 2015.

[15] D. Hambling, Success for Andrea Rossi's E-Cat cold fusion system, but mysteries remain, Wired, October 29, 2011.

[16] G. Levi et al., Indication of anomalous heat energy production in a reactor device containing hydrogen loaded nickel powder, Cornell University Library arXiv, 2013.

[17] G. Levi et al., Observation of abundant heat production from a reactor device and of isotopic changes of fuel, Unpublished manuscript, October 2014.

[18] PRWire, PRNewswire, Industrial heat has acquired Andrea Rossi's E-Cat technology, Research Triangle Park, January 24, 2014.
[19] A. Parkhomov and E. Belousova, Researches of the heat generator similar to high temperature Rossi reactor, Poster Presented at ICCF-19. Padua, Italy, April 2015.
[20] M. McKubre, A Russian experiment: high temperature, nickel, natural hydrogen, *Infinite Energy,* Issue 120, March/April. 9–11, 2015.
[21] Jiang, S., 2015. New Result of Anomalous Heat Production in Hydrogen-Nickel Metals at High Temperature. Presentation Posted at E-Cat World. www.e-catworld. May 30.
[22] D. Nagel, Scientific and commercial overview of ICCF18, *Infinite Energy,* Issue 112, November/December 2013.
[23] JET Energy Inc., undated. JET Energy NANOR©-type LANR Technology. Online available: http://world.std.com/~mica/nanortechnology.htm.
[24] P. Hagelstein, On theory and science generally in connection with the Fleischmann–Pons experiment, *Infinite Energy,* Issue 108, . March/April 2013, pp. 5–12.
[25] R. George et al., About Brillouin Corp, Online2014, a vailable: http://brillouinenergy.com/ ?page=home_page.
[26] Brillouin Energy, HHT^{TM} reactor and results, Poster at ICCf-19 Conference, Padua, Italy, April 2015.
[27] T. Grimshaw, Evidence-based public policy toward cold fusion: rational choices for a potential alternative energy source, Austin, TX, Lyndon B. Johnson School of Public Affiars. Unpublished Professional Report, December 2008.
[28] T. Grimshaw, Evidence-based public policy for support of cold fusion (LENR) development, Poster Presented at *17th Int. Conf. on Cold Fusion*, Daejeon, South Korea, August 2012.
[29] T. Grimshaw, Public policy planning for broad deployment of cold fusion (LENR) for energy production, Paper FrM1-1, *17th Int. Conf. on Cold Fusion*, Daejeon, South Korea, August 2012.
[30] T. Grimshaw, Cold fusion public policy: rational – and urgent – need for change, Presentation at 2014 Cold Fusion/LANR Colloquium at MIT, Cambridge, MA. March 2014.
[31] White, L. Irvin et al., Energy from the West: summary report, US environmental protection agency, Prepared by Science and Public Policy Program, University of Oklahoma, Prepared for Office of Research and Development, EPA (600/9-79-027), August 1979.
[32] L. Johns et al., A technology assessment of coal slurry pipelines. office of technology assessment, NTIS order #PB-278675, March 1978.

J. Condensed Matter Nucl. Sci. 19 (2016) 98–109

Research Article

Current Status of the Theory and Modeling Effort based on Fractionation

Peter L. Hagelstein*
Massachusetts Institute of Technology, Cambridge, MA, USA

Abstract

The theoretical problems associated with excess heat in the Fleischmann–Pons experiment were once viewed as insurmountable; nevertheless, some progress has been made in the past quarter century. Conceptually the problem can be split into one area involving new physics which is needed to address the microscopic physics of the reaction process; and a second area involving known physics which allows one to connect with the different practical issues involved in the experiments and observations. We review the ideas and progress first in the area of new physical mechanisms, in which models that describe the down-conversion of the large nuclear quantum allowing for coherent energy exchange of the nuclear energy into vibrational energy. These ideas provide a connection between excess heat experiments with Pd and Ni, with D or a mixture of H and D; experiments where tritium or low-level nuclear emission is seen, and other experiments with collimated X-ray emission. In the area of conventional physics mechanisms, we are interested in the basic physics of PdH and PdD; modeling cathode loading and understanding why some cathodes loaded very poorly in the early days; understanding active sites which we attribute to monovacancies in PdD and NiH; figuring out how active sites are created; triggering excess heat; and removing the helium made in the new reactions.

Keywords: Excess heat, Fleischmann–Pons experiment, Karabut experiment, Mechanism, Theory

1. Introduction

Ever since the announcement of excess heat in the Fleischmann–Pons experiment [1,2], theorists have generally struggled with the seemingly impossible theoretical ramifications. The majority of theorists in the scientific community generally regard the theoretical problems as impossible, concluding that the experiments which show anomalies simply must be in error. Even though a non-nuclear explanation appears to be ruled out by the absence of chemical products commensurate with the energy produced, and also by the very large estimates for the energy produced per unit volume, mass or atom; one still encounters a surprisingly wide variety of proposed explanations in terms of various chemical reactions, the water reaction, the release of stress energy, stored chemical energy, or mechanical energy. For those who accept that the origin of the energy must be nuclear, there is no agreement as to the nature of the microscopic

*E-mail: plh@mit.edu

© 2016 ISCMNS. All rights reserved. ISSN 2227-3123

mechanism; whether it be fusion, fission, some kind of neutron or nuclear fragment transfer, beta decay, more general nuclear disintegration, or perhaps some new kind of process involving exotic particles such as dark matter. Some regard the effect as implying some basic failure of modern physical theory, and have proposed alternatives to quantum mechanics, to field theory, and to the standard model. Quite a few theorists have focused on microscopic reaction mechanisms involving d+d and p+d fusion reactions; in this case a key issue has been finding some way around the Coulomb barrier in order to arrive at a predicted rate that is relevant to experiment.

2. Helium and Absence of Neutrons

^{4}He was observed in the gas phase by Bush, Miles and coworkers [3–6], and was found to be correlated with the energy produced [7]. Although there remains uncertainty in the energy produced per unit ^{4}He, a picture consistent with the majority of the measurements is one in which the energy release is due to the mass energy difference between D_2 and ^{4}He, and not all of the ^{4}He produced makes it into the gas phase unless steps are taken to scrub it out.

If we consider a microscopic reaction in which two deuterons react to make ^{4}He (independent of the many issues as to how such a reaction might happen), a natural question might be to ask how much of the 24 MeV reaction energy the alpha particle is born with. A knowledge of this energy could be helpful in clarifying the microscopic reaction mechanism. However, a direct measurement of the alpha energy seems problematic, as the introduction of a silicon surface barrier detector, or other conventional detector, would likely impact the experiment and inhibit excess heat production. It was proposed to view the PdD and D_2O present in the experiment as nuclear detectors [8,9] since energetic alphas would be expected to produce primary neutrons from deuteron disintegration, and secondary neutrons from inelastic collisions with deuterons (which could gain sufficient energy for subsequent dd-fusion reactions).

Experiments were found in which excess heat bursts were observed at the same time neutron measurements were carried out. In general there is no correlation between excess heat bursts and neutron emission, with an upper limit of about 1 neutron emitted per 100 J of energy produced. The conclusion from this analysis is that the alpha particles measured subsequently as ^{4}He have less than 0.1% of the 24 MeV reaction energy at birth, which rules out all plausible conventional Rutherford picture reactions. Energetic alphas if present would also lead to X-ray production [10], which is not seen in experiments. One could imagine the reaction energy somehow going into many making fast deuterons [11]; however, in such a scenario the energy would need to be distributed among more than 25,000 deuterons for the resulting neutron emission rate to be consistent with experiment [12].

3. Models Based on Down-conversion

For us the absence of energetic nuclear radiation in amounts commensurate with the energy produced was the most fundamental theoretical issue raised by the Fleischmann–Pons experiment, and was generally the focus of our early research effort. Our sense was that if the effect was real, then it implied the existence of a mechanism to down-convert a large nuclear quantum into a great many low energy quanta. In recent years we have made use of the term "fractionation" to indicate strongly enhanced down-conversion via the new mechanism reviewed in what follows. We began analyzing quite a few specific models of candidate systems seeking a down-conversion mechanism with no luck for many years. The first glimmer of success finally came in 2002 when we understood that down-conversion in the spin-boson model was limited by a destructive interference effect, and that the introduction of loss could remove this destructive interference [13]. The ideas, models, and progress on this problem have been reviewed several times over the years [14–17]. More general reviews are available in the form of video lectures based on IAP courses given at MIT in the past few years [18].

Although the early modeling focused on specific systems, it became evident that associated mechanisms could be studied more efficiently using "toy" models where specific transitions were abstracted into equivalent two-level systems, and specific low energy condensed matter degrees of freedom (such as a vibrational mode) could be abstracted as an oscillator. The problem that results is one involving many two-level systems coupled to an oscillator. This problem is known in the literature, and falls under the general heading of a "spin-boson" model. We were interested in coherent energy exchange between the two-level systems and oscillator under conditions that the transition energy of the two-level systems is large, and the oscillator energy is small. Coherent energy exchange in the multi-photon regime of the spin-boson model is known, and we have contributed to the analysis of this problem and related issues [19–22].

The generalization of the model augmented with loss is complicated in general [17], and we focused initially on a restricted version where loss of infinite strength was modeled [23,24]. It had been found earlier that when the destructive interference effect was eliminated, the resulting coherent energy exchange rate was very fast, but depended relatively weakly on the details of the loss model. Hence we wanted to construct a simple but crude model which would give the maximum coherent energy exchange rate; this would be helpful in evaluating the model, and could provide a limit for more sophisticated models to be worked on later. With some work this model was analyzed systematically, and numerical and analytic results were established quantifying the coherent energy exchange rate and associated scaling laws [25–29]. The scaling in the limit where a very large quantum is down-converted to many small energy quanta was found to be algebraic, a gentle result which suggested a real possibility for the fractionation of an MeV quantum down to the eV range. Fractionation in this kind of model proceeds by the sequential exchange of many quanta one at a time, while overall coherence is maintained. The coherent energy exchange rate under conditions of fractionation is faster the stronger the individual coupling, the fewer small energy quanta involved, the more two-level systems present, and the stronger the oscillator excitation.

To connect with the Fleischmann–Pons experiment, interest focused on a transition from an initial D_2 state to a final ^{4}He state, with the energy going into a highly excited optical phonon mode. Since the Coulomb interaction inhibits tunneling, the resulting coupling matrix element is very small; hence this transition could never fractionate a large quantum. A solution was to transfer the excitation to other transitions strongly coupled to the oscillator, which would be able to fractionate the large quantum [30]. A key feature of this kind of model is that the associated coherent dynamics [31] ends up predicting a maximum coherent energy exchange rate linear in the coupling matrix, so that the Gamow factor associated with tunneling only comes in once [30]; as opposed to incoherent transitions where the Gamow factor comes into the rate quadratically. There has been much discussion over the past quarter century of how to avoid the Coulomb barrier [32], which in our view is impossible; instead, the impact of the Coulomb barrier on a coherent process is very much reduced. Estimates for the maximum reaction rate from this kind of model come out to be in the same range as experimental observations with no requirement for heroic screening effects.

Although the focus has been on the D_2/^{4}He transition for excess heat in connection with the Fleischmann–Pons experiment, it is clear that the same mathematical models would apply equivalently to HD/^{3}He transitions which are candidate reactions for light water experiments. Unimportant so far, but also similar is the HT/^{4}He transition. We have in the past proposed that a D_2/HT coherent reaction could be responsible for tritium production in PdD experiments. We worry about the D_2/n^{3}He channel which we fear might have been responsible for effects seen by Kevin Wolf [33]. In the case of light water systems, perhaps tritium can be produced by the beta decay of HD under conditions of incomplete fractionation.

Fractionation of the large quantum would be accomplished by nuclear transitions strongly coupled to the highly-excited oscillator. In the case of PdD systems, we contemplate the possibility that transitions in the deuteron may be responsible for optical phonon mode excitation, and transitions in the host Pd nuclei responsible in the case of acoustic mode operation. For Ni gas loaded systems we consider transitions in the host Ni nuclei as candidates for the fractionation. Note that fractionation in the host lattice nuclei has the potential to lead to elemental anomalies [34,35] through a coherent disintegration process [36].

4. Interaction and Collimated X-ray Emission in the Karabut Experiment

Once a mechanism was found that could down-convert a large nuclear quantum into a large number of smaller quanta, it was thought that a theory for excess heat in the Fleischmann–Pons experiment would quickly follow. Unfortunately, there were more difficulties. The rate at which excess heat was produced in experiment seemed consistent with the maximum rate predicted in the donor and receiver model [30] if the associated MeV-level quantum were fractionated; and it was clear that the scaling laws were sufficiently gentle that an MeV-level quantum could be fractionated down to quantum below an eV; however, try as we might we were unable for years to identify a specific transition capable of the required fractionation. The basic problem was that the weak second-order coupling leading to phonon exchange in connection with a nuclear transition was far too weak to mediate the needed down-conversion.

Progress on this problem next came from an unexpected quarter. Since there are not energetic nuclear products commensurate with the energy produced in the Fleischmann–Pons experiment, it does not seem possible to prove the reaction mechanism under consideration in this review by direct measurements. On the other hand, if the down-conversion mechanism is involved, we should be able to study it in isolation in some other application. We pursued the design of such an experiment which focused on up-converting many low energy vibrational quanta to produce a large quantum needed for nuclear excitation. The led to consideration of an experiment focusing on nuclear excitation in ^{201}Hg, where the 1565 eV transition is the lowest among stable nuclei starting from the ground state [15]. Subsequently came the interpretation of collimated X-ray emission near 1.5 keV in the Karabut experiment [37–39] as arising from the up-conversion of vibrational energy leading to phased array collimated emission from impurity Hg nuclei as contaminants on the surface.

It became clear that we should be able to apply the new model to analyze the Karabut experiment; and when done initially we concluded that the weak second-order interactions which we had been working with were just not strong enough to produce the effects observed by Karabut. Whatever interaction produced nuclear excitation and exchanged phonons needed to be orders of magnitude stronger. Ultimately this led to the identification of the first-order interaction Hamiltonian from a relativistic description of the nuclei in the lattice [40,41] as the only possible interaction capable of producing the requisite fractionation power. Still, there remained problems in making a connection between theory and experiment. The basic picture being contemplated for the Karabut experiment involved vibrations in the small metal cathode as responsible for the up-conversion; however, we were not able to get quantitative agreement between theory and experiment [42,43]. Incredibly, within this picture the coupling seemed still to be too weak. The basic model was re-analyzed, and a new regime of the model was found where the coupling was much stronger [44], which seemed to provide a resolution.

Meanwhile, tests for X-ray emission and charge emission from a vibrating copper foil were ongoing at SRI and later at MIT [45,46]. These experiments did not produce X-ray emission under conditions where one might have expected if the above interpretation of the Karabut experiment were correct. In the anomalous regime of the model there could be sufficient up-conversion strength in the copper foil, as long as sufficiently strong transitions were present to get into the anomalous regime in the first place. For a copper foil, the lowest energy excited states are above 500 keV (too high to be useful), so the only candidate transitions strong enough to reach the anomalous regime in principle are those connecting to negative energy states. Hence, the tentative interpretation of a negative result from the SRI and MIT experiments with copper is that transitions with negative energy states do not mediate up-conversion and down-conversion [47]. This suggested a revised interpretation of the Karabut experiment, where MHz vibrations in the much more massive vacuum chamber was responsible for the up-conversion, with the 14.4 keV transition in ^{57}Fe mediating the conversion. Such an interpretation brings into alignment the water jet experiment of Kornilova et al. [48] where collimated X-rays are reported observed from a nearby steel plate.

We note the possibility of up-conversion to produce coherent disintegration [36] in connection with the experiments of Carpinteri and co-workers [49]. If correct, this suggests the possibility of the up-conversion of vibrational energy to produce disintegration in some high-Z element (such as lead) for excess heat production by cold fission.

So far there have been calculations of two matrix elements based on the new phonon-exchange interaction. One is for coupling with deuterons in the lattice [50]; which leads to selection rules which appear to be consistent with magnetic field effects having a substantial effect on excess heat production in PdD systems [51,52]. The other is for phonon exchange in D_2/^{4}He transitions [53], which results in a matrix element sufficiently large to be relevant to excess heat production experiments [54].

5. Physics of PdD and D/Pd Loading

Deuterium is loaded into palladium electrochemically in the Fleischmann–Pons experiment, so to model the Fleischmann–Pons experiment we need to understand some of the associated (conventional) physics. One can find reviews of metal hydrides in general and palladium hydride in particular in the literature [55–61]. Over the years people have used a variety of theoretical methods to model hydrogen in Pd, starting with the important analysis of [62]. We pursued density functional calculations [63,64] in an effort to understand the phase diagram from first principles; from this effort we understood that even though the accuracy of such methods is very good, one needs better accuracy to model the phase diagram. From our perspective, the best approach to develop the models needed is an empirical one. We recently made use of empirical models including both octahedral and tetrathedral site occupation to compare against experimental data for α-phase PdH_x and PdD_x, with the result that it is possible to estimate quantitatively model parameters for both the O-site and T-site energies and other model parameters [65]. A similar model was used to compare with experiment in the case of high loading [66], and once again we were able to extract estimates for the model parameters. The resulting models are useful in describing PdH and PdD in the β-phase up to higher than stoichiometric loading.

Hydrogen and deuterium evolution reaction models are used to model the electrochemical loading of the Pd surface and bulk. We were interested in these models in order to understand why there was so much difficulty in obtaining high D/Pd ratios in the early years following the announcement of cold fusion in 1989. We focused on a subset of the reactions in the Volmer-Tafel regime, and noticed that there was a big variation in the resulting model parameters when different specific experiments were considered. Our conclusion was that it was likely that the rate of internal D_2 leaks in the different cathodes varied by as much as three orders of magnitude [67,68].

For modeling the solubility and also the loading of D and H in Pd we need a good model for the fugacity of H_2 and D_2 gas, especially under conditions of high loading. We examined different empirical models that have been developed over the years [69] and concluded that the best available model at this point is a recent one due to Joubert [70]. In this study we were also concerned with the difference between the fugacity of H_2 and the fugacity of D_2, concluding that the small difference between them was not accurately dealt with over the full range currently by any set of models.

6. Monovacancies and Co-deposition

It was recognized early on that getting two deuterons close together in PdD was problematic due to the high background electron density due to the Pd [71]. It was also noted early on that the repulsion between hydrogen or deuterium atoms was reduced in vacancies [72] since the background electron density is reduced. Another early paper focuses on the possibility of close encounters between deuterium atoms in a monovacancy with 6–9 deuterium atoms [73].

Sigma-bonded dihydrogen complexes involve the bonding of molecular H_2 as a molecule; they were discovered in the 1980s [74], and have been the focus of much subsequent research. For many years we have been interested in the possibility of dideuterium formation in a monovacancy in PdD [75–77]. The basic idea is that deuterons would be in close proximity in such a configuration, and the high background electron density is helpful for screening (enhanced screening is thought to play a role in low-energy deuteron beam experiments [78]). A dihydrogen complex similar to what we proposed has been found on the surface of highly loaded PdH [79].

In the past few months a modified version of this scenario has become of interest to us. We note that the O-site and T-site states around a monovacancy in Pd and Ni are nearly degenerate [81,80], so that when H or D atoms occupy a monovacancy they might occupy either O-sites or T-sites democratically until repulsion effects become important. Hence with an occupation of six deuterium atoms, a stable configuration could be where all the O-sites are occupied; and with eight deuterium atoms, all the T-sites could be occupied; in both cases there are no O–T nearest neighbors. When deuterium atoms occupy neighboring octahedral and tetrahedral sites, there is a repulsion energy [82] on the order of 150 meV. Current thinking is that the deuterium chemical potential becomes sufficiently high above a bulk D/Pd loading of 0.83 for neighboring O–T occupation in monovacancies, where tunneling might be enhanced by screening due to background electrons. This kind of configuration would be similar in Pd and Ni monovacancies, and probably for other FCC lattices as well. It seems likely something similar can happen in vacancies in BCC metal deuterides.

Probably a large number of monovacancies is required for excess heat production in the Fleischmann–Pons experiment, so the issue of how they are made becomes relevant. The vacancy formation energy in Pd is near 1.6 eV, which implies a low vacancy population in thermal equilibrium. However, the vacancy formation energy is decreased for every neighboring H or D atom, and at sufficiently high loading the vacancies become favored in thermal equilibrium. This was exploited by Fukai and Okuma [83,84] who created vacancy phase PdH and NiH which had 25% host metal atom vacancies. High loading is known to be important in the Fleischmann–Pons experiment. We have interpreted the requirement that Pd cathodes must reach a D/Pd loading of 0.95 [85] for excess heat to be seen as consistent with where vacancies are stabilized near room temperature. The problem is that even if stabilized, vacancy diffusion is sufficiently slow at room temperature that one would expect negligible formation of superabundant vacancies in a Fleischmann–Pons cell due to diffusion alone. We proposed that superabundant vacancy formation should occur in new material created through co-deposition, so that the monovacancy active sites are near the surface. We interpreted the early Szpak co-deposition experiment [86] as producing excess heat with no need for an extended loading period due to superabundant vacancy creation due to co-deposition under conditions of high loading so that vacancies are thermodynamically preferred. Support for this point of view comes from the Letts high-current density co-deposition protocol [87], which was confirmed by Miles [88]. This approach was exploited for cathode preparation in the cryogenic experiments done recently at SRI [89,90]. Co-deposition at lower current density would be expected to make fewer vacancies, resulting in insufficient active sites for sufficient gain for excess heat production, but which result in low-level nuclear emissions [91].

In the case of the NiH gas loading experiments of the Piantelli group [92,93] the question might be asked as to how monovacancies might be created since hydrogen solubility is so low in bulk Ni. In this case, if the H/Ni ratio is sufficiently high so that the loading is in the mixed phase region of the miscibility gap, then H is present in regions of α-phase and β-phase NiH_x. This is interesting as there is the possibility that the hydrogen density in the regions of clumped β-phase material to be sufficiently high to stabilize the vacancies. At the elevated temperature of the Piantelli experiment, there is the possibility of some vacancy diffusion. This potentially provides a mechanism for the development of increasing levels of vacancies with cycles of H loading and de-loading near the surface. If the Ni contains impurities, there is the possibility that the sites near the impurities will look attractive the the H, and also that the vacancy formation energy for the vacancy will start out a bit lower than for Ni. This may be relevant to the claims made for the recent experiment of Parkhomov (discussed at ICCF19), which involves impure Ni and quite high temperature.

In light of the comments above, it seems clear that the Pd and Ni structures produced in experiments to make superabundant vacancies would be attractive candidates for experimentation, since the monovacancy concentration in maximized. We are not aware of such experiments having been carried out at this point, even though we have advocated this approach for two decades.

7. Other Issues

Suppose that a candidate material has been produced with a large number and density of monovacancies, and that the sample has been loaded with whatever mixture of D and/or H is desired. Of interest then is to stimulate the sample in order to trigger excess heat production. Within the approach under discussion there are some theoretical issues to be considered. Vibrations in the lattice involve a large number of modes; to maximize the coherent dynamics of the models we are interested in an ideal system in which a single highly-excited vibrational mode is present, preferably one which interacts with D_2 or HD equivalently at all of the active sites. In PdD the Γ-point optical phonon mode works this way, and also has the added advantage that the group velocity is zero, so that vibrational energy generated at the Γ-point will remain largely localized in space. This leads us to consider Letts' two-laser experiment [94,97], in which two visible lasers tuned so that the difference frequency is in the THz range is scanned. Maximum excess heat is found at three frequencies; the lowest near 8.2 THz, identified with the optical phonon modes near the Γ-point; a second near 15.1 THz, identified with optical phonon modes near the L-point (also where the group velocity is zero); and a third near 20.8 THz. The latter peak does not have a confirmed identification; we have speculated that it might be due to the L-point associated with H impurities. We note that a confirmation of this experiment has not yet been obtained. Of interest in these experiments is that once excess heat has been stimulated in the two laser experiment, it is observed to remain on, or to decay slowly following shutting off the lasers. An interpretation of this effect is that the two lasers introduce an initial excitation in the optical phonon mode locally causing nuclear energy from the coherent process to be drawn into the excited modes (analogous to what happens in a laser); probably the area where the reactions occur expands over several hour time scale. When the lasers are turned off, the thought is that the nuclear energy continues supporting excitation of the vibrational modes, providing indirect support for the conjecture that nuclear energy is down-converted in this experiment. Direct support for the conjecture would come if large amplitude optical phonon vibrations were seen directly in these experiments, either with optical Raman type measurements, or with inelastic neutron scattering.

This suggests that triggering excess heat could be accomplished simply by providing excitation in high frequency vibrational modes, which could be done through interstitial hydrogen or deuterium diffusion, ion bombardment, electrical stimulation, or electromagnetic stimulation (excess heat has been observed under conditions which could be interpreted involving THz vibrational excitation by each of these mechanisms). In the event that the vibrational mode loss is low, as in the two-laser experiment, it may be that the vibrational energy can be maintained so that no further stimulation is needed. If the group velocity is high so that the vibrational energy leaves the area where the monovacancy active sites are, or if the vibrational modes involves some complicates superposition of local plane wave modes excited by interstitial diffusion, then one can imagine that the loss of vibrational energy is sufficiently high that the coherent process can only be maintained by continued vibrational excitation. In the first case, we might identify sustained excess heat production with what has been termed "Mode B" operation (in which the excess heat in a Fleischmann–Pons cell is nearly independent of the current density once initiated), while the second case would correspond with the more typical "Mode A" operation (where the excess heat is strongly correlated with the electrochemical current density).

If the system is working well, and D_2 is converted to ^{4}He (which is born stationary with all of the reaction energy going into vibrations), then one might expect ^{4}He accumulation in the monovacancies. This helium accumulation would be expected to clog up the monovacancy active sites, limiting the rate of excess power production. This picture could account for the observation that power levels seem to increase as the temperature gets higher, as helium diffusion away from the active sites is more efficient as the temperature increases. In some experiments the temperature dependence of the excess power is consistent with the helium diffusion coefficient [7]. One could imagine conditions under which the bottleneck might be the initial removal of the helium from the monovacancy which would involve a higher barrier energy [97]. We have proposed that the issue of helium removal may account for the advantage of nano-scale excess heat systems, such as described in [98].

According to the models, the ability of an equivalent lossy spin-boson system to fractionate a large quantum depends on the microscopic coupling strength, as well as other factors including the Dicke number [27], which can be larger if there are more two-level systems. There are a number of implications for this. One is that the HD/^{3}He transition is one that is "easier" to work with since the large quantum is 5.5 MeV which is less than the 23.9 MeV of the D_2/^{4}He transition. One could imagine a cold fusion system which runs with the D_2/^{4}He transition when it works well, but if it were running less well so to be unable to fractionate the large 24 MeV quantum, one could imagine it switching over to the D_2/HT transition. One could also contemplate a frustrated version of the system only able to fractionate a much smaller quantum, which perhaps would lead to low-level nuclear emissions as seen in the SPAWAR experiments [91].

References

[1] M. Fleischmann, S. Pons and M. Hawkins, *J. Electroanal Chem.* **201** (1989) 301; errata, **263** (1990) 187.

[2] M. Fleischmann, S. Pons, M.W. Anderson, L.J. Li and M. Hawkins, *J. Electroanal. Chem.* **287** (1990) 293.

[3] B.F. Bush, J.J. Lagowski, M.H. Miles and G.S. Ostrom, Helium production during the electrolysis of D_2O in cold fusion, *J. Electroanal. Chem.* **304** (1991) 271.

[4] M.H. Miles and B. Bush, Search for anomalous effects during D_2O electrolysis using palladium cathodes, *Proc. Third Int. Conf. on Cold Fusion*, Nagoya, Japan, 1992, p. 189.

[5] M.H. Miles, R.A. Hollins, B.F. Bush, J.J. Lagowski and R.E. Miles, Correlation of excess power and helium production during D_2O and H_2O electrolysis using palladium cathodes, *J. Electroanal. Chem.* **346** (1993) 99.

[6] M.H. Miles, B. Bush and J.J. Lagowski, Anomalous effects involving excess power, radiation and helium production during D_2O electrolysis using palladium cathodes, *Fusion Technol.* **25** (1994) 478.

[7] P.L. Hagelstein, M.C.H. McKubre, D.J. Nagel, T.A. Chubb and R.J. Hekman, New physical effects in metal deuterides, *Proc. 11th Int. Conf. on Cold Fusion*, Marseilles, France, 2004, p. 23.

[8] P.L. Hagelstein, Constraints on energetic particles in the Fleischmann–Pons experiment, *Naturwissenschaften* **97** (2010) 345.

[9] P.L. Hagelstein, Secondary neutron yield in the presence of energetic alpha particles in PdD, *J. Condensed Matter Nucl. Sci.* **3** (2010) 41–49.

[10] P.L. Hagelstein, On the connection between K_α X-rays and energetic alpha particles in Fleischmann-ŰPons experiments, *J. Condensed Matter Nuclear Science* **3** (2010) 50–58.

[11] Y. Kim, Conventional nuclear theory of low-energy nuclear reactions in metals: Alternative approach to clean fusion energy generation, *J. Condensed Matter Nucl. Sci.* **13** (2014) 264–276.

[12] P.L. Hagelstein, Neutron yield for energetic deuterons in PdD and in D_2O, *J. Condensed Matter Nucl. Sci.* **3** (2010) 35–40

[13] P.L. Hagelstein, A unified model for anomalies in metal deuterides, *Proc. 9th Int. Conf. on Cold Fusion*, Beijing, China, 2002, p. 121.

[14] P.L. Hagelstein, Unified phonon-coupled SU (N) models for anomalies in metal deuterides, *Proc. 10th Int. Conf. on Cold Fusion*, Cambridge, USA, 2003, pp. 837–870.

[15] P.L. Hagelstein, Bird's eye view of phonon models for excess heat in the Fleischmann-ŰPons experiment, *J. Condensed Matter Nucl. Sci.* **6** (2011) 169.

[16] P.L. Hagelstein, On the phonon model in cold fusion/LENR, *Infinite Energy* **112** (2013) 12–17.

[17] P.L. Hagelstein and I.U. Chaudhary, Phonon models for anomalies in condensed matter nuclear science, *Current Sci.* **108** (2015) 507–513.

[18] G. Verner, M. Swartz and P. Hagelstein, Summary report: Introduction to Cold Fusion–IAP course at the Massachusetts Institute of Technology, *Current Sci.* **108** (2015) 653–654.

[19] P.L. Hagelstein and I.U. Chaudhary, Level splitting in association with the multiphoton Bloch–Siegert shift, *J. Phys. B: At. Mol. Opt. Phys.* **41** (2008) 035601.

[20] P.L. Hagelstein and I.U. Chaudhary, Multiphoton Bloch–Siegert shifts and level-splittings in spin-one systems, *J. Phys. B: At. Mol. Opt. Phys.* **41** (2008) 035602.

[21] P.L. Hagelstein and I.U. Chaudhary, Multiphoton Bloch–Siegert shifts and level splittings in a three-level system, *J. Phys. B: At. Mol. Opt. Phys.* **41** (2008) 105603.

[22] P.L. Hagelstein and I.U. Chaudhary, Excitation transfer in two two-level systems coupled to an oscillator, *J. Phys. B: At. Mol. Opt. Phys.* **41** (2008) 135501.

[23] P.L. Hagelstein and I.U. Chaudhary, Energy exchange using spin–boson models with infinite loss, *J. Condensed Matter Nucl. Sci.* **4** (2011) 202–212.

[24] P.L. Hagelstein and I.U. Chaudhary, Energy exchange in the lossy spin–boson model, *J. Condensed Matter Nucl. Sci.* **5** (2011) 52–71.

[25] P.L. Hagelstein and I.U. Chaudhary, Second-order formulation and scaling in the lossy spin–boson model, *J. Condensed Matter Nucl. Sci.* **5** (2011) 87–101.

[26] P.L. Hagelstein and I.U. Chaudhary, Local approximation for the lossy spin–boson model, *J. Condensed Matter Nucl. Sci.* **5** (2011) 102–115.

[27] P.L. Hagelstein and I.U. Chaudhary, Coherent energy exchange in the strong coupling limit of the lossy spin–boson model, *J. Condensed Matter Nucl. Sci.* **5** (2011) 116–139.

[28] P.L. Hagelstein and I.U. Chaudhary, Errata and comments on a recent set of papers in *Journal of Condensed Matter in Nuclear Science*, *J. Condensed Matter Nucl. Sci.* **7** (2012) 1–7.

[29] P.L. Hagelstein and I.U. Chaudhary, Pulse and amplitude approximation for the lossy spinŰboson model, *J. Condensed Matter Nucl. Sci.* **9** (2012) 30–49.

[30] P.L. Hagelstein and I.U. Chaudhary, Generalization of the lossy spin–boson model to donor and receiver systems, *J. Condensed Matter Nucl. Sci.* **5** (2011) 140–154.

[31] P.L. Hagelstein and I.U. Chaudhary, Dynamics in the case of coupled degenerate states, *J. Condensed Matter Nucl. Sci.* **5** (2011) 72–86.

[32] T.A. Chubb and S.R. Chubb, Overcoming the Coulomb barrier in cold fusion, *J. Condensed Matter Nucl. Sci.* **2** (2009) 51–59.

[33] T.O. Passell, Radiation data reported by Wolf at Texas A&M as transmitted by T. Passell, *EPRI Report*.

[34] V. Nassisi, G. Caretto, A. Lorusso, D. Manno, L. Famà, G. Buccolieri and U. Mastromatteo, Modification of Pd–H_2 and Pd–D_2 thin films processed by He–Ne laser, *J. Condensed Matter Nucl. Sci.* **5** (2011) 1–6.

[35] E. Campari, S. Focardi, V. Gabbani, V. Montalbano, F. Piantelli, S. Veronesi, S. and I.U. Siena, Surface analysis of hydrogen loaded nickel alloys, *Proc. 11th Int. Conf. on Cold Fusion*, Marseilles, France, 2004, pp. 414–420.

[36] P.L. Hagelstein and I.U. Chaudhary, Anomalies in fracture experiments and energy exchange between vibrations and nuclei, *Meccanica* **50** (2015) 1189–1203.

[37] A.B. Karabut, Research into powerful solid X-ray laser (wave length is 0.8–1.2 nm) with excitation of high current glow discharge ions, *Proc. 11th Int. Conf. on Emerging Nuclear Energy Systems*, 29 September–4 October 2002, Albuquerque, New Mexico, USA, pp. 374–381.

[38] A. B. Karabut, Experiments characterizing the X-ray emission from a solid state cathode using a high-current glow discharge, *Proc. 10th Int. Conf. on Cold Fusion*, August 24–29, 2003, Cambridge, MA, USA 585–596.

[39] A.B. Karabut, E.A. Karabut and P.L. Hagelstein, Spectral and temporal characteristics of X-ray emission from metal electrodes in a high-current glow discharge, *J. Condensed Matter Nucl. Sci.* **6** (2012) 217–240.

[40] P.L. Hagelstein and I.U. Chaudhary, Including nuclear degrees of freedom in a lattice hamiltonian, *J. Condensed Matter Nucl. Sci.* **7** (2012) 35–50.

[41] P.L. Hagelstein and I.U. Chaudhary, Phonon–nuclear coupling for anomalies in condensed matter nuclear science, *J. Condensed Matter Nucl. Sci.* **12** (2014) 105–142.

[42] P.L. Hagelstein and I.U. Chaudhary, Lossy spinŰboson model with an unstable upper state and extension to N-level systems, *J. Condensed Matter Nucl. Sci.* **11** (2012) 59–92.

[43] P.L. Hagelstein and I.U. Chaudhary, Born-ŰOppenheimer and fixed-point models for second-order phonon exchange in a metal, *J. Condensed Matter Nucl. Sci.* **12** (2011) 69–104.

[44] P.L. Hagelstein and I.U. Chaudhary, Models for phonon–nuclear interactions and collimated X-ray emission in the Karabut experiment, *J. Condensed Matter Nucl. Sci.* **13** (2014) 177–122.

[45] F.L. Tanzella, P.L. Hagelstein, J. Bao and M.C.H. McKubre, Progress report on an experiment to clarify collimated X-ray emission in Karabut's experiment, *X-ray Lasers 2014: Proc. the 14th Int. Conf. on X-ray Lasers*, Fort Collins, Colorado, USA, 2015.

[46] P.L. Hagelstein, Directional X-ray and gamma emission in experiments in condensed matter nuclear science, *Current Sci.* **108** (2015) 601–607.

[47] F.L. Tanzella, P.L. Hagelstein, J. Bao and M.C.H. McKubre, Seeking X-rays and charge emission from a copper foil driven at MHz frequencies, submitted to *Proc. ICCF19.*

[48] A.A. Kornilova, V.I. Vysotskii, N.N. Sysoev, N.K. Litvin, V.I. Tomak and A.A. Barzov, Generation of intense X-rays during ejection of a fast water jet from a metal channel to atmosphere, *J. Surface Investigation. X-ray, Synchrotron and Neutron Techniques* **4** (2010) 1008–1017.

[49] A. Carpinteri, G. Lacidogna, A. Manuello and O. Borla, Piezonuclear fission reactions from earthquakes and brittle rocks failure: evidence of neutron emission and non-radioactive product elements, *Exp. Mechanics* **53** (2013) 345–365.

[50] P.L. Hagelstein and I.U. Chaudhary, Coupling between a deuteron and a lattice, *J. Condensed Matter Nucl. Sci.* **9** (2012) 50–63.

[51] J. O'M. Bockris, R. Sundaresan, Z. Minevski and D. Letts, Triggering of heat and sub-surface changes in Pd–D systems, *Trans. Fusion Technol.* **26** (1994) 267–290.

[52] M. Swartz, G. Verner, J. Tolleson, L. Wright, R. Goldbaum and P.L. Hagelstein, *J. Condensed Matter Nucl. Sci.* **15** (2015) 66–80.

[53] P.L. Hagelstein and I.U. Chaudhary, Central and tensor contributions to the phonon-exchange matrix element for the $D_2/^4He$ transition, *J. Condensed Matter Nucl. Sci.* **11** (2013) 15–58.

[54] P.L. Hagelstein and I.U. Chaudhary, Models for excess heat in PdD and NiH, *Proc. ICCF17*, Seoul, South Korea, 2013.

[55] W. Mueller, J. P. Blackedge and G. G. Liebowitz, *Metal Hydrides*, Academic Press, New York, 1968.

[56] E. Wicke, H. Brodowsky and H. Züchner, Hydrogen in palladium and palladium alloys, in *Hydrogen in Metals*, Vol. 2, Springer, Berlin, 1978, pp. 73–155.

[57] T.B. Flanagan, Thermodynamics of metal–hydrogen systems, In *Metal Hydrides*, Springer, Berlin, 1981, pp. 361–377.

[58] L. Schlapbach, *Hydrogen in Intermetallic Compounds I.* Springer, New York, 1988.

[59] T.B. Flanagan and W.A. Oates, The palladium–hydrogen system, *Ann. Rev. Materials Sci.* **21** (1991) 269–304.

[60] Y. Fukai, *The Metal–Hydrogen System*, Springer, New York, 1991.

[61] F.D. Manchester, A. San-Martin and J.M. Pitre, The H–Pd (hydrogen–palladium) system, *J. Phase Equilibria* **15** (1994) 62–83.

[62] J.R. Lacher, A theoretical formula for the solubility of hydrogen in palladium, *Proc. Roy. Soci. London. Ser. A, Math. and Phy. Sci.* **161** (1937) 525–545.

[63] P.O. Orondo, A theoretical model of interstitial hydrogen: pressure–composition–temperature, chemical potential, enthalpy and entropy, MIT PhD Thesis, Cambridge, USA, 2012.

[64] P. Orondo and P.L. Hagelstein, Basic physics model for PdH thermodynamics, *J. Condensed Matter Nucl. Sci.* **13** (2014) 149–164.

[65] P.L. Hagelstein, O-site and T-site occupation of α-phase PdH_x and PdD_x, *J. Condensed Matter Nucl. Sci.*, submitted.

[66] P.L. Hagelstein, Empirical models for octahedral and tetrahedral occupation in PdH and in PdD at high loading, *J. Condensed Matter Nucl. Sci.*, in press.

[67] P.L. Hagelstein, M.C.H. McKubre and F.L. Tanzella, Electrochemical models for the Fleischmann–Pons experiment, *Proc. ICCF15*, Rome, Italy, 2009, p. 16.

[68] P.L. Hagelstein, Deuterium evolution reaction model and the Fleischmann–Pons experiment, *J. Condensed Matter Nucl. Sci.* **16** (2015) 46–63.

[69] P.L. Hagelstein, Equation of state and fugacity models for H_2 and for D_2, *J. Condensed Matter Nucl. Sci.* **16** (2015) 23–45.

[70] J.-M. Joubert, A Calphad-type equation of state for hydrogen gas and its application to the assessment of Rh–H system, *Int. J. Hydrogen Energy* **35** (2010) 2104–2111.

[71] O.B. Christensen, P.D. Ditlevsen, K.W. Jacobsen, P. Stoltze, O.H. Nielsen and J.K. No, HH interactions in Pd, *Phy. Rev. B* **40** (1989) 1993–1996.

[72] P. Nordlander, J.K. No, F. Besenbacher and S.M. Myers, Multiple deuterium occupancy of vacancies in Pd and related metals,

Phy. Rev. B **40** (1989) 1990–1992.

[73] V. Rosato and F. Cleri, Deuterium clusters in a strained palladium lattice, *J. Materials Res.* **5** (1990) 2094–2099.

[74] G.J. Kubas, *Metal Dihydrogen and s-Bond Complexes: Structure, Theory and Reactivity*, Springer, Berlin, 2001.

[75] P.L. Hagelstein and I.U. Chaudhary, Arguments for dideuterium near monovacancies in PdD, *Proc. ICCF15*, Rome, Italy, 2009, p. 282.

[76] J.H. He, L.F. Dechiaro, D.L. Knies, G.K. Hubler, K.S. Grabowski, A.E. Moser, D.D. Dominguez, D.A. Kidwell and P.L. Hagelstein, Stability of a hydrogen molecule in a vacancy of palladium hydrides, *Int. J. Hydrogen Energy* **37** (2012) 12351–12357.

[77] P.L. Hagelstein, Molecular D_2 near vacancies in PdD and related problems, *J. Condensed Matter Nuclear Sci.* **13** (2014) 138–148.

[78] K. Czerski, A. Huke, P. Heide and G. Ruprecht, The ^{2}H(d, p)^{3}H reaction in metallic media at very low energies, *Europhys. Lett.* **68** (2004) 363–369.

[79] P.K. Schmidt, K. Christmann, G. Kresse, J. Hafner, M. Lischka and A. Gross, Coexistence of atomic and molecular chemisorption states: H_2/Pd (210), *Phy. Rev. Lett.* **87** (2001) 096103.

[80] F. Besenbacher, B.B. Nielsen, J.K. Norskov, S.M. Myers and P. Nordlander, Interaction of hydrogen isotopes with metals: Deuterium trapped at lattice defects in palladium, *J. Fusion Energy* **9** (1990) 257–261.

[81] F. Besenbacher, J.K. Nørskov, M.J. Puska and S. Holloway, Excitation of hydrogen motion inside a nickel vacancy, *Phy. Rev. Lett.* **55** (1985) 852–855.

[82] R. Nazarov, T. Hickel and J. Neugebauer, Ab initio study of H-vacancy interactions in fcc metals: Implications for the formation of superabundant vacancies, *Phy. Rev. B* **89** (2014) 144108.

[83] Y. Fukai and N. Okuma, Evidence of copious vacancy formation in Ni and Pd under a high hydrogen pressure, *Japanese J. Appl. Phys.* **32** (1993) L1256–L1259.

[84] Y. Fukai and N. Okuma, Formation of superabundant vacancies in Pd hydride under high hydrogen pressures, *Phy. Rev. Lett.* **73** (1994) 1640–1643.

[85] M.C.H. McKubre, F.L. Tanzella and V. Violante, What is needed in LENR/FPE studies?, *J. Condensed Matter Nuclear Sci.* **8** (2012) 187–197.

[86] S. Szpak, P.A. Mosier-Boss and J.J. Smith, Reliable procedure for the initiation of the Fleischmann–Pons effect, *Second Ann. Conf. on Cold Fusion*, Lake Como, Italy, 1991, 87–91.

[87] D. Letts and P.L. Hagelstein, Modified Szpak protocol for excess heat, *J. Condensed Matter Nuclear Sci.* **6** (2012) 44–54.

[88] M. Miles, Co-deposition of palladium and other transition metals in H_2O and D_2O solutions, *J. Condensed Matter Nucl. Sci.* **13** (2014) 401–410.

[89] F. Tanzella, J. Bao, M. McKubre and P.L. Hagelstein, Stimulation of metal deuteride wires at cryogenic temperatures, *J. Condensed Matter Nucl. Sci.* **8** (2012) 176–186.

[90] M. McKubre, J. Bao, F. Tanzella and P.L. Hagelstein, Calorimetric studies of the destructive stimulation of palladium and nickel fine wires, *J. Condensed Matter Nucl. Sci.* **13** (2014) 337–345.

[91] P.A. Mosier-Boss, L.P. Forsley, F.E. Gordon, D. Letts, D. Cravens, M.H. Miles, M. Swartz, J. Dash, F. Tanzella, P.L. Hagelstein, M. McKubre and J. Bao, Condensed matter nuclear reaction products observed in Pd/D co-deposition experiments, *Current Sci.* **108** (2015) 656–659.

[92] S. Focardi, R. Habel and F. Piantelli, Anomalous heat production in Ni–H systems, *Il Nuovo Cimento A* **107** (1994) 163–167.

[93] S. Focardi, V. Gabbani, V. Montalbano, F. Piantelli and S. Veronesi, Large excess heat production in Ni–H systems, *Nuovo Cimento A* **111** (1998) 1233–1242.

[94] D. Letts, D. Cravens and P.L. Hagelstein, Dual laser stimulation and optical phonons in palladium deuteride, in low-energy nuclear reactions and new energy technologies, *Low-Energy Nuclear Reactions Sourcebook*, Vol. 2, American Chemical Society, Washington DC, 2009, pp. 81-Ű93.

[95] P.L. Hagelstein, D. Letts and D. Cravens, Terahertz difference frequency response of PdD in two-laser experiments, *J. Condensed Matter Nuclear Sci.* **3** (2010) 59–76.

[96] P.L. Hagelstein and D.G. Letts, Analysis of some experimental data from the two-laser experiment, *J. Condensed Matter Nuclear Sci.* **3** (2010) 77–92.

[97] P.L. Hagelstein and D.G. Letts, Temperature dependence of excess power in two-laser experiments, *J. Condensed Matter Nuclear Sci.* **13** (2014) 165–176.

[98] M.R. Swartz, G.M. Verner, J.W. Tolleson and P.L. Hagelstein, Dry, preloaded NANOR®-type CF/LANR components, *Current Sci.* **108** (2015) 595–596.

J. Condensed Matter Nucl. Sci. 19 (2016) 110–118

Research Article

Seeking X-rays and Charge Emission from a Copper Foil Driven at MHz Frequencies

F.L. Tanzella, J. Bao and M.C.H. McKubre

SRI International, Menlo Park, CA, USA

P.L. Hagelstein*

Massachusetts Institute of Technology, Cambridge, MA, USA

Abstract

The absence of energetic nuclear particles in the Fleischmann–Pons experiment has been interpreted as indicating that a large nuclear quantum can be down-converted into a great many low energy vibrational quanta. Models that describe this also suggest that low energy vibrational quanta can be up-converted to produce nuclear excitation. Karabut's collimated X-ray emission has been interpreted as being due to the up-conversion of vibrational energy in a small cathode to produce excitation in ^{201}Hg. To test this, we developed a new experiment to vibrate a copper foil with and without surface Hg, and we looked for X-ray emission and charge emission from the surface. Signals were observed in the detectors in both cases; however, the signals in the X-ray detector did not respect the absorption edge of the Be window and are artifacts; and the large current signals associated with charge emission did not charge a capacitor in a simple configuration. We conclude that both are artifacts. The absence of collimated X-ray emission in this case is interpreted as due to the absence of strong low energy nuclear transitions in the copper, and also as ruling out a candidate theoretical model involving up-conversion due to interactions with negative energy transitions. A new interpretation of the Karabut experiment focuses now on lower frequency vibrations in the massive steel cathode holder and vacuum chamber as responsible for the up-conversion, and transfer of the up-converted energy to surface ^{201}Hg to produce the collimated X-ray emission.

Keywords: Charge emission, Collimated X-ray emission, Karabut experiment, Up-conversion, Vibrations

1. Introduction

The most significant claim made by Fleischmann and Pons in 1989 was for their observation of an unexpected excess heat effect in electrochemical experiments where palladium was loaded with deuterium in a D_2O/LiOD electrolyte [1,2]. No commensurate chemical products were found, which motivated Fleischmann to conjecture that the origin of the energy produced must be nuclear. However, one would expect under normal conditions to observe energetic

*Corresponding author. E-mail: plh@mit.edu

nuclear radiation in amounts commensurate with the energy, and this is not seen in Fleischmann–Pons experiments [3]. For the scientific community, this absence of commensurate nuclear radiation has been sufficient to conclude the experiments must be in error [4]; for us it is a potential advantage to be able to produce clean nuclear energy in large amounts for possible commercial applications.

This situation leaves us with a number of scientific questions. For example, how might it be possible for nuclear energy to be released without a matching amount of nuclear radiation. A question often asked back in 1989 was: where are the neutrons? However, we recognize now that this absence of a commensurate amount of neutrons also extends to protons, alphas, other nuclear fragments, electrons, and gammas. Another important question has to do with what specific reaction mechanism is involved. For example, in a conventional exothermic nuclear reaction one has the possibility of detecting the reaction products as energetic particles. Based on the measurements of the products one normally can understand most everything important about how the reaction works, and in some cases make use of computation for a theoretical understanding. In the Fleischmann–Pons experiment the reaction mechanism cannot be similarly studied, since no commensurate energetic reaction products are created in the first place. ^{4}He has been detected in the off-gas correlated with the energy produced [5–8]; but due to the absence of secondary neutrons we know that the ^{4}He nucleus is born with almost none of the reaction energy [9,10].

All of this in the end leads to a tremendously puzzling situation. We do not know by what mechanism energy is produced in the Fleischmann–Pons experiment, and we cannot make use of normal nuclear diagnostics to study it since the primary reaction mechanism does not produce energetic nuclear products. If a bright theorist managed to deduce the microscopic reaction mechanism responsible, based on what we know it would not be possible to prove it directly (as with conventional nuclear reactions) by performing a Fleischmann–Pons experiment.

A theoretical approach has been developed over many years that is based on the down-conversion of the large nuclear quantum to low energy condensed matter degrees of freedom, such as optical phonons [11–15]. In an experiment which works this way, there should be large amplitude vibrations which could be detected by optical measurements, or using low-energy inelastic neutron scattering. Models developed to describe this effect also predict that vibrational energy should be able to be up-converted to produce nuclear excitation. This raises the possibility that we might be able to study the underlying mechanism responsible for excess heat in the Fleischmann–Pons experiment by developing new experiments that provide vibrational stimulation which is up-converted to produce nuclear excitation, and then detect the subsequent nuclear decay [16].

This line of thought motivated us to seek stable nuclei which have the lowest energy excited states among the stable nuclei, resulting in the identification of candidates such as ^{201}Hg (1565 eV), ^{181}Ta (6240 eV) and ^{181}Tm (8410 eV). For example, the up-conversion of vibrational energy could produce excitation of the 1565 eV transition in ^{201}Hg, which could be diagnosed by looking for electron or X-ray emission. If the nuclei are excited in-phase, it would be possible for the output radiation to be collimated. These considerations focus our attention on the high-current density glow discharge experiments of Karabut [17–22]. In 2002 Karabut reported observations of collimated X-ray emission near 1.5 keV from a variety of different cathodes and a selection of different discharge gases [17]. It was conjectured that strong vibrations in the cathode by voltage spikes in the discharge; that the nuclei in the cathode work together to up-convert the vibrational energy; that the excitation is transferred to a small number of ^{201}Hg nuclei on the surface; and finally that this in-phase excitation of the nuclei results in collimated X-rays through a phased array emission effect [16,23]. The associated line shape is broad, consistent with the broad line shape from the models, suggesting that perhaps Karabut's glow discharge implements the scheme under discussion.

2. The experiment

To make progress on this line of argument, we decided to develop an experiment in which a copper foil could be vibrated over a range of frequencies in the MHz regime, and with different amplitudes; with the possibility of coating a surface with mercury, and looking for electron and X-ray emission from the coated surface [24]. We chose to work with a copper foil since Hg binds well to the surface (and subsequently diffuses slowly into the bulk); and also because Karabut reported positive X-ray emission results with copper cathodes.

There are no measurements of vibrational frequencies excited in Karabut's experiments. From theory we would expect up-conversion to be easier the fewer phonons required in the up-conversion, which focused our attention on the fundamental cathode resonances which were estimated to be in the range of 50–250 MHz. Constraints imposed by the available electronics and supply of copper foils led to the selection of rolled copper foils roughly 73 μm thick for our experiment, with vibrational resonances above 20 MHz.

One surface was capacitively coupled to a thick driver; first through air, and later through PVDF to avoid air breakdown. No DC bias was used so that the frequency of the force on the foil surface was at twice the frequency of the voltage applied to the driver. The applied voltage and frequency was computer controlled through an Agilent 8648A signal generator and amplified with an ENI 603L linear amp. In early experiments we used measurements of the input impedance to try to detect vibrational energy storage and energy loss from the foil. Broad resonances were seen in some cases, which suggested the possibility that vibrational energy might leak into the holder. This motivated us to implement a low-loss cylindrical resonator to hold the foil (Fig. 1).

The 1565 eV excited state of ^{201}Hg decays primarily through internal conversion [25], so we implemented charge detection using a Keithley 617 electrometer, as illustrated in the schematic of Fig. 2. We made use of an Amptek X-123 Si-PIN detector to look for X-ray emission. We were able to measure for charge, or for X-ray emission, but only one at a time.

Figure 1. Close-up of driver, foil and resonator.

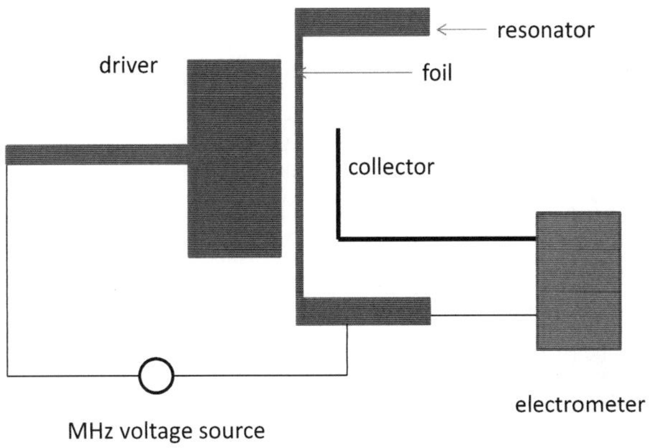

Figure 2. Simplified schematic of foil and resonator, driver, drive circuit and charge detection.

3. Measurement Results and Noise Issues

During the experimental campaign we saw signals on the electrometer that seemed to be consistent with charge emission, and which appeared to be correlated in frequency with some of the expected vibrational resonances; an example is shown in Fig. 3. The sign of the current was negative, consistent with electron emission. A bias on the order of 5 V results in a weak modification of the signal, and a bias near 100 V had a much stronger impact on it. The magnitude of the current signal was somewhat larger with no surface Hg, which suggests that whatever the origin of the effect, it

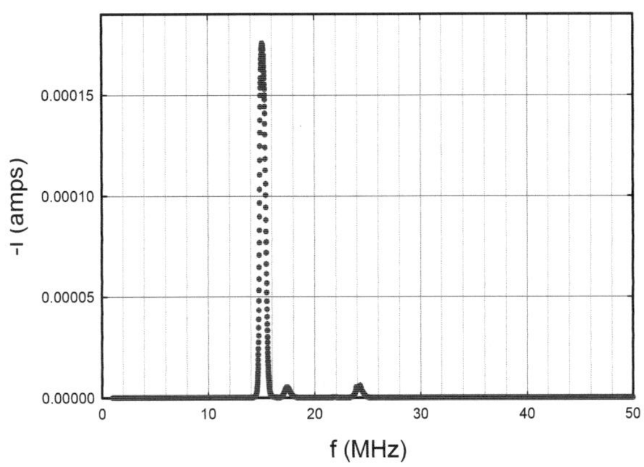

Figure 3. Example of a signal from the electrometer; I is the equivalent current and f is the drive frequency (the vibrational frequency is twice this drive frequency).

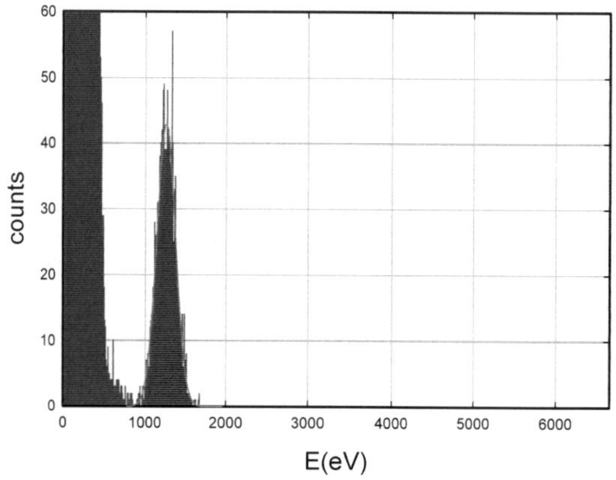

Figure 4. Example of a spectrum from the X-123 detector.

was not a result of electron emission from excited ^{201}Hg. In connection with the current signal we considered the possibility that up-converted vibrational energy was promoting electrons which could be picked up by air molecules and transported to the collector. In later tests it was found that there was no light evident near the foil surface in connection with the current signal, and also no obvious odor which would be expected if energetic electrons were ejected.

When the current signal was first seen, we were motivated to attempt measurements of X-rays. We saw features above 1 keV which appeared to be correlated with some of the expected vibrational resonances when Hg was on the surface; an example is shown in Fig. 4. It was possible to obtain clean spectrum with no surface Hg with no counts above 1 keV.

Perhaps it is understandable that results of this type could stimulate excitement. However, when we attended to writing up the experiment and results for publication, it became clear that there were unresolved issues associated with the calibration of the X-ray spectra. New calibration spectra were taken, and all of the data was systematically recalibrated and reviewed. In the process it became clear that the low energy feature near 1.5 keV that looks so promising in Fig. 4 would in some of the spectra show up at lower energies, and even at sufficiently low energy that it should have been absorbed by the 0.5 mil (12.5 μm thick) Be window (we would expect little X-ray transmission through this window below 1 keV). From this study it became clear that the detector was responding to something other than X-rays [24].

From discussions of this experiment with colleagues at MIT and elsewhere, it became clear that there was concern that the charge emission signals might also be artifacts. A similar experiment was set up at MIT and studied for some months, and in some cases signals showed up that were candidates for charge emission. With additional testing it was determined that these signals were noise. Near the end of this first MIT campaign we decided to develop a stringent test for charge emission at where the collector was connected through a capacitor to ground; in this case charge from the foil should charge up the capacitor and produce a measurable voltage. No significant voltage above a roughly 40 mV noise level was seen on a 5 pF capacitor at MIT.

Analogous experiments with a capacitor were carried out at SRI. In one set of experiments, a microfarad capacitor was used and monitored in real time with the electrometer running in voltage mode with a high impedance ($10^{14}\,\Omega$).

Strong signals were seen that appeared to provide a confirmation of the earlier current signals, showing a similar dependence on the drive frequency, and consistent with the sign and magnitude of the current signal. This appeared to provide confirmation of the charge emission result, and resulted in our submitting an abstract on our results for ICCF19.

A modified version of this test done subsequently led to different results, and has called into question the earlier current measurements. When a microfarad capacitor was again used for the test, but not monitored by the electrometer when the foil was driven, then no significant voltage was seen on the capacitor when measured after the excitation was turned off. Voltages up to about 10 V were seen with a 6 pF capacitor in this configuration, which would have been significant if real; but subsequent tests suggested that such signals were consistent with noise.

4. Discussion

It is clear from these studies that there is a learning curve associated with this kind of experiment. One issue is that using a linear amp to drive a low impedance reactive capacitor load ends up generating harmonics (at more than 10th order in experiments at MIT). The X-123 detector is very sensitive to noise; although at this point it has not been clarified whether the noise is due to RF, ultrasound, or due to something else. Measuring for charge emission is problematic in the presence of noise associated with the capacitive coupling excitation of the foil. The capacitor measurements suggest at this point that no charge emission has been seen; however it remains to be explained why monitoring the voltage on a microfarad capacitor with an electrometer that has an input impedance rated at 10^{14} ohms should lead to a false positive result.

Independent of how painful the learning curve associated with this new experiment might be, we are interested in how to implement a better version of the experiment generally, including reducing the noise, but also one which provides a better test. We are also concerned with the impact that a negative result so far on collimated X-ray emission has on the underlying theoretical picture.

It will be convenient to focus on the theoretical issue first. In the Introduction above we discussed briefly that a model had been developed over the years which describes both down-conversion relevant to the Fleischmann–Pons experiment, and up-conversion relevant to the Karabut experiment; it will be useful here to take this discussion a bit further. The early work on the models which focus on up-conversion and down-conversion required only that there be some coupling between the equivalent two-level systems standing in for the nuclear transitions, and the highly-excited oscillator which in the present discussion so far refers to the vibrational mode of the foil. When Karabut's experiment was analyzed making use of the relatively weak second-order coupling that would be expected between nuclear transitions involving phonon exchange, there was no way to obtain quantitative agreement between model and experiment. Subsequently, it became clear that agreement between theory and experiment could only come about if there were a much stronger coupling, which led to the adoption of a new fundamental Hamiltonian based on a relativistic description of the nuclei [26]. In this case there is a very strong first-order interaction between internal nuclear transitions and the center of mass momentum of the nuclei involved in the vibrations. Under normal conditions this first-order coupling is rotated out leaving a weak second-order relativistic correction that is usually ignored. However, we know from the analysis of the lossy spin-boson model (which is closely related algebraically) that the rotation becomes unhelpful for this kind of problem when the oscillator is strongly excited and when loss is present [27,28].

Once the new coupling became available, the theoretical up-conversion in Karabut's collimated X-ray experiments was revisited, multiple times. There were difficulties in the results from the modeling, which are worth considering here in light of the experimental results so far. As explained in the Introduction, the basic picture under consideration involved the up-conversion of vibrational quanta in the range of 50–250 MHz to produce nuclear excitation near 1.5 keV. In the normal up-conversion regime of the model relevant to this problem, the up-conversion power increases with a higher vibrational frequency, with stronger vibrations, with more nuclei involved, and with a strong low-energy nuclear transition. In the course of the campaign it became clear that the copper foil was not a particularly good candidate, since there are no strong low-energy nuclear transitions, and the number of nuclei present in a thin foil is limited. A response to this observation was to add the copper resonator to the SRI experiment in the hope of increasing the number of nuclei vibrating.

However, there remained headaches in the analysis. It became clear that even with the much stronger relativistic coupling, it was not possible to obtain agreement between theory and Karabut's experiment in the normal region of the model, assuming that a copper cathode was responsible for the up-conversion. Subsequently another pass was made through the analysis of the model which turned up an anomalous regime of the model, where the scaling laws were very different [29]. This anomalous regime of the model could be accessed when the coupling was extremely strong, or else if the nuclear transition energy is very low; and within this regime up-conversion is very much enhanced. Consequently, there seemed to be the possibility that a small cathode could provide the up-conversion needed to go from MHz to keV as long as it operated in the anomalous regime. For a copper cathode, since there are no relevant low energy nuclear transitions (the lowest energy excited state in ^{63}Cu is at 670 keV, and in ^{65}Cu is at 771 keV), this would imply that the very strong coupling required involved transitions with negative energy states. An analysis has not yet been done specific for negative energy state transitions, but we note that this is very much not an attractive model; we would only want to pursue this kind of approach if forced by experimental results that required it as the only possible option.

Given the unattractiveness of a model which relies on strong transitions to negative energy states for up-conversion in the anomalous regime of the model, it would be tempting to argue that a negative result with a copper foil rules out the scenario. Probably this would attribute too much weight to the present negative result. Nevertheless, a consideration of the different scenarios and issues suggests that we should re-think the model, the interpretation of the Karabut experiment, and we should also re-think the vibrating foil experiment. From this new perspective, our attention should be focused on stable isotopes with low-energy excited states, and we might expect these to be involved in the up-conversion process. Above we noted that ^{201}Hg and ^{181}Ta were candidates with the lowest energy transitions (at 1.56 and 6.24 keV); however, further down the list is ^{57}Fe with a transition energy of 14.4 keV [16]. This is significant due to the report of Kornilova et al. of the observation of collimated X-ray emission from a steel plate close to a water jet [30]. We recall that in Karabut's experiment the target chamber is made of steel.

This suggests that we should contemplate a revised interpretation for Karabut's collimated X-ray experiment. In this case, there are short (sub-ns) large amplitude (more than 50 kV) voltage spikes under normal operation in the high-current discharge which would be expected to couple to the cathode surface across the cathode fall region, and produce vibrations in the MHz regime that would vibrate a large number of ^{57}Fe nuclei in the roughly 80 pounds of steel target chamber. The vibrational frequency range of interest in this case is probably in the MHz region (note that the fundamental compressional mode in the 3 mm thick steel pate of Kornilova et al. is near 1 MHz). The conjecture is that vibrations in the large steel structure produces the up-conversion which is transferred to a small number [$O(10^{10})$] of impurity ^{201}Hg contaminant nuclei on the cathode surface, which then produces directional X-ray emission via a phased array emission effect.

In light of this new picture, a next generation version of the experiment should work with a larger sample that contains a suitable low-energy transition. Based on the interpretation of Karabut's collimated X-ray experiment, and also the interpretation of the water jet experiment of Kornilova et al., the focus should probably be on a steel sample; however also very much of interest would be a sample made of tantalum. We would like to drive this larger sample much harder, perhaps using a transducer matched to a resonance. To reduce noise we would like not to drive a reactive load with a linear amplifier. Given that the X-123 is so sensitive to noise, we would like to make use of X-ray film (keeping in mind that film has noise issues as well). We are continuing the experimental effort along these lines now at MIT.

References

[1] M. Fleischmann, S. Pons and M. Hawkins, *J. Electroanal Chem.* **201** (1989) 301; errata, **263** (1990) 187.

[2] M. Fleischmann, S. Pons, M.W. Anderson, L.J. Li and M. Hawkins, *J. Electroanal. Chem.* **287** (1990) 293.

[3] P.L. Hagelstein, M.C.H. McKubre, D.J. Nagel, T.A. Chubb and R.J. Hekman, New physical effects in metal deuterides, *Proc. ICCF11* (2004) 23.

[4] J.R. Huizenga, *Cold Fusion. The Scientific Fiasco of the Century*, Oxford University Press, Oxford, 1994.

[5] B.F. Bush, J.J. Lagowski, M.H. Miles and G.S. Ostrom, Helium production during the electrolysis of D_2O in cold fusion, *J. Electroanal. Chem.* **304** (1991) 271.

[6] M.H. Miles and B. Bush, Search for anomalous effects during D_2O electrolysis using palladium cathodes, *Proc. Third Int. Conf. on Cold Fusion*, 1992, Vol. 189.

[7] M.H. Miles, R.A. Hollins, B.F. Bush, J.J. Lagowski and R.E. Miles, Correlation of excess power and helium production during D_2O and H_2O electrolysis using palladium cathodes, *J. Electroanal. Chem.* **346** (1993) 99.

[8] M.H. Miles, B. Bush and J.J. Lagowski, Anomalous effects involving excess power, radiation, and helium production during D_2O electrolysis using palladium cathodes, *Fusion Technology* **25** (1994) 478.

[9] P.L. Hagelstein, Constraints on energetic particles in the Fleischmann–Pons experiment, *Naturwissenschaften* **97** 345 (2010).

[10] P.L. Hagelstein, Secondary NeutronYield in the presence of energetic alpha particles in PdD, *J. Cond. Mat. Nucl. Sci.* **3** (2011) 41.

[11] P.L. Hagelstein and I.U. Chaudhary, Energy exchange in the lossy spin-boson model, *J. Cond. Mat. Nucl. Sci.* **5** (2011) 52.

[12] P.L. Hagelstein and I.U. Chaudhary, Second-order formulation and scaling in the lossy spin-boson model, *J. Cond. Mat. Nucl. Sci.* **5** (2011) 87.

[13] P.L. Hagelstein and I.U. Chaudhary, Local approximation for the lossy spin-boson model, *J. Cond. Mat. Nucl. Sci.* **5** (2011) 102.

[14] P.L. Hagelstein and I.U. Chaudhary, Coherent energy exchange in the strong coupling limit of the lossy spin-boson model, *J. Cond. Mat. Nucl. Sci.* **5** (2011) 116.

[15] P.L. Hagelstein and I.U. Chaudhary, Errata and comments on a recent set of papers in Journal of Condensed Matter in Nuclear Science, *J. Cond. Mat. Nucl. Sci.* **7** (2012) 1.

[16] P.L. Hagelstein, Bird's eye view of phonon models for excess heat in the Fleischmann–Pons experiment, *J. Cond. Mat. Nucl. Sci.* **6** (2011) 169.

[17] A.B. Karabut, Research into powerful solid X-ray laser (wave length is 0.8–1.2 nm) with excitation of high current glow discharge ions, *Proc. 11th Int. Conf. on Emerging Nuclear Energy Systems*, 29 September–4 October 2002, Albuquerque, New Mexico, USA, pp. 374–381.

[18] A.B. Karabut, Experimental research into characteristics of X-ray emission from solid-state cathode medium of high-current glow discharge, *Proc. 10th Int. Conf. on Cold Fusion*, August 24–29, 2003, Cambridge, MA, USA.

[19] A.B. Karabut, Research into characteristics of X-ray emission laser beams from solid-state cathode medium of high current glow discharge, *Proc. 11th Int. Conf. on Cold Fusion*, 31 October-Ǘ5 November, 2004, France, pp. 253–257.

[20] A.B. Karabut, Study of energetic and temporal characteristics of X-ray emission from solid state cathode medium of high current glow discharge, *Proc. 12th Int. Conf. on Cold Fusion*, December 2–7, 2006, Japan, pp. 344–350.

[21] A.B. Karabut, E.A. Karabut, Research into energy spectra of X-ray emission from solid cathode medium during the high current glow discharge operation and after the glow discharge current switch off, *Proc. 14th Int. Conf. on Cold Fusion*, August 10–15, 2008, USA.

[22] A.B. Karabut, E.A. Karabut and P.L. Hagelstein, Spectral and temporal characteristics of X-ray emission from metal electrodes in a high-current glow discharge, *J. Cond. Mat. Nucl. Sci.* **6** (2012) 217.

[23] P.L. Hagelstein, Directional X-ray and gamma emission in experiments in condensed matter nuclear science, *Current Science* **108** (2015) 601.

[24] F.L. Tanzella, P.L. Hagelstein, J. Bao and M.C.H. McKubre, Progress report on an experiment to clarify collimated X-ray emission in Karabut's experiment, *Proc. Int. Conf. on X-ray Lasers ICXRL2014* (in press).

[25] V. Meot, J. Aupiais, P. Morel, G. Gosselin, F. Gobet, J. N. Scheurer and M. Tarisien, Half-life of the first excited state of Hg 201, *Phys. Rev. C* **75**(6)(2007) 064306. Half-life of the first excited state of ^{201}Hg, *Phys. Rev. C* **75** (2007) 064306.

[26] P.L. Hagelstein and I.U. Chaudhary, Including nuclear degrees of freedom in a lattice hamiltonian, *J. Cond. Mat. Nucl. Sci.* **7** (2011) 35.

[27] P.L. Hagelstein and I.U. Chaudhary, Phonon-nuclear coupling for anomalies in condensed matter nuclear science, *J. Cond. Mat. Nucl. Sci.* **12** (2014) 105.

[28] P.L. Hagelstein and I.U. Chaudhary, Phonon models for anomalies in condensed matter nuclear science, *Current Science* **108** (2015) 507.

[29] P.L. Hagelstein and I.U. Chaudhary, Models for phonon-nuclear interactions and collimated X-ray emission in the Karabut experiment, *J. Cond. Mat. Nucl. Sci.* **13** (2014) 177.

[30] A.A. Kornilova, V.I. Vysotskii, N.N. Sysoev, N.K. Litvin, V.I. Tomak and A.A. Barzov, Generation of intense X-rays during ejection of a fast water jet from a metal channel to atmosphere, *J. Surface Investigation. X-ray, Synchrotron and Neutron Techniques* **4** (2010) 1008–1017.

J. Condensed Matter Nucl. Sci. 19 (2016) 119–126

Research Article

The Launch of a New Plan on Condensed Matter Nuclear Science at Tohoku University

Yasuhiro Iwamura *, Jirohta Kasagi and Hidetoshi Kikunaga
Condensed Matter Nuclear Reaction Division, Research Center for Electron Photon Science, Tohoku University, Japan

Hideki Yoshino, Takehiko Itoh and Masanao Hattori
CLEAN PLANET Inc., Japan

Tadahiko Mizuno
Hydrogen Engineering Application & Development Company, Japan

Abstract

A new division devoted to Condensed Matter Nuclear Reaction (CMNR) was established at the Research Center for Electron Photon Science of Tohoku University in April 2015. This division consist of researchers from Tohoku University, Mitsubishi Heavy Industries and Clean Planet Inc., who have been actively engaged in the field of CMNR. In this division, fundamental research on condensed matter nuclear reaction, R&D on energy generation and nuclear waste decontamination will be performed.

Keywords: Collaborative research division, Condensed matter nuclear reaction, Electron screening, Nano-structured Pd–CaO complex, Permeation, Transmutation

1. Introduction

The Research Center for Electron Photon Science of Tohoku University and CLEAN PLANET Inc. agreed to the establishment of the collaborative research division: the Condensed Matter Nuclear Reaction Division. This is the first official research division for condensed matter nuclear science and its application in Japan.

In this paper, the research activities of participating members are briefly described. This includes the organization and research plan of the division, research background and objectives.

*E-mail: iwamura@Lns.tohoku.ac.jp

© 2016 ISCMNS. All rights reserved. ISSN 2227-3123

2. Activities of Participating Members

2.1. Tohoku university

J. Kasagi and his co-workers of Tohoku University have been investigating the effects of metal environments acting on nuclear reactions. They succeeded in measuring the reaction rates at lower bombarding energies, and were the first to deduce the screening energy of the d + d reaction in metal, which is a quantitative scale of the enhancement at very low energies [1–4].

The excitation functions of the yield of protons emitted in the D(d,p)T reaction in Ti, Fe, Pd, PdO and Au were measured for bombarding energies between 2.5 and 10 keV. It was found that the reaction rate at lower energies varies greatly with the host materials. The most strongly enhanced DD reaction occurs in PdO at E_d = 2.5 keV; it is enhanced by factor of fifty from the bare deuteron rate and the screening energy deduced from the excitation function, which amounts to 600 eV. Figure 1 shows the relative yield of protons emitted in the D(d,p)T reaction in the five hosts as a function of the bombarding energy of deuterons; (a) two independent measurements for PdO, (b) for Pd and Fe and (c) for Au and Ti. In the upper sections, the data normalized to the yield at 10 keV are plotted. In the lower sections, the experimental yields divided by those presented with the dotted curve are shown. The dotted curves correspond to the relative yields calculated for the bare DD reactions without screening. Solid and dashed curves correspond to calculations with the screening energy indicated in each section [4].

An enhancement of this size cannot be explained by electron screening alone but suggests the existence of an additional and important mechanism of the screening in solids. This is the first clear evidence that the fusion reaction rate can be enhanced in the metal environment during keV deuteron bombardment.

They also investigated on Li + d reaction in Pd [5], Li + d and Li + p reactions in Liquid Lithium [6], d + d and Li + d reactions in metal lithium acoustic cavitation [7] under low energy condition.

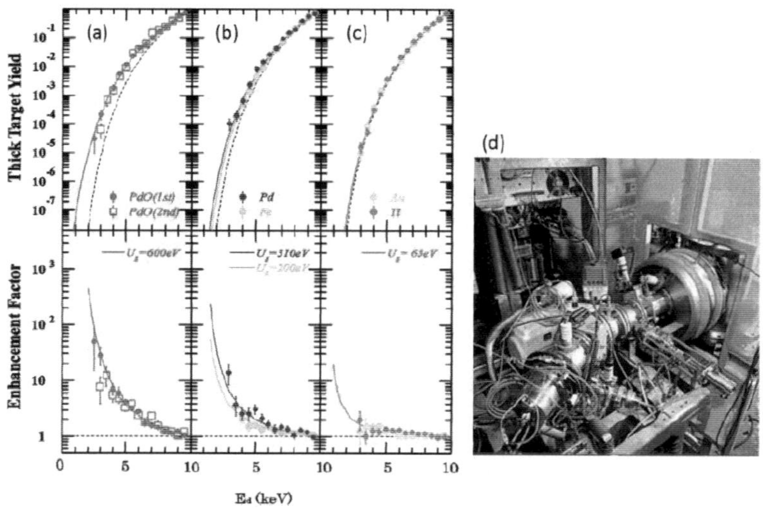

Figure 1. Excitation functions of the yield of protons emitted in the D(d,p)T reaction in Ti, Fe, Pd, PdO and Au [4].

2.2. Mitsubishi heavy industries Ltd.

Y. Iwamura, T. Itoh and their co-workers of the Advanced Technology Research Center of Mitsubishi Heavy Industries have been investigating the condensed matter nuclear transmutation reactions observed in the nano-sized Pd complexes induced by D_2 gas permeation. They first observed transmutation reaction of Cs into Pr using a very simple and compact setup shown in Fig. 2 [8].

They fabricated nano-structured Pd–CaO complexes with an element that is specifically targeted to be transmuted. The sample is a Pd complex composed of bulk Pd on the bottom, alternating CaO and Pd layers, and a Pd thin film on top. The multi-layers of CaO and Pd were obtained by alternately sputtering 2-nm-thick CaO and 20-nm-thick Pd layers five times. The top surface was Pd 40 nm thick. These processes were performed by the Ar ion beam sputtering method or the magnetron sputtering method. After fabricating the Pd–CaO complex, Cs was deposited on the surface of the thin Pd layer.

Elemental changes in the Pd complexes were measured by the X-ray Photoelectron Spectroscopy (XPS) without taking the complexes out of the vacuum chamber. This prevented contamination from outside of the chamber. The first and second results are plotted in Fig. 2(b). There was no Pr at the beginning of the experiments. The number of Cs atoms decreased while Pr atoms increased over time during the experiment. The number of atoms was evaluated by XPS spectra. The amount of deuterium permeation was proportional to the elapsed time. Pr was confirmed by the Inductively Coupled Plasma Mass Spectrometry (ICP-MS) and the (Time of Flight Secondary Ion Mass Spectrometry (TOF-SIMS) [9,10, 12] and the in-situ X-ray Fluorescence analysis (XRF) measurement [11,12]. As the control experiments:

(1) H_2 gas permeation experiments using the same Pd multilayer samples (Pd/CaO/Pd),
(2) D_2 gas permeation using the same Pd multilayer samples (Pd/CaO/Pd) without Cs,
(3) D_2 gas permeation using the Pd sample without CaO were performed.

These control experiments were performed under the same temperature and pressure as foreground experiments.

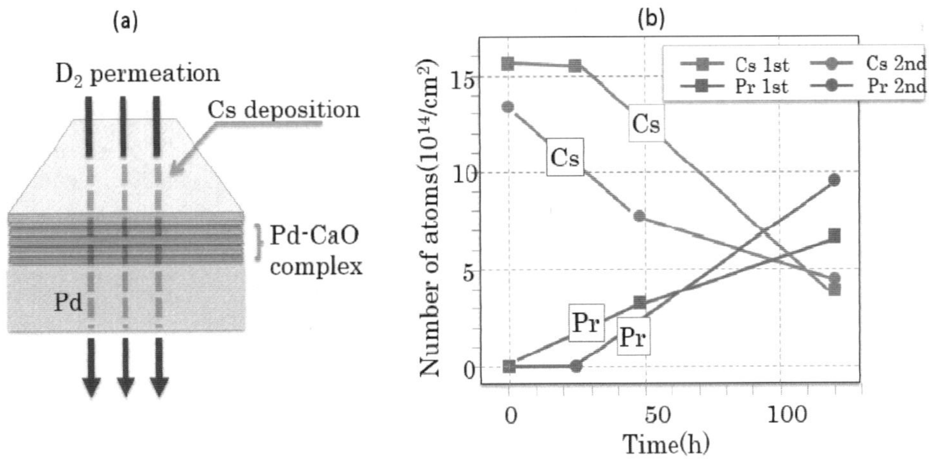

Figure 2. Schematic representation of experimental method; (a) Condensed matter nuclear transmutation method with the nano-sized Pd complexes induced by D_2 gas permeation and (b) experimental results on transmutation of Cs into Pr [8].

No Pr was detected in any of the control experiments. This suggests that both deuterium and nano-structured Pd multilayer with CaO are necessary factor to observe transmutation reactions.

Replication experiments have been performed in some universities and institutes, mainly in Japan. T. Higashiyama et al. of Osaka University observed transmutation of Cs into Pr in 2003 [13]. H. Yamada et al. performed similar experiments [14] using Cs and detected increase of mass number 137 by TOF-SIMS. They used a couple of nano-structured Pd multilayer thin film and observed the increase of mass number 141 (corresponding to Pr) only when ^{133}Cs was present on the Pd sample. T. Hioki and N. Takahashi et al., the researchers of Toyota Central R&D Labs detected Pr using ICP-MS from the permeated Pd sample. They presented these findings at ICCF-17 and published their results in the Japanese Journal of Applied Physics [15]. They performed careful experiments and concluded that the amount of Pr could not be explained by a contamination process.

These replications provide important information concerning about the nature of this phenomenon. Other than Osaka University, the samples were independently fabricated with different Pd dimensions or fabrication techniques. Nevertheless, similar results were observed. These experimental evidences strongly suggest to us that transmutation reactions of Cs into Pr occurred using this simple method.

The D_2 gas permeation transmutation reactions were observed for the other elements. Table 1 summarizes the typical transmutation reactions observed so far [12]. If we deposit ^{138}Ba on the nano-structured Pd film, we obtain ^{150}Sm by D_2 gas permeation. And if we deposit ^{137}Ba on the Pd complex, we obtain ^{149}Sm. The observed transmutation reactions of Ba into Sm belong to a reaction category in which the increase of mass number is 12 and the increase of atomic number is 6. Nuclear transmutation induced by this method is not limited to the category in which the increase of mass number is 8 and the increase of atomic number is 4 (Cs$\rightarrow$Pr).

The obtained experimental results so far suggest that a certain rule seems to exist for this deuterium permeation transmutation. We can notice that 2d, 4d or 6d appear to be reacting with the deposited elements. From another point of view, it might be considered that α-capture reactions occur in deuterium permeation experiments. At present, this is just speculation; however, it is important that a certain rule seems to exist.

To apply this deuterium permeation transmutation phenomenon to practical use it will be necessary to increase the mass of the transmutation products. This phenomenon might be used as an innovative nuclear transmutation method for radioactive waste, or as a new energy source if we can solve many technical problems that would prevent commercialization. Iwamura et al. have been trying to increase the amount of transmutation products for these years. They applied an electrochemical method to increase the local deuteron density near the surface of the nano-structured Pd multilayer film and increased the amount of transmutation products up to $\sim$1 μg/cm^2 [16,17]. Furthermore, they developed a solution circulation type of experimental apparatus aiming consecutive transmutation of Cs in solution [18].

2.3. Clean Planet Inc.

T. Mizuno, H. Yoshino and their colleagues have been developing excess heat generation methods and devices for the past few years. Figure 3 shows the experimental reactor setup and the example of excess heat generation [19]. Their experimental process is as follows. First, Ni metal was heated under vacuum condition and D_2 gas was introduced

Table 1. Typical transmutation reactions observed in Mitsubishi heavy industries Ltd. [12].

Elements		Apparent reactions
Cs	4d	${}^{133}_{55}\mathrm{Cs} \xrightarrow{4d(2\alpha)} {}^{141}_{59}\mathrm{Pr}$
Ba	6d	${}^{138}_{56}\mathrm{Ba} \xrightarrow{6d(3\alpha)} {}^{150}_{62}\mathrm{Sm}$, ${}^{137}_{56}\mathrm{Ba} \xrightarrow{6d(3\alpha)} {}^{149}_{62}\mathrm{Sm}$
W	4d or 2d	${}^{182}_{74}\mathrm{W} \xrightarrow{4d(2\alpha)} {}^{190}_{78}\mathrm{Pt}$? ${}^{186}_{74}\mathrm{W} \xrightarrow{2d(\alpha)} {}^{190}_{76}\mathrm{Os}$?

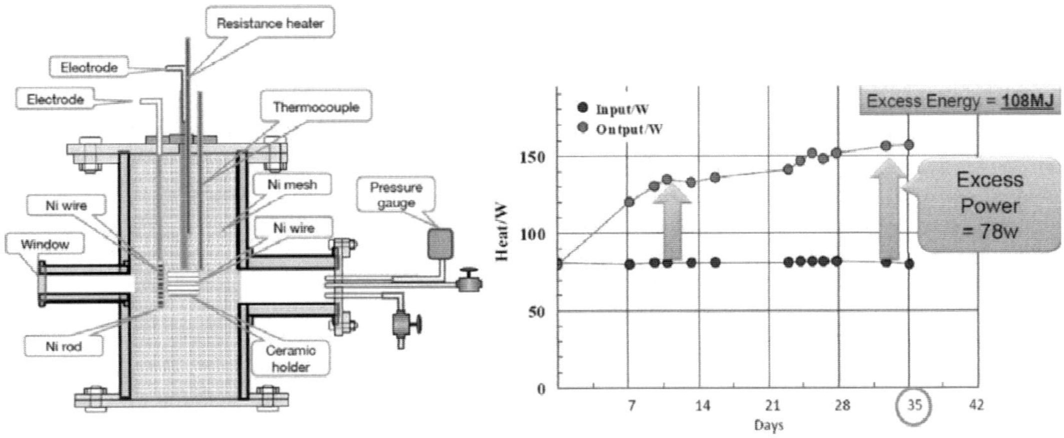

Figure 3. Experimental reactor setup and an example of excess heat generation by Clean Planet Inc. [19].

into the chamber. The Ni mesh was subjected to the plasma discharge and cooled down. This process was repeated 4 or 5 times to create the nano-structured Ni surface. After this activation process, D_2 gas, at about 100–300 Pa, was introduced at 200°C, and then excess heat generation was observed as shown in Fig. 3. In this case, they observed about 78 W excess power that is almost the same as input power.

T. Mizuno is now improving the method of measuring excess heat measurement and analysis, in order to obtain more precise results of excess heat generation [20].

3. Condensed Matter Nuclear Reaction Division (April 2015 –)

The Condensed Matter Nuclear Reaction Division of Research Center for Electron Photon Science at Tohoku University has established in April 2015. It is the first official research division created for condensed matter nuclear reaction (CMNR) and its application in Japan. In United States, the Sidney Kimmel Institute for Nuclear Renaissance (SKINR) at University of Missouri was established in 2012 and the Center for Emerging Energy Sciences (CEES) at Texas Tech. University has recently started. Below, we describe the purpose of the division, organization of the division and the outline of our research plan.

3.1. Purpose of the division

With the aim of creating revolutionary innovation in the energy industry, the Research Center for Electron Photon Science at Tohoku University and Clean Planet Inc. have established a Condensed Matter Nuclear Reaction Division. Through this new joint research collaboration, we will perform the following:

(1) Fundamental Research on Condensed Matter Nuclear Reaction (CMNR),
(2) Development of a New Energy Generation Method,
(3) Development of a New Nuclear Waste Decontamination Method.

Experimental data that indicates the presence of CMNR have been accumulated and experimental conditions for inducing CMNR are gradually becoming clear, although systematic experimental study is still insufficient. So we will

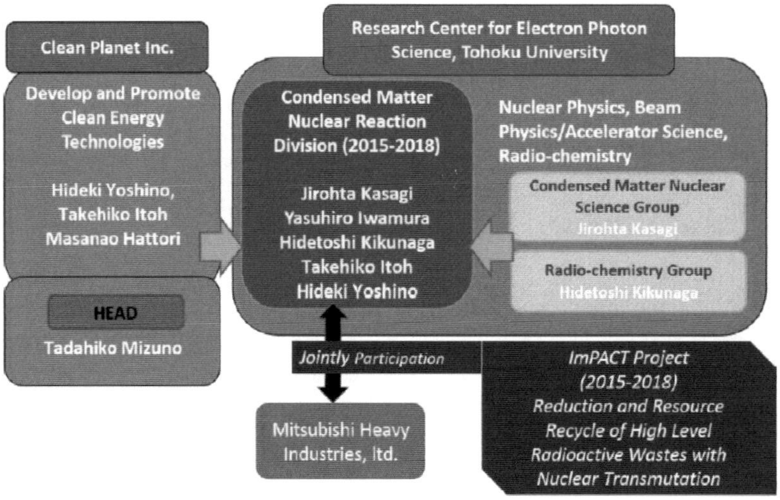

Figure 4. Organization of Condensed Matter Nuclear Reaction Division of Research Center for Electron Photon Science at Tohoku University.

obtain more systematic data and improve the reliability of measurement on CMNR. It leads to better understanding of ultra-low-energy nuclear reactions in condensed matter. We will also work on application development research aimed at commercializing new clean energy devices and new nuclear waste decontamination methods. We hope to bring major changes to Japan's energy industry, through the conceptual change of conventional nuclear reaction.

3.2. Organization of the division

The organization of the new division is illustrated in Fig. 4. Jirohta Kasagi, Yasuhiro Iwamura, Hidetoshi Kikunaga. Takehiko Itoh and Hideki Yoshino participate in the division.

Iwamura and Itoh were investigating nuclear transmutation reactions observed in the nano-sized Pd complexes induced by D_2 gas permeation and left Mitsubishi Heavy Industries, ltd. at the end of March 2015 to join the division. Now Iwamura is a research professor of Tohoku University and Itoh is a visiting associate professor. Itoh is also a director of Clean Planet Inc. J. Kasagi is a professor emeritus and has been investigating the electronic and ionic screening effects on low-energy nuclear reactions in condensed matter as described in the former section. H. Kikunaga is an associate professor and has been engaged in the field of radiochemistry. H. Yoshino is a visiting researcher in this division and also a CEO of Clean Planet Inc.

The nuclear waste decontamination research is supported by the ImPACT Program "Reduction and Resource Recycle of High Level Radioactive Wastes with Nuclear Transformation", which is a Japanese national research project. The Electron Photon Science of Tohoku University and Mitsubishi Heavy Industries jointly participate in this project.

3.3. Outline of Research Plan

The outline of our research plan is shown in Fig. 5. As for the excess heat generation, we will seek the most probable method for energy generation for 2 years. We will try new methods of excess heat generation based on the Mizuno's

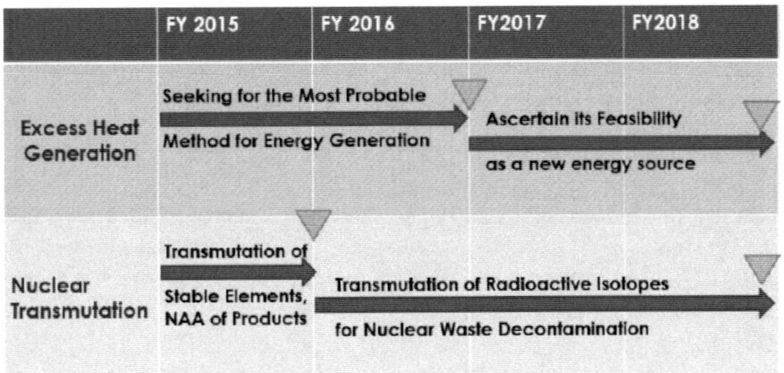

Figure 5. Outline of our research plan.

method and the transmutation method induced by D_2 gas permeation. Measurement of excess heat on the Mitsubishi transmutation method have not been performed. If we measure heat release, this will confirm excess energy. It is very important to measure the released heat precisely as well as to evaluate correct input power. We will develop an elementary method for energy generation until FY2016, and we will ascertain its feasibility as a new energy source. The focus in this stage will be on controllability and reliability of energy production reactions, as well as economical issues.

Nuclear transmutation research will performed with funding from the ImPACT program. Pr, which is transmuted from Cs, will be re-confirmed by other methods such as Nuclear Activation Analysis (NAA) or Rutherford Backscattering (RBS). (However, if we do not get operational resumption of Japan's nuclear power reactors during FY2015, then we may not be able to use NAA.) We will confirm that Pr is transmuted from Cs by RBS. Stable Zr, Se and Pd transmutation will be tried using the transmutation method in the nano-sized Pd complexes induced by D_2 gas permeation. If we obtain the positive results, we will be able to go to the next stage. We will make transmutation experiments using radioactive isotopes for nuclear radioactive waste decontamination after FY2016.

Fundamental Research on Condensed Matter Nuclear Reaction (CMNR) will be performed through these two research activities. We hope to clarify what is happening during CMNR by obtaining systematic experimental data.

References

[1] J. Kasagi et al., *J. Phys. Soc. Jpn.* **64** (1995) 777.
[2] H. Yuki et al., *J. Phys. Soc. Jpn.* **66** (1997) 73.
[3] H. Yuki et al., JETP Lett. **68** (1998) 823,
[4] J. Kasagi et al., *J. Phys. Soc. Jpn.* **71** (2002) 2881.
[5] J. Kasagi et al., *J. Phys. Soc. Jpn.* **73**, (2004) 608.
[6] K. Fang et al., *J. Phys. Soc. Jpn.* **80** (2011) 084201.
[7] Y. Toriyabe et al., *Phys. Rev. C* **85** (2012) 054620.
[8] Y. Iwamura et al., *Jpn. J. Appl. Phys.* **41** (2002) 4642.
[9] Y. Iwamura et al., *Condensed Matter Nuclear Science*, P.L. Hagelstein and S. Chubb (Eds.), World Scientific, Singapore, 2006, p. 435.
[10] Y. Iwamura et al., *Condensed Matter Nuclear Science*, J.-P. Biberian (Ed.), World Scientific, Singapore, 2004, p. 339.
[11] Y. Iwamura et al., *Condensed Matter Nuclear Science,* A. Takahashi et al. (Eds.), World Scientific, Singapore, 2006, p. 178.
[12] Y. Iwamura et al., *J. Condensed Matter Nucl. Sci.* **4** (2011) 132.

[13] T. Higashiyama et al., *Condensed Matter Nuclear Science*, P.L. Hagelstein and S. Chubb (Eds.), World Scientific, Singapore, 2006, p. 447.
[14] H. Yamada et al., *Condensed Matter Nuclear Science,* A. Takahashi et.al. (Eds.), World Scientific, Singapore, 2006, p. 196.
[15] T. Hioki et al., *Jpn. J. Appl. Phys.* **52** (2013) 107301.
[16] Y. Iwamura et al., *J. Condensed Matter Nucl. Sci.* **10** (2013) 63.
[17] Y. Iwamura et al., *J. Condensed Matter Nucl. Sci.* **13** (2014) 242.
[18] Y. Iwamura et al., *Current Sci.* **108** (2015) 628.
[19] H. Yoshino et al., Presentation at 2014 CF/LANR Colloquium at MIT, March. 21–23, 2014, Massachusetts Institute of Technology, Cambridge, MA, USA.
[20] T. Mizuno et al., *Proc. JCF15*, November 1–2, 2014, Sapporo, Japan, pp. 33–62.

J. Condensed Matter Nucl. Sci. 19 (2016) 127–134

Research Article

Screening Energy of the d+d Reaction in an Electron Plasma Deduced from Cooperative Colliding Reaction

J. Kasagi* and Y. Honda

Research Center for Electron Photon Science, Tohoku University, Sendai, Japan

Abstract

We have measured protons and tritons from the d (d,p) t reaction in liquid In, Sn, Pb and Bi during the D_3^+ molecular deuterium beam bombardment for $15 < E < 60$ keV. Observed energy spectra and yield excitation functions are anomalous and suggest that the target deuteron is also in motion. It is found that the reaction mechanism is unique to the molecular beam in such a way that one deuteron in a molecule is elastically scattered by a host metal atom and then it collides with the other to cause the d+d reaction. We call it as the cooperative colliding mechanism (CCM). Experimental data are compared with the CCM calculation and values of the screening energy of the d+d reaction are deduced. They are 380 ± 80, 570 ± 80, 670 ± 100 and 490 ± 100 eV for In, Sn, Pb and Bi, respectively. It is concluded that the screening energy of the d+d reaction in an electron plasma in metal is really large: the averaged value of 530 eV is more than ten times larger than the simple Thomas–Fermi screening prediction.

Keywords: Cooperative colliding mechanism, DD reaction in metal, Liquid metal, Low-energy deuteron beam, Screening potential

1. Introduction

Conduction electrons in metal behave as a Fermi plasma with low temperature ($T<10^3$ K) and high density ($\rho \sim 10^{22}/\mathrm{cm}^3$). The interaction between two deuterons under such an electron plasma is very much different from that of naked deuterons. Obviously, the interaction potential between the two nuclei may be characterized by a large reduction of the repulsive Coulomb part. Consequently, the nuclear reaction rate in low energies may be enhanced by many orders of magnitude. Thus, it is very important to study low-energy d+d reactions in a metallic electron environment in order to predict the reaction rate of nuclear fusion. Such environments include cold fusion in metal, nuclear fusion in high-density plasma as well as nuclear synthesis and energy productions in stars.

Up to the present, the screening potential of the d+d reaction in the various metals has been measured [1–3]. However, the deduced values are anomalously larger than expected; at least several times larger than theoretical predictions using the Thomas–Fermi approximation. Huke et al. [4] discussed the experimental difficulties in deducing

*E-mail: kasagi@lns.tohoku.ac.jp

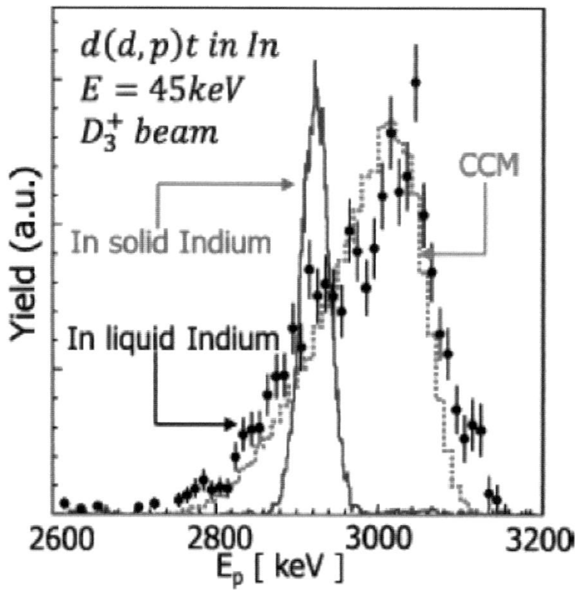

Figure 1. Energy spectra of the d (d,p) t proton measured at θ = 142° for 45 keV D_3^+ beam bombardments. Solid circles show the one for the liquid In, while the solid blue line for the solid In. the dotted red line shows calculated one based on CCM.

the screening potential of the d+d reaction in metal: most of the uncertainty originates from the target deuteron density distribution as well as the surface cleanness. Although experiments have been performed to minimize such systematic errors, it is highly desirable to know the target deuteron density more accurately.

We have developed nuclear reactions in liquid metal in order to explore another environment which enhances the reaction rate more. In the present work, we bombarded liquid In, Sn, Pb and Bi with a D_3^+ molecular beam and found a new reaction mechanism, which is unique to the molecular beam and is called the cooperative colliding mechanism (CCM). It turned out that the CCM is very useful for determining the screening potential of the d+d reaction without the uncertainty of the target deuteron density. The screening potential in the liquid metals is determined by using the CCM, for the first time.

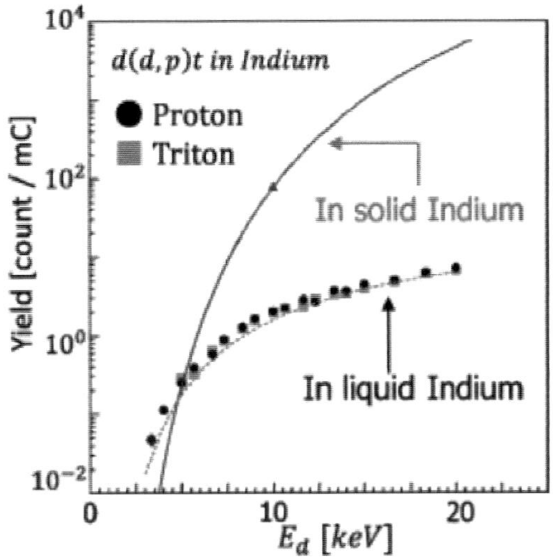

Figure 2. Excitation functions of the protons and the tritons by D_3^+ beam bombardment at θ = 142°. Solid circles and solid squares show the yield of the protons and that of the tritons for the liquid In, solid blue line shows the yield of the protons for the solid In, a dotted red line shows calculated one based on CCM, respectively.

2. Experimental Setup

The experiment was performed by using a low-energy deuteron beam from the high-current ion beam generator [5] at the Research Center for Electron Photon Science of Tohoku University. The experimental procedure is almost the same as described in our previous work [6], and will be briefly mentioned here.

The low-energy D_3^+ beam of several 10 μA was mostly used with bombarding energies ranging from 10 to 60 keV, corresponding to 3.3–20 keV for single deuteron. The beam was vertically bent by 60° with respect to the horizontal and injected into the target chamber from an upper port. The pressure in the target chamber is kept 10^{-5}–10^{-6} Pa. Inside the chamber, the beam passes through a 6-mm-diameter aperture and collides with a liquid target which is placed in the target holder equipped with a heater. For the liquid metal targets, In, Sn, Pb and Bi were, respectively, heated up to 200–380°C (T_m: 156.6°C), 360–390°C (T_m: 231.9°C), 380–410°C (T_m: 327.5°C) and 325–355°C (T_m: 211.5°C). During the measurement, the beam current was measured by inserting a Faraday cup for periodic sampling at a rate that was controlled by a data acquisition program (3-second sampling in every 10 s). The charged particles emitted during beam bombarding were detected with a silicon semiconductor detector (SSD). The SSD had an active area of 450 mm^2 and thickness of 100 μm, and was placed at 124° and 142° with respect to the beam direction and 30 mm away from the target. A 2-μm-thick Al foil was placed in front of the SSD to stop elastically scattered deuterons and sputtered particles. The energy resolution for this setup was about 55.1 keV in FWHM for 2.96 MeV protons.

3. Experimental Results

Protons and tritons emitted from the d (d,p) t reaction were measured in various conditions: with liquid and solid target; with two detection angles at 124° and 142°, bombarding energies from 10 to 60 keV by 2.5 keV steps with D_3^+ beams (i.e., from 3.3 to 20 keV for a single deuteron), and beam currents from a few μA to about 100 μA. In addition, D_2^+

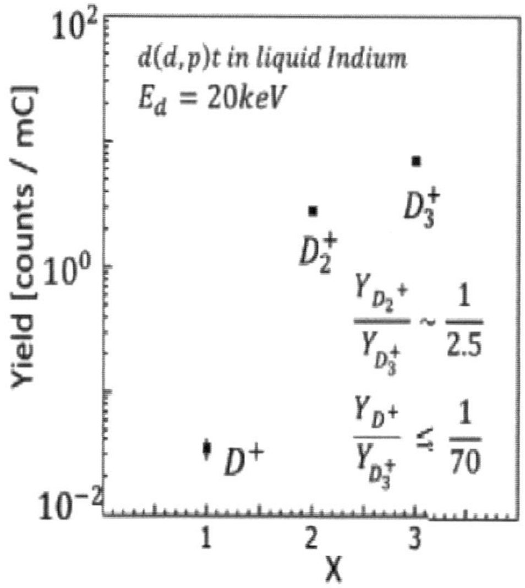

Figure 3. The proton yields versus the number of deuterons (X) in a deuterium molecular of the beam (D_X^+) bombarding the liquid In.

and D^+ beams were also used to bombard the liquid In.

The energy spectrum of protons from the d (d,p) t reaction measured for the liquid In by D_3^+ beam bombardment is plotted in Fig. 1 with solid circles, which is compared with the one for the solid In plotted with a solid blue line. As seen, the spectrum for the liquid In target is quite different from the one for the solid. Usually, deuterons impinging upon the material play a double role, i.e., projectiles and targets of the d+d reaction. In this case, target deuterons are usually at rest and distributed from the surface.

For the solid In, the spectrum having a sharp peak structure can be well reproduced by a kinematical calculation for such a condition. Thus, we can say that the spectrum for the liquid is quite different from the usual one. The observed shape is very broad and largely skewed. Moreover, the peak position shifts to the higher energy side as compared with the normal spectrum for the solid In. The triton spectrum for the liquid In also shows the same characteristics.

The excitation function measured for the liquid In are shown in Fig. 2, where solid circles and squares show the yield of protons and tritons, respectively. A solid line shows a function $Y \propto \exp(-B/\sqrt{E_d})$ with $B = 44.39$, which is known to approximate the energy dependence of the thick-target yield of the d (d,p) t reaction. The solid line which is normalized to the experimental yield at E_d = 10 keV well reproduces the other one at $E_d = 5$ keV. Thus, the excitation function of the yield for the solid In can be well explained as the normal d+d reaction, while the excitation function for the liquid In cannot. The yield for the liquid decreases much more slowly with decrease of incident energy than the one for the solid does.

Figure 3 shows the proton yields versus the number of deuterons (X) in a deuterium molecular beam (D_X^+) bombarding the liquid In. The yield depends on X very strongly: the yield of the d+d reaction almost diminishes to nothing when bombarded by an atomic beam (D^+). This suggests that the anomalous behavior of the d (d,p) t reactions in the liquid In should be related with the use of the molecule beam.

4. d+d Cooperative Collision in Liquid Metal

The anomalous features seen in energy spectra as well as in the excitation function strongly indicate that the target deuteron of the d (d,p) t reaction is not at rest but is in motion with a finite momentum. As a result of spreading of the collision energy, the energy of emitted particles spreads widely and the excitation function becomes less steep. Since these phenomena are observed only for the molecular beams, we have inferred that two deuterons in the same molecule collide to cause the d (d,p) t reaction. Figure 4 shows a schematic presentation of this process. In this case, the reaction proceeds by two steps: First, one deuteron in a molecule is elastically scattered by an In atom, and then, it collides with the other deuteron. Since the reaction occurs with the partner in the same molecule, kinematical variables of the secondary d (d,p) t reaction depends only on the scattering angle of the elastic scattering In(d,d)In. This process is called the cooperative colliding mechanism (CCM). We do not see the normal d (d,p) t reaction in the liquid In for the reason that deuterons diffuse quickly in the liquid In due to the lack of the lattice structure so that target deuterons for the subsequent deuterons scarcely exist at the surface region of the liquid.

The reaction yield is essentially a product of two factors: the probability of the elastic scattering In(d,d)In and that of the d (d,p) t reaction. When the deuteron A scattered by In within the small solid angle $d\Omega_A$ at the angle of (θ_A,ϕ_A), the proton yield measured at the angle of $(\theta_{lab},\phi_{lab})$ is expressed as

$$dY(\theta_{lab},\phi_{lab}) = I_d\rho_{In}dx\frac{d\sigma_R}{d\Omega_A}(E_d,\theta_A,\phi_A)\,d\Omega_A \times N_d\frac{d\sigma_{dd}}{d\Omega_{CM}}(E_{CM},\theta_{CM},\phi_{CM})\frac{d\Omega_{CM}}{d\Omega_{lab}}\Delta\Omega_{lab}. \quad (1)$$

where I_d is the number of beam deuterons, ρ_{In} the number density of In, dx the thickness of the liquid In (x the distance from the surface), $d\sigma_R/d\Omega_A$ the cross section of the first In(d,d)In scattering, E_d the deuteron incident energy, N_d the number of the target deuterons/(unit area), $d\sigma_{dd}/d\Omega_{CM}$ the cross section of the d (d,p) t reaction at the energy of E_{CM} in the center of mass (CM) system, θ_{CM} and ϕ_{CM} stand for the angle of emitted protons in the CM system, $d\Omega_{CM}/d\Omega_{lab}$ the ratio of the solid angle, and $\Delta\Omega_{lab}$ is the solid angle subtended by the detector. It should be noted that the following relations can be easily deduced because of the CCM; $E_{CM} = E_d(1 - \cos\theta_A)$ (by neglecting the recoil of In) and $N_d = (X-1)/2\pi R_{dd}^2$ (X is the number of deuterons in a molecule and R_{dd} is the bond length of the molecule ion).

5. Screening Energy deduced from the CCM d+d Reaction

For the calculations, we use the screened Rutherford scattering cross section for $d\sigma_R/d\Omega_A$ and $d\sigma_{dd} = d\Omega_{CM}$ is calculated from the astrophysical S factor and angular distributions reported in [7]. In Eq. (1), E_d and I_d are not constants but decrease as x increases. Thus, we divide the liquid target into thin layers and calculate the energy and the

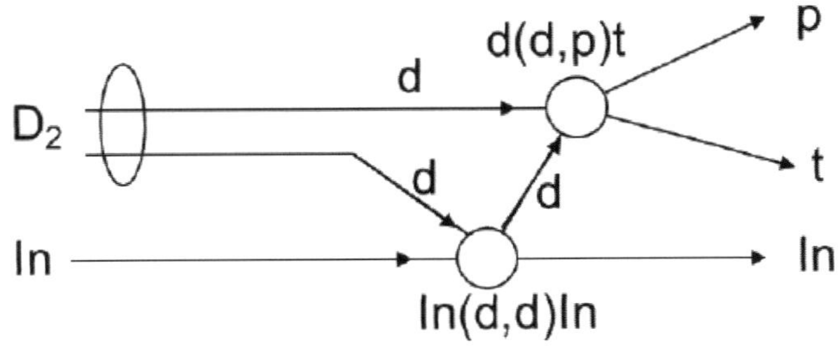

Figure 4. Schematic presentation of CCM causing the d+d reaction. Both deuterons are from the same molecule.

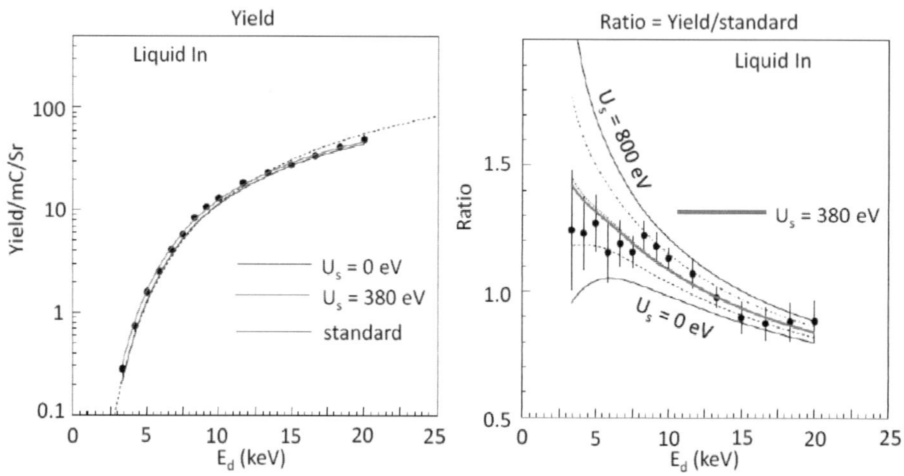

Figure 5. Excitation function of proton yields measured for the liquid In target. The left-hand side figure shows raw yields vs incident energy per deuteron. On the right-hand side figure, the yields are divided by a standard analytical function $S(E_\mathrm{d}) = A\exp\left(-B/E_\mathrm{d}^{1/2}\right)$. The CCM calculations are shown by several lines corresponding to different screening energies of from 0 to 800 eV.

flux loss for each layer. The calculation based on the CCM is compared with the experimental data. In Fig. 1, we show the calculated proton spectrum with the red dotted line. The anomalous features of the spectrum are well reproduced by the calculation. The excitation function of the proton yield for the liquid In target is compared in Fig. 5.

In the liquid metals, the d (d,p) t reaction occurs in a sea of the conduction electrons of the liquid metal. Thus, we introduce the screening energy U_s and simply replace E_d in

$$\frac{\mathrm{d}\sigma_\mathrm{dd}}{\mathrm{d}\Omega_\mathrm{CM}}(E_\mathrm{d}) \text{ as } E_\mathrm{d}+U_\mathrm{s}, \text{ i.e., } \frac{\mathrm{d}\sigma_\mathrm{dd}}{\mathrm{d}\Omega_\mathrm{CM}}(E_\mathrm{d}+U_\mathrm{s}).$$

In the left-hand side figure, the excitation function of raw yields are plotted. Although the calculation with $U_\mathrm{s} =$ 380 eV gives the best fit to the data, it is not so clear in this plot. In the right-hand side figure, we show the yields divided by a standard analytical function $S(E_\mathrm{d}) = A\exp(-B/E_\mathrm{d}^{1/2})$; $A = 2500$ (In and Sn), 3700 (Pb and Bi), and $B = 17$ for all. They are compared with calculations for various values of U_s. The curves shown are the CCM calculations with U_s from 0 to 800 eV by 200 eV steps. It is clear that the experimental yields are enhanced over the calculation without the screening, and $U_\mathrm{s} = 380 \pm 50$ eV is determined for the d+d reaction in conduction electrons in liquid In.

Plots of the ratio vs E_d of the present work are shown in Fig. 6. Obtained values of the screening energy are 480 ± 80, 620 ± 80, 700 ± 100 and 540 ± 100 eV for In, Sn, Pb and Bi, respectively. It should be noticed that the values of the screening energy deduced from the CCM d (d,p) t reaction are more reliable than those from the normal d (d,p) t reaction in solid metals, because the number of the target deuterons can be known more precisely. Raiola et al. [9] reported the values of the screening energy for the same metals; they are 520 ± 50 (In), 130 ± 20(Sn), 480 ± 50 (Pb) and 540 ± 60 (Bi) eV. Except for In, the presently deduced values are larger than the previously reported ones. Thus, it is concluded that the screening energy of the d+d reaction in an electron plasma in metal is really large. However, this large screening effect cannot be explained by the Thomas–Fermi screening model; the present averaged value $U_\mathrm{s} = 570$ eV is more than ten times larger than the Thomas–Fermi screening prediction. Czerski et al. [8]

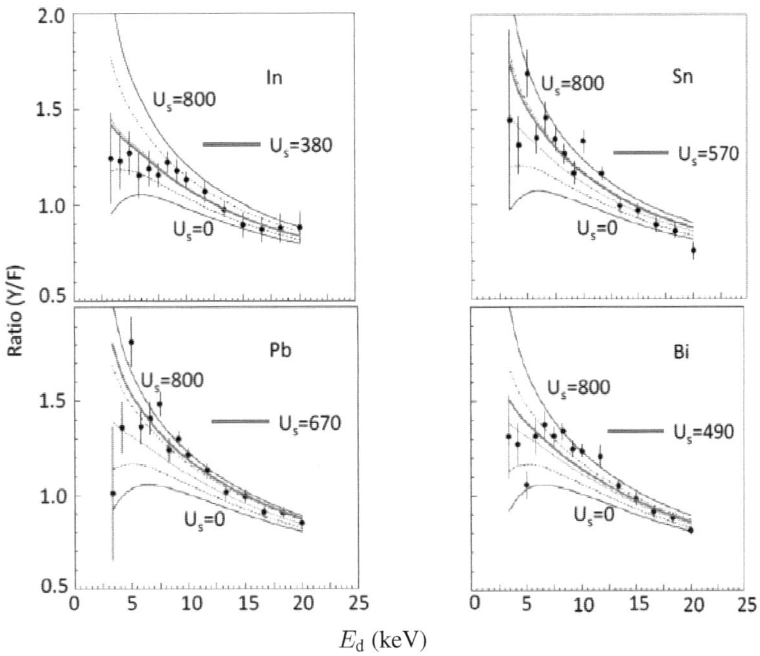

Figure 6. Plots of ratio (yield/$S(E_d)$) vs deuteron energy for In, Sn, Pb and Bi. Curves are the CCM calculations. The red curve corresponds to the deduced U_s value.

calculated the screening energy including the cohesive effect, but their reported values are still 400 eV smaller than the experimental values.

6. Conclusions

We have found a new reaction process called the cooperative colliding d+d reaction in liquid metal bombarded with low-energy D_3 molecular beam. The CCM d+d reaction has turned out to be useful in measuring the screening energy of the d (d,p) t reaction in conduction electrons in liquid metals. Using the CCM reaction, we have established that d+d reactions are really enhanced in conduction electrons; obtained values of the screening energy are 480 ± 80, 620 ± 80, 700 ± 100 and 540 ± 100 eV for In, Sn, Pb and Bi, respectively. It should be emphasized that the values of the screening energy presently deduced are more reliable than those from the normal d (d,p) t reaction in solid metals.

The obtained large screening energy of $U_s \sim 570$ eV would give huge reaction cross section of about 100 pb at the room temperature in metals, if a simple extrapolation could be made. This enables the formation of ^{4}He* with the rate of $\sim 10^{10}$/cm^3/s. (The energy dissipation mechanism from ^{4}He* is another issue.)

The mechanism to give such a large U_s is unknown. If conduction electrons in metals behave as classical gas, then the screening energy can be 600 eV for 2×10^{23} electrons/cm^3. However, they are low-temperature Fermi particles following the Fermi statistics. In this case, we need 10^6 larger electron density than the normal one, in order to explain the large screening energy. Cohesive effects (d+d)→(dd) in dense electrons should be understood more deeply.

Acknowledgement

This work was supported by JSPS KAKENHI Grant Numbers 23540333 and 26400288.

References

[1] H. Yuki et al., *JETP Lett.* **68** (1998) 823.
[2] K. Czerski et al., *Euro. Phys. Lett.* **54** (2001) 449.
[3] F. Raiola et al., *Euro. Phys. J.* A **13** (2002) 377.
[4] A. Huke et al., *Phys. Rev.* C **78** (2008) 015803.
[5] H. Yuki et al., *J. Phys. Soc. Jpn.* **66** (1997) 73.
[6] Y. Toriyabe et al., *Phys. Rev.* C **85** (2012) 054620.
[7] A. Krauss et al., *Nucl. Phys.* A **465** (1987) 150.
[8] K. Czerski et al., *Euro. Phys. Lett.* **68** (2004) 363.
[9] F. Raiola et al., *Euro. Phys. J.* **19** (2004) 283.

J. Condensed Matter Nucl. Sci. 19 (2016) 135–144

Research Article

Effect of Minority Atoms of Binary Ni-based Nano-composites on Anomalous Heat Evolution under Hydrogen Absorption

A. Kitamura* †, A. Takahashi‡, R. Seto and Y. Fujita
Technova Inc., 1-1 Uchisaiwaicho 1-chome, Chiyoda-ku, Tokyo 100-0011, Japan

A. Taniike and Y. Furuyama
Kobe University, Kobe 658-0022, Japan

Abstract

Ni-based binary nano-composite samples supported by mesoporous silica and single-component nickel samples were subjected to hydrogen absorption runs at various temperatures up to 350°C. The former include $Pd_{0.016}Ni_{0.070}/SiO_2$ (PNSII) and $Cu_{0.011}Ni_{0.077}/SiO_2$ (CNS2), and the latter NiO/Ni (NN) and Ni/SiO_2 (NS). Only the binary nano-composite samples, PNSII and CNS2, showed excess power reaching 8 W/g-Pd and 1 W/g-Ni with integrated excess energy of 3.8 keV/Pd (6.5 keV/atom-H) and 0.6 keV/Ni (10 keV/atom-H), respectively, in elevated temperature runs, implying a catalytic effect of the minority atoms on the phenomena. These excess heat values were observed after the saturation of H-absorption to metal, which implies a non-chemical source of energy.

Keywords: $Cu{\cdot}Ni/SiO_2$ nano-composite, Excess energy, Hydrogen gas absorption, Oil-flow calorimetry, Mesoporous silica, 10 keV/atom-H

1. Introduction

Phenomena of anomalous heat evolution from hydrogen-isotope-loaded metal-oxide samples have been studied using a twin absorption system A_1A_2, e.g., [1,2], and a scaled-up system C_1 with a ten-times larger reaction chamber equipped with a flow calorimeter employing an oil coolant with a boiling point of 390°C [3–5]. The former was used for experiments at room temperature mainly for Pd-based and Ni-based nano-composite samples supported by or mixed with zirconia to show the effectiveness of metal-oxide nano-particles for the production of anomalously large hydrogen uptake and associated heat. The samples include Pd/ZrO_2 (PZ), $Pd_{0.08}Ni_{0.35}/ZrO_2$ (PNZ), $Pd_{0.04}Ni_{0.29}/ZrO_2$ (PNZ2B),

*E-mail: kitamuraakira3@gmail.com
†Also at: Kobe University, 5-1-1 Fukaeminami, Higashinada, Kobe 558-0022, Japan.
‡Also at: Osaka University, 1-1 Yamadaoka, Suita, Osaka 565-0871, Japan.

Ni/ZrO_2 (NZ), $Cu_{0.081}Ni_{0.36}/ZrO_2$ (CNZ) and $Cu_{0.21}Ni_{0.21}/ZrO_2$ (CNZIV). The system was also used for samples supported by mesoporous silica (mp-silica), Pd/SiO_2 (PS) and $Pd_{0.011}Ni_{0.062}/SiO_2$ (PNS), to reveal the effectiveness of the mp-silica as a substrate for the nanoparticles.

On the other hand, the latter system, C_1, was fabricated to make precise measurements at elevated temperatures, mainly for Ni-based binary nano-composite samples, $Cu_{0.02}Ni_{0.083}/SiO_2$ (CNS), $Cu_{0.076}Ni_{0.36}/ZrO_2$ (CNZ4), $Pd_{0.016}Ni_{0.070}/SiO_2$ (PNSII) and $Cu_{0.011}Ni_{0.077}/SiO_2$ (CNS2). References [3–5] describe the results of the absorption runs yielding some excess heat up to 30–100 eV/atom-Ni or 5 W/g-Ni.

The above results have made us infer that surface adatoms of the minor component have a catalytic effect of enabling hydrogen absorption of the host crystalline bulk and release of enhanced anomalous heat. To confirm this point, a single-component sample supported by mp-silica, Ni/SiO_2 (NS), has been subjected to the absorption runs. The present paper describes the performance of the NS sample in comparison with those of the binary nano-composite samples [5] to discuss effect of minority atoms of the binary nano-composites on anomalous heat evolution.

2. Description of the System

A schematic of the experimental system C_1 is shown in Fig. 1. For detailed description of the system including the results of calibration of the flow calorimetry with a flow rate of 10 cm^3/min, refer to the previous paper [5]. From

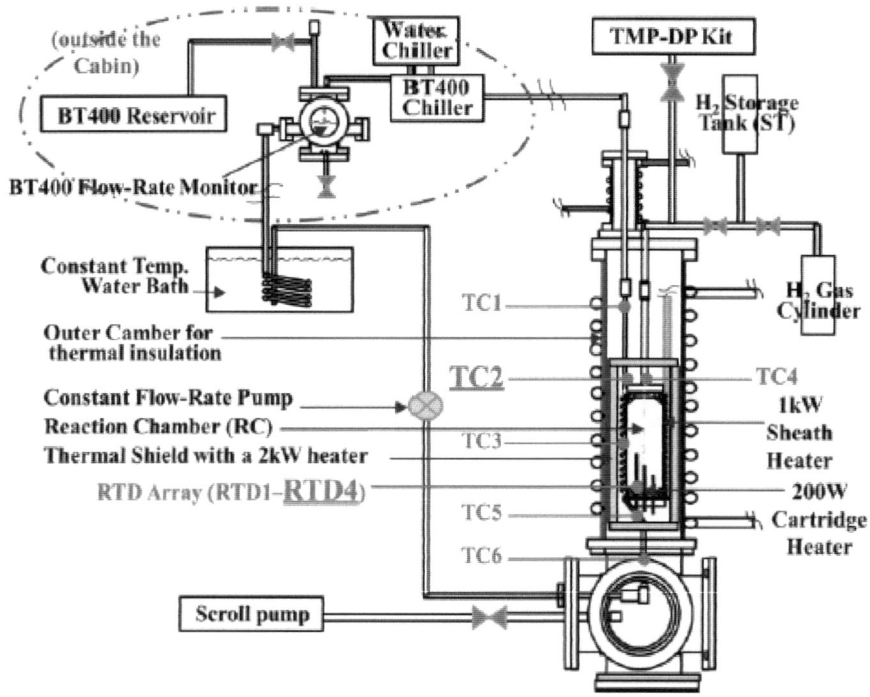

Figure 1. Schematic of the C_1 absorption system after modification of the flow-rate-monitor and addition of the 200-W cartridge heater inside the reaction chamber.

the calibration runs using an Al_2O_3 powder A-12, Showa Denko K.K. [6], the heat conversion coefficient, dT_{C2}/dW = 2.7°C/W or 1.33°C/W, was obtained at room temperature or at the temperature range from 200 to 300°C, respectively. The heat recovery rate was calculated by

$$R_h = F\rho C\,(T_{C2} - T_{C6})/(W_1 + W_2), \tag{1}$$

where F, ρ and C are the flow rate, the mass density and the specific heat capacity, respectively, of the coolant BarrelTherm-400 (BT400), Matsumura Oil Co. Ltd., and W_1 and W_2 the outer (#1) and the inner (#2) cartridge heater power, respectively. The mean value of $R_h = 0.8$ is a little lower than that in the case of the C_1 system before the addition of the heater #2 due to increased loss of heat through the lead wire. The values of dT_{C2}/dW = 2.7°C/W or 1.33°C/W will be used to calculate the excess power observed in room temperature runs or elevated temperature runs, respectively. The calibration run also serves as a control run for foreground runs using the Ni-based samples.

3. Hydrogen Absorption Run for Single-component Nickel Sample NS

The single-component nickel nanoparticle samples include the commercially available pure nickel nanoparticle sample, "QSI-Nano Nickel" powder, QuantumSphere Inc. [7], and a home-synthesized nickel nano-particles supported by the mp-silica, which we call the "NS" sample. Performance of the former, which we call the "NN" sample, has been described in [5]. In the present paper, the absorption runs for the NS sample and its performance are described.

The NS sample was synthesized from a nickel chloride solution, according to the recipe provided by Admatechs Inc. [8]. The solution contains the mp-silica powder as a suspended material. After filtration, the mp-silica was annealed at 800°C for 3 h, and milled with a ball mill. Since we prepared sufficient amount of the sample to occupy 500 cm^3 volume of the reaction chamber, it was not necessary to add any filler like alumina used in the cases of CNS, CNZ4 and PNSII samples. The 144 g of the sample supporting 16.7 g (0.28 mol) of nickel filled the reaction chamber. Although we have not yet done XRD measurements for this sample, we expect similar results to those for the CNS2 sample described in [5], showing that almost all nickel atoms in the virgin sample are oxidized and perfectly deoxidized during the hydrogen absorption runs. We also expect results of STEM/EDS similar to those for the CNS2 sample described in the next section, showing that nickel nano-particles are trapped in pores of mp-silica. These points would be confirmed by measurements of XRD and STEM/EDS for the NS sample.

Protium (H) absorption runs, H-NS#1 through H-NS#4, were performed after vacuum baking for 45 h at temperatures 230–290°C with the heater power of $(W_1 + W_2)$ = (69+20) W. The temperature history in the H-NS#1 run, e.g., is shown in Fig. 2. Each time the heater power was varied, the phase number is advanced; #1–1 for (0 + 0) W, #1–2 for (10 + 5) W, #1–3 for (20 + 10) W, and so on. In the figure also shown are the pressures at the reaction chamber and at the ST, P_r and P_s, respectively, and (H/Ni), the number of H atoms lost from the gas phase relative to the number of nickel atoms, which is calculated from the values of P_r and P_s, and volumes of the reaction chamber and the ST with a correction for the temperature based on the Boyle–Charles' law.

The most important variable among the temperatures at TC1 through TC6 and RTD1 through RTD4 is that at TC2, monitoring the oil-outlet temperature T_{C2}, while the oil-inlet temperature being monitored by TC6 always remained near room temperature. The TC2 temperature in the H-NS runs, T_{C2}(NS), are compared in Fig. 3 with that in the calibration/control run, H-$Al_2O_3$3, $T_{C2}(Al_2O_3)$. As is seen, T_{C2}(NS) is somewhat higher than T_{C2},(Al_2O_3) in almost all elevated temperature phases.

In the course of these runs, fluctuation of the flow rate of BT400 was recorded in the flow-rate monitor. A correction for the flow rate fluctuation was made to $T_{C2}(Al_2O_3)$ by adding the deviation ΔT_{C2} due to the fluctuation ΔF,

$$\Delta T_{C2} = (dT_{C2}/dF)\Delta F = (-\Delta F/F)(W_1 + W_2)(dT_{C2}/dW)\alpha, \tag{2}$$

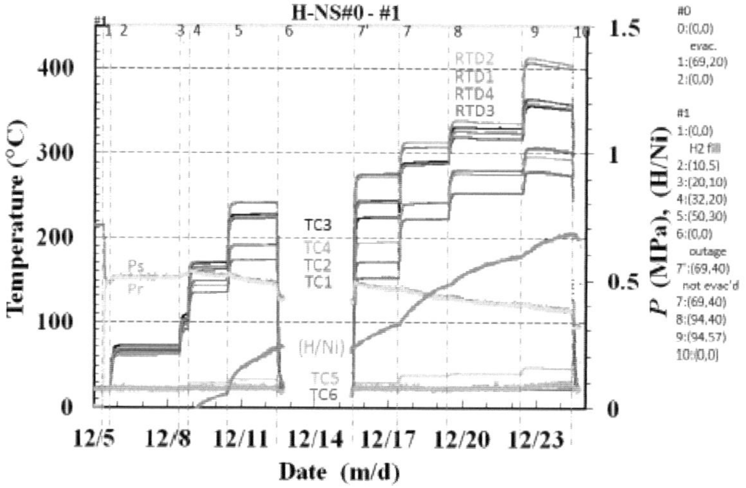

Figure 2. Temperature history in the H-NS#1 run.

where α is a fitting parameter. Essentially α has to be 1.0, which is confirmed by deriving Eq. (2) from Eq. (1), or Eq. (1′) below. However, α is determined to be 0.8 from a blank run which is supposed to show no excess heat at all, Ar-CNS2#6 to be described in Section 4. This is consistent with our experience that $(T_{C2}–T_{C6})$ varied with a flow-rate (F) dependence weaker than the linear one, that is $\alpha < 1$ in the modified equation;

$$T_{C2} - T_{C6} = (W_1 + W_2)R_h/(F^{\alpha}\rho C). \qquad (1')$$

The corrected reference temperature, $T_{C2}(Al_2O_3) + \Delta T_{C2}$, is expressed by the blue chain line in Fig. 3. When compared with this corrected reference with the systematic error of about 10°C taken into account, we see no meaningful increase in TC2 temperature T_{C2}(NS) in all H-NS runs. We conclude that in the NS powder is induced no anomalous effect, or no excess heat in the temperature range from room temperature to 400°C.

In Fig. 3 the values of (H/Ni) are plotted again. In addition to absorption/adsorption (collectively called 'sorption'), the variable (H/Ni) could involve hydrogen consumption due to oxygen pickup reaction,

$$NiO + H_2 \rightarrow Ni + H_2O + 0.033 \text{ eV}, \qquad (3)$$

if the H_2O molecules are either in the liquid phase or in the solid phase due to adsorption. The latter would be possible, if strong adsorption sites happen to be created on the surface of the NS particles or on the mp-silica. The H_2O molecules would remain stuck on the surface throughout the run before the succeeding baking-absorption runs.

The former, i.e., condensation into liquid phase, is also possible in the present system. The virgin sample has 0.28 mol of NiO. If all of NiO molecules are deoxidized, the same amount of H_2O is produced, consuming 0.28 mol of H_2, i.e., (H/Ni) $\sim$ 1.0. The H_2O gas would take on a pressure of 1.5 MPa in the part downstream of the SuperNeedle valve including the reaction chamber with a free volume of 420 cm^3, if it can remain in the gas phase. However, since the vapor pressure of water at room temperature is 2.3 kPa, almost all of the H_2O molecules will be condensed in the tube outside the reaction chamber at room temperature. The liquid water will be vaporized and removed by subsequent

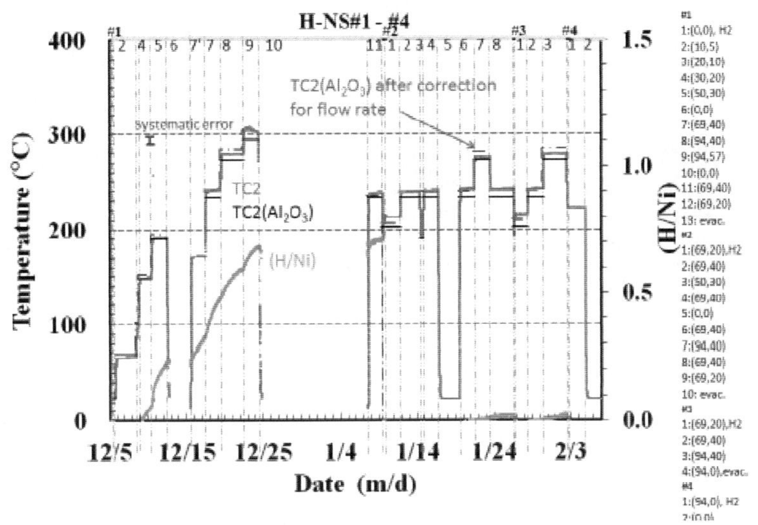

Figure 3. Comparison of the temperature at TC2 (*red line*) for NS-sample runs, H-NS/Al_2O_3, with that for TC2(Al_2O_3) in the calibration/control run (*black line*), H-$Al_2O_3$3, and with the TC2(Al_2O_3) corrected for the flow rate fluctuation (*blue chain line*) to see nonexistence of excess temperature in the H-NS/Al_2O_3 runs. Heater powers (W_1,W_2) in W are indicated on the right. "H2" means H_2 gas fill, and "evac." means evacuation.

evacuation prior to the next run. The value of (H/Ni) observed at the final phase before re-baking, #1-12, is 0.7. Then about 1/4 of NiO might have been emitted directly and exhausted during the rebaking phase #1-13.

We observe very small values of (H/Ni) in the succeeding runs, #2, #3 and #4. This means that little reaction (absorption/adsorption) takes place on the single-component nickel nanoparticles supported by the mp-silica at temperatures from room temperature up to 400°C. This is in contrast to the nickel composite samples described in the following section.

4. Brief Description of Performance of the CNS2 Sample

The hydrogen absorption runs for the binary nano-composite samples supported by mp-silica have been presented in detail in [3] for the CNS sample and [5] for the PNSII and CNS2 samples. Here results of STEM/EDS characterization and deuterium and argon exposure runs for the CNS2, D-CNS2#5 and Ar-CNS2#6, are described, while the results of the hydrogen absorption runs of the CNS2 samples are only briefly mentioned for comparison. For the detailed description of PNSII and CNS2 samples including XRD characterization, refer to [5].

The 160-g CNS2 sample synthesized in our laboratory with a composition of $Cu_{0.011}Ni_{0.077}/SiO_2$, containing 12 g (0.21 mol) of nickel and 1.9 g (0.03 mol) of copper, occupied the 500 cm^3 volume of the reaction chamber, leaving no void to be filled with Al_2O_3 powder. A STEM/EDS photo is shown in Fig. 4. It appears that nickel and copper atoms are included in the same pores of the mp-silica with a density ratio approximately equal to the mixing ratio, and diameter of 5–20 nm.

The D and Ar absorption runs, D-CNS2#5 and Ar-CNS2#6, were performed after vacuum baking at temperatures higher than 240°C similarly to other runs. The temperature evolution (the red line) compared with that in the calibra-

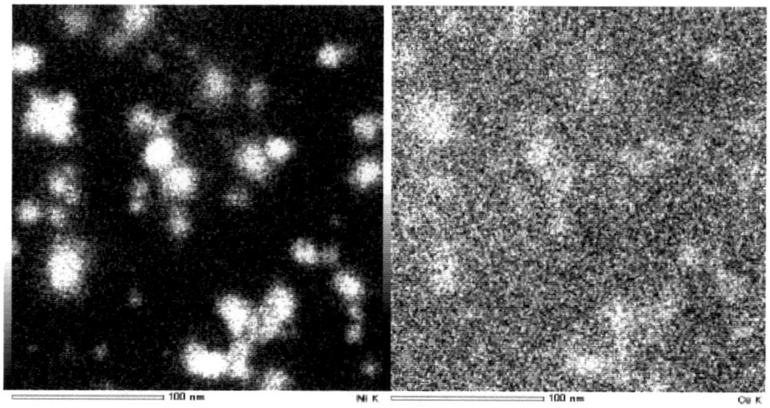

Figure 4. STEM/EDS photo of the CNS2 sample showing Cu/Ni particles confined in the nano-pores.

tion/control run H-$Al_2O_3$3 (*the black line*) and the corrected one for flow rate (the blue chain line) is shown in Fig. 5 together with (H/Ni) (*orange line*). Here it is assumed that the Cu atoms are acting as a catalyst for D absorption of nickel lattice, and do not form a lattice to absorb themselves. The correction is based on Eq. (2), with the fitting parameter $\alpha = 0.8$ determined so as to best fit the observed temperature in the Ar-CNS2#6-2 phase, which is supposed to show no excess heat.

This selection gives also null excess in the deuterium absorption run D-CNS2#5, i.e., the differences between

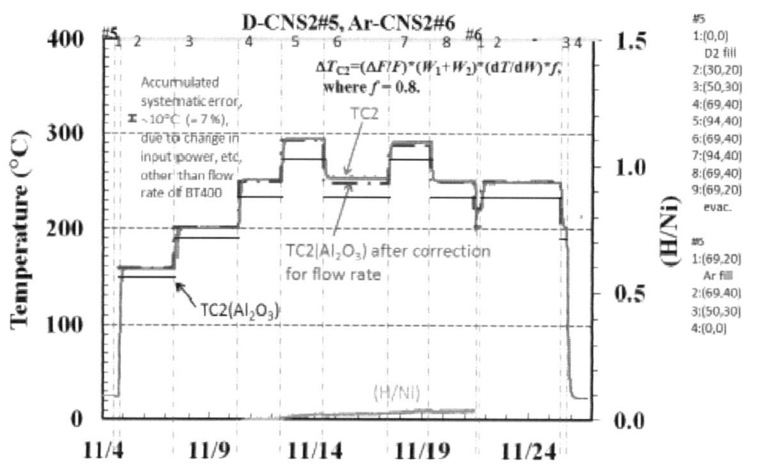

Figure 5. Evolution of the temperature at TC2 in the D-CNS2#5 and Ar-CNS2#6 runs compared with that in the calibration/control run H-$Al_2O_3$3. The corrected reference temperature TC2(Al_2O_3) is calculated by adding ΔT_{C2} due to deviation of the flow rate of BT400 to the latter.

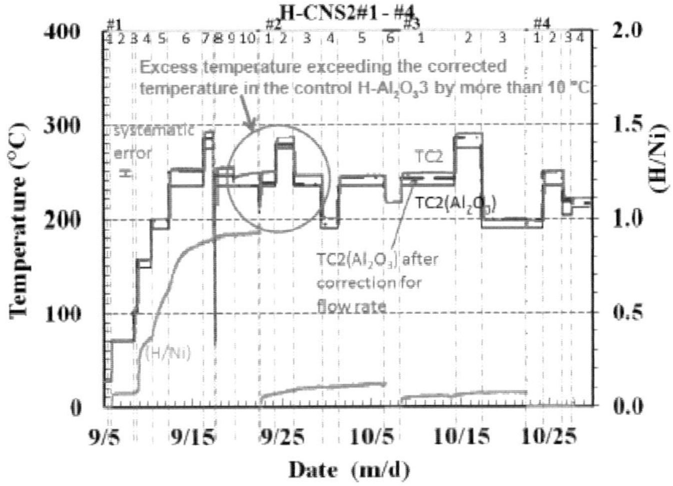

Figure 6. An example of the excess heat runs: evolution of the temperature at TC2 in the H-CNS2 runs compared with that in the calibration/control run H-$Al_2O_3$3. Heater powers (W_1,W_2) in W are: **#1**-1; (0,0) H_2 fill, -2;(10,5), -3;(20,10), -4;(30,20), -5;(50,30), -6;(69,40), -7; (94,40), -8;(0,0), -9;(69,40), -10;(69,40) flow rate readjusted, -11;(69,20) evac.; **#2**-1;(69,40) H_2 fill, -2;(94,40), -3;(69,40), -4;(50,30), -5;(69,40), -6; (69,20) evac.; **#3**-1;(69,40) H_2 fill, -2;(94,40), -3;(50,30), -4;evac.; **#4**-1; (50,30) H_2 fill (0.095 MPa), -2;(69,40), -3;(69,20), -4;(69,20) evac.

TC2 and the corrected TC2(Al_2O_3) are smaller than the systematic error in all phases. This means that the excess heat phenomenon could be isotope dependent in the case of nickel, since the excess heat was clearly observed in H-CNS2#1 through #4 runs. This is consistent with the common belief that nickel works with hydrogen, but not with deuterium. To confirm this point, the temperature evolution in the runs is re-presented in Fig. 6.

In most phases of a variety of the input power, the measured TC2 temperature agrees with the corrected control temperatures (blue chain line) within the systematic error of 10°C, which is comprised of estimated uncertainties in applied heater voltages, location of the thermocouples, and the position / distribution of the heat sources. However, during the period from the phase #1-9 through #2-3, substantial increase beyond the error range is observed. After this period, the excess temperature shows a decline below the systematic error to become null in the phase #3-3 and later.

During this period the maximum excess temperature corresponds to the excess power of 10 W, or 0.8 W/g-Ni, over the input power of 109–134 W. The phase-averaged sorption energy, $\eta_{\mathrm{av},i}$ (eV/atom-H), and the total excess energy per unit absorbent atom, $E_{\mathrm{a,ex}}$ (eV/atom-Ni), are calculated with the following:

$$\eta_{\mathrm{av},i} = \frac{\int_{t-T_i}^{t} W_{\mathrm{a,ex}}\,\mathrm{d}t}{L(t) - L(t-T_i)}, \tag{4}$$

$$E_{\mathrm{a,ex}} = \int_0^t W_{\mathrm{a,ex}}\,\mathrm{d}t, \tag{5}$$

where $W_{\mathrm{a,ex}}$ is the specific excess energy (eV/atom-Ni),$L \equiv$ (H/Ni) is the time-dependent loading ratio, and T_i is the period of the phase i. They are plotted at the end of each phase in Fig. 7.

The maximum values are $\eta_{\mathrm{av,max}}$ = 1.5 × 10^4 eV/atom-H and $E_{\mathrm{a,ex}}$ = 3.8 × 10^2 eV/atom-Ni. The former and the latter are more than three orders of magnitude and two orders of magnitude, respectively, larger than the largest

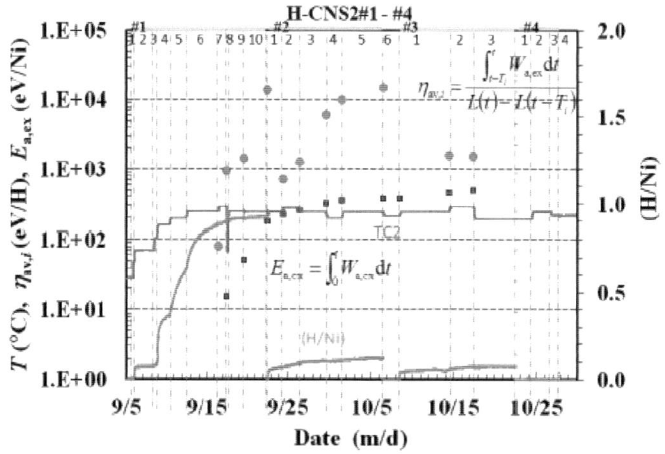

Figure 7. Excess energy per H atom, $\eta_{av,i}$ (•), and the integrated excess per nickel atom, $E_{a,ex}$ (■), both plotted at the end of each phase.

possible value of the conventional chemical reaction energy (which is a few eV/atom). It should be emphasized here that the excess heat emerges in the phases with the loading ratio (H/Ni) approaching saturation and changing only slightly. Especially, $\eta_{av,max}$ values are estimated by assuming that the small change of the H/Ni ratio contributed totally to the excess heat reaction. However, we do not know what portion of this change of hydrogen transfer was responsible

Table 1. Comparison of the samples in regard to excess heat.

	Sample name	NN/Al_2O_3	NS	$PNSII/Al_2O_3$	CNS2
	M	Ni	Ni	Pd (Pd·Ni)	Ni
Transient phase at room temperature in #1 run	*Loading* ratio (H/M)	0.02	∼ 0	3.2	∼ 0
	Specific power (W/g-M)	0.26	∼ 0	3.9	∼ 0
	Sorption energy (eV/atom-M)	0.17	∼ 0	3.0	∼ 0
Transient phase at room temperature in #2 or later	Loading ratio (H/M)	< 0.02	–	0.87–1.0	–
	Specific power (W/g-M)	< 0.004	–	0.35–0.43	–
	Sorption energy (eV/atom-M)	< 0.003	–	0.21–0.19	–
Saturation phase at *elevated temperature*	*Loading* ratio (H/M) in #1 run	0.23	> 0.73	7.6 (1.4)	0.9
	Loading ratio (H/M) in #2 or later	0.06	0.02	2.0–2.5 (0.4)	0.12–0.08
	Max. excess power (W)	∼ 0	∼ 0	11 – 8	10
	Max. specific excess power $W_{a,ex}$ (W/g-M)	∼ 0	∼ 0	7.9–5.9 (2.0)	0.8
	Specific excess energy $E_{a,ex}$ (keV/atom-M)	∼ 0	∼ 0	3.8 (0.68)	0.38
	Phase-averaged sorption Energy $\eta_{av,i}$ (keV/atom-H)	∼ 0	∼ 0	2.2–6.5	15

* 'Loading' in #1 runs includes hydrogen atoms spent for deoxidation of PdO or NiO.
** 'H' stands for either H or D, and 'M' stands for either nickel or palladium.

for the excess energy. The portion may be much smaller. These points make us infer that the excess energy may be of nuclear reaction origin.

5. Summary and Concluding Remarks

Another Ni-based nano-composite sample, PNSII, and another single-component nickel sample, NN, were also subjected to hydrogen isotope absorption tests using the same system, and described in detail in [5]. A comparison of the performances of the samples including these two is shown in Table 1. The NN and NS samples showed little heat evolution with no anomaly. At room temperature only the PNSII sample containing palladium absorbed substantial amount of hydrogen with positive heat evolution. The virgin PNSII sample exposed to D_2 gas at room temperature showed evolution of heat much larger than the energies emerging from the bulk sorption in addition to oxygen pickup reactions, while in the #2 and later runs the loading ratio and the heat evolution had the values a little larger than or nearly equal to the values characteristic of the bulk palladium sample.

At elevated temperatures both in the PNSII runs and the H-CNS2 runs, excess power was observed in time intervals where the (H/M) ratios varied very slowly, while little excess power was recorded in the phases with rapidly increasing (H/M) in the #1 runs. It appears that the anomalous excess heat at elevated temperatures has some origins other than chemical energy of H-absorption/adsorption. Anyway, it is emphasized that only the binary nano-composite samples showed the excess heat at elevated temperatures. This suggests that the anomalous heat evolution is caused by some catalytic effect of minority atoms, probably the ones making up the conditioned near-surface sites.

Finally, we have to mention that any noticeable change both in the γ-ray and the neutron counting rates was not

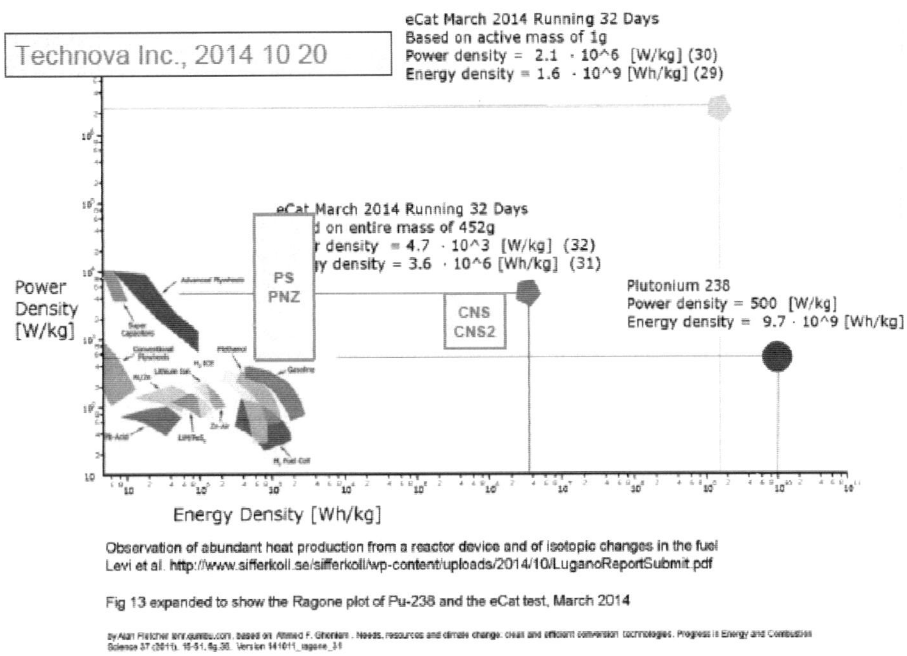

Figure 8. Extended Ragone plot. The data from Kobe University–Technova group are added to the original [9].

coincident with the excess power evolution. The range of the specific excess power and the specific excess energy obtained in our laboratory so far are shown in Fig. 8, the extended Ragone plot originally presented in Ref. [9], where the power density vs the energy density obtained by various energy sources are plotted. The sources include the ^{238}Pu battery and 'eCat' device in addition to conventional ones such as gasoline, H_2 fuel cells, Li-ion batteries, advanced flywheels, etc. It is obvious that the present device of ours as well as the 'eCat' has the parameters far superior to conventional energy sources. The energy density of the CNS and CNSII sample runs is as high as 1000 times that of gasoline. Thus, the present apparatus could hopefully be used as a distributed high-energy-density source without hard radiation or radioactive waste.

Acknowledgement

The authors thank Mr. S. Nagai, Kobe University, for his help in performing the STEM/EDS analysis.

References

[1] Akira Kitamura, Takayoshi Nohmi, Yu Sasaki, Akira Taniike, Akito Takahashi, Reiko Seto and Yushi Fujita, *Phys. Lett. A* **373** (2009) 3109–3112.
[2] A. Kitamura, Y. Miyoshi, H. Sakoh, A. Taniike, A. Takahashi, R. Seto and Y. Fujita, *J. Condensed Matter Nucl. Sci.* **5** (2011) 42–51.
[3] A. Kitamura, A. Takahashi, R. Seto, Y. Fujita, A. Taniike and Y. Furuyama, *J. Condensed Matter Nucl. Sci.* **15** (2015) 231–239.
[4] A. Kitamura, A. Takahashi, R. Seto, Y. Fujita, A. Taniike and Y. Furuyama, *Proc. JCF14* (2014) 1–13.
[5] A. Kitamura, A. Takahashi, R. Seto, Y. Fujita, A. Taniike and Y. Furuyama, *Proc. JCF15* (2015) 1–19.
[6] http://www.sdk.co.jp/english/products/140/142/2046.html.
[7] http://qsinano.com/wp-content/uploads/2014/05/qsi_nano_nickel_ni_5_oct_09.pdf.
[8] Y. Ookawauti; Private communication.
[9] http://www.networkworld.com/article/2824558/.

J. Condensed Matter Nucl. Sci. 19 (2016) 145–154

Research Article

High-energetic Nano-cluster Plasmoid and its Soft X-ray Radiation

A. Klimov*, A. Grigorenko, A. Efimov, N. Evstigneev, O. Ryabkov, M. Sidorenko, A. Soloviev and B. Tolkunov

Limited Liability Company – New Inflow, Osenny blv. bld.6, apt.348, 121609 Moscow, Russian Federation

Abstract

Artificial stable microwave (MW) plasmoids were obtained and studied by Kapitsa in swirl gas flow. This MW plasmoid had unusual physical properties close to those of *natural ball lightning*. We studied the physical parameters and properties of a longitudinal heterogeneous plasmoid (plasma formation with erosive nano-clusters) created by capacity coupled high-frequency (HF) discharge in high-speed swirl flow in our previous papers. These nano-clusters were created by an electrode's erosion in a plasma reactor. Measurement of the power balance of this heterogeneous plasmoid is very important for aluminium–hydrogen power energetics, using a new plasma generator with high coefficient of performance (COP). This work is a continuation of our previous work. We determined that there is extra power released in heterogeneous plasmoids created by combined discharge (HF discharge + DC discharge). The measured COP in this plasmoid is about 2–10. We suppose that this extra power release in a heterogeneous plasmoid is connected with LENR. The obtained experimental results (COP, optical spectra, soft X-ray spectra, chemical composition of dusty particles) prove our suggestion. It was revealed that a heterogeneous non-equilibrium plasmoid creates intensive soft X-ray radiation (with quantum energy about 1–10 keV). We determined that excited and charged cluster particles are responsible for this soft X-ray radiation creation, to be exact. Parameters of heterogeneous non-equilibrium plasmoid (N_e, T_e, T_V, T_R and others) were measured in the experimental plasma vortex reactor (PVR). *The main goals of this work are the following:* (1) Creation of plasma-chemical vortex reactor (PVR) with high COP. (2) Obtaining of the key experimental results in this reactor for the future theoretical LENR model creation based on the PVR's operation.

Keywords: Plasmoid, plasma-chemical reactor, Soft X-ray radiation, Swirl flow

1. Experimental Setup PVR

A schematic of the experimental setup of the PVR is shown in Fig. 1. Test section (1) 60 mm diameter and 60 cm length of this reactor is manufactured from Quartz glass tube. There are tangential injection and axial injection of testing gas mixture into the swirl generator [2]. (2) So, it is possible to change the swirl flow parameter over a wide

*E-mail: klimov.anatoly@gmail.com

© 2016 ISCMNS. All rights reserved. ISSN 2227-3123

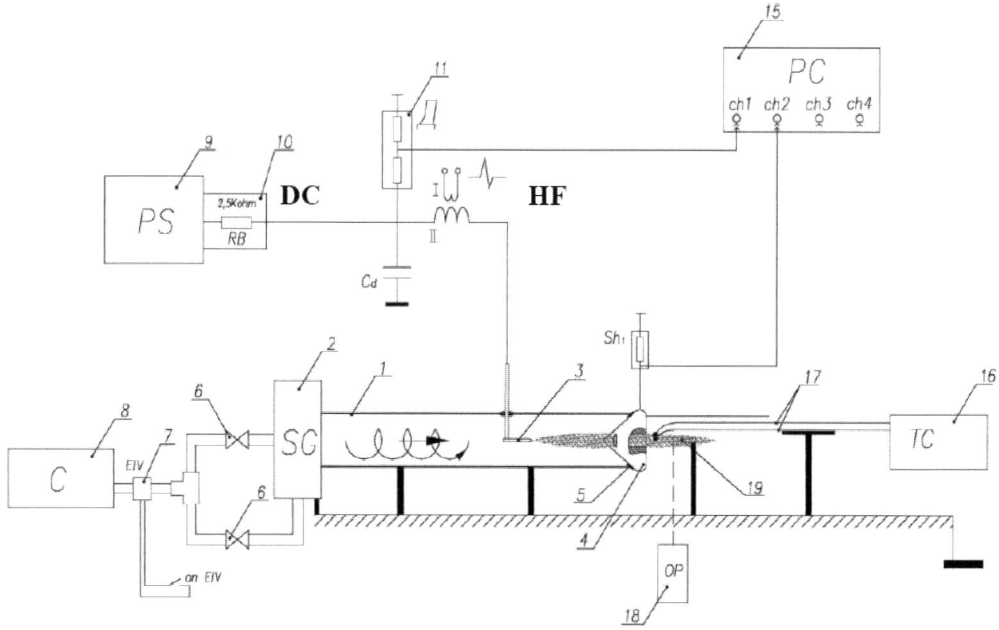

Figure 1. Schematic of experimental setup PVR. 1 – Quartz test section; 2 – swirl generator; 3 – anode; 4 – cathode- nozzle; 5 – rubber gaskets; 6,7 – valves; 8(C) – compressor, 9-PS - power supply DC, 10; 11 – high voltage probe 1:1000; 12 - current probe; 15 – oscilloscope TDS 2014B; 16,17 – thermocouples (CENTER 304); 18 – X-ray spectrometer and optical spectrometer.

range. The maximum mass flow rate is about 30 g/s in this swirl generator. The nozzle (4) is connected with PVR by rubber gaskets hermetically. Combined discharge is created in the PVR by DC power supply (PS DC) and HF power supply (HF). Mean electric power consumption used in this reactor is about 0.2–3 kW.

The following diagnostic instrumentation is used in our experiments: IR pyrometer with laser (MS6550A), thermocouple probes (CENTER 304), optical PIV device, optical spectrometer (AvaSpec 2048), X-ray spectrometer (X-123), ion mass spectrometer, high-speed video camera (Motion Pro), current probes and voltage probes, mass gas flow meter (Inflow).

The following flow and plasma parameters are measured in this work: mass flow rate of different gases (argon, water steam and others), initial gas mixture temperature, final gas mixture temperature behind the reactor nozzle, volt–ampere signals, mean electrical power input in plasma, optical spectra and X-ray spectra, high-speed video of plasmoid dynamics and discharge evolution, chemical analysis of LENR chemical species.

A heterogeneous plasmoid created by combined discharge in Ar–H_2O mixture swirl flow in the PVR is shown in Figs. 1 and 2.

2. Heterogeneous Plasmoid in Ar–H_2O Mixture Swirl Flow

Power losses in PVR are measured by electric power of nickel–chrome wire heater in pure argon flow (or known electric power consumption in argon plasma) in the beginning of each experiment (regime -PVR calibration). It was shown in the work [2–9] that there is no extra power release in pure argon flow when plasma is on.

Figure 2. Heterogeneous plasmoid in PVR. 1 – swirl generator, 2 – water steam injector-anode, 3 – plasmoid near cathode and 4 – cathode.

Gas flow mixture Ar–H_2O = 1 : 1 is used in the experiments considered in this work. Typical images of the heterogeneous plasmoid in PVR obtained in this mixture are shown in Figs. 2 and 3. One can see that the plasmoid's length is close to the electrode's gap length of 80 mm. We determined that many tiny metal droplets are created (with diameters 0.01–0.1 mm) with plasma halos around them (diameters about of 10 mm) in the discharge zone. These metal droplets are created by hot cathode erosion. So, a large heterogeneous plasmoid consists of many small plasmoids (metal droplet kernel and plasma halo). These small plasmoids are charged formations. We measured the typical plasmoid's potential at about of $\phi \sim$ - 2–3 kV. It is important that these plasmoids can move to the anode in an external electric field at constant propagation velocity $V \sim$ 2–5 m/s (against the incoming gas flow). These droplets touch and cover the anode's surface. It is very simple to estimate electric charge and electric potential of these droplets by using of mechanic impulse equation and the Coulomb force. Note that this force equals the drag force at constant

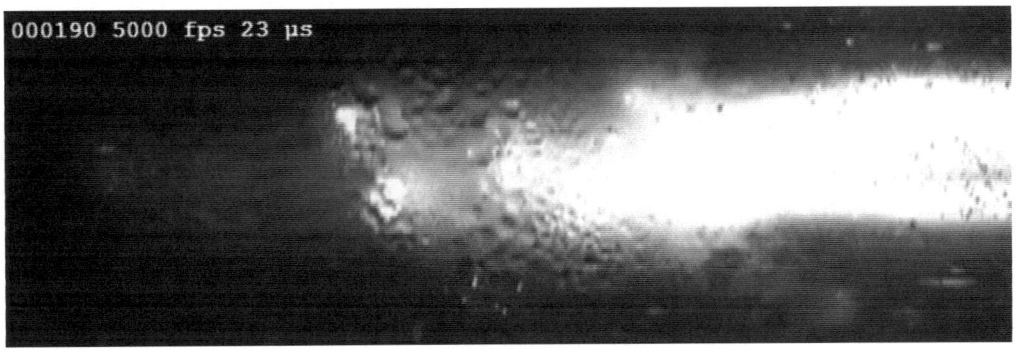

Figure 3. Frame of high-speed video. Exposure time is 20 μs. Right -cathode. Argon mass flow rate is 2 g/s and water vapour mass flow rate is 2 g/s.

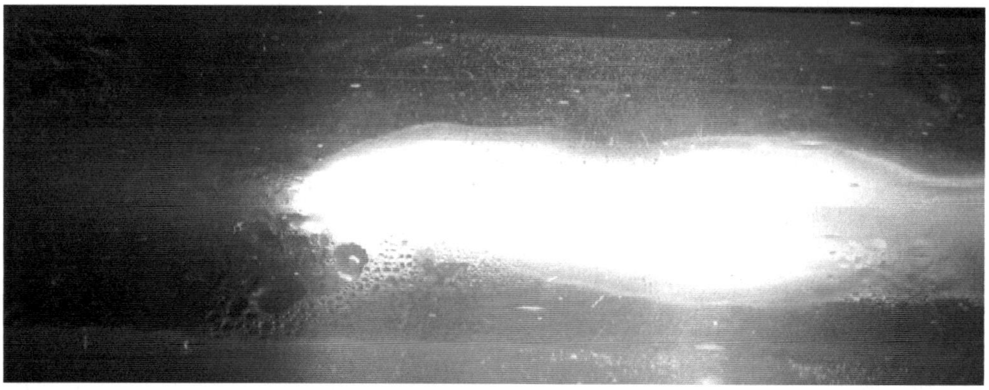

Figure 4. Integral photo of heterogeneous plasmoid 80 mm length in PVR. Exposure time 500 μs. Right -anode. Argon mass flow rate is 2 g/s and water vapour mass flow rate is 2 g/s .

propagation velocity V of these particles.

The optical spectra obtained in this experiment are shown in Figs. 4 and 5. One can see that there are the optical atomic lines Ni, H, O, Ar and the molecular band OH in these spectra. There is continuous spectrum in stable homogeneous plasmoid also, as shown in Fig. 5. This continuous spectrum is connected with hot cluster radiation (black body radiation). It is revealed that amplitudes of the hydrogen lines $I_{\alpha-\delta}$ are decreased considerably in this regime. A

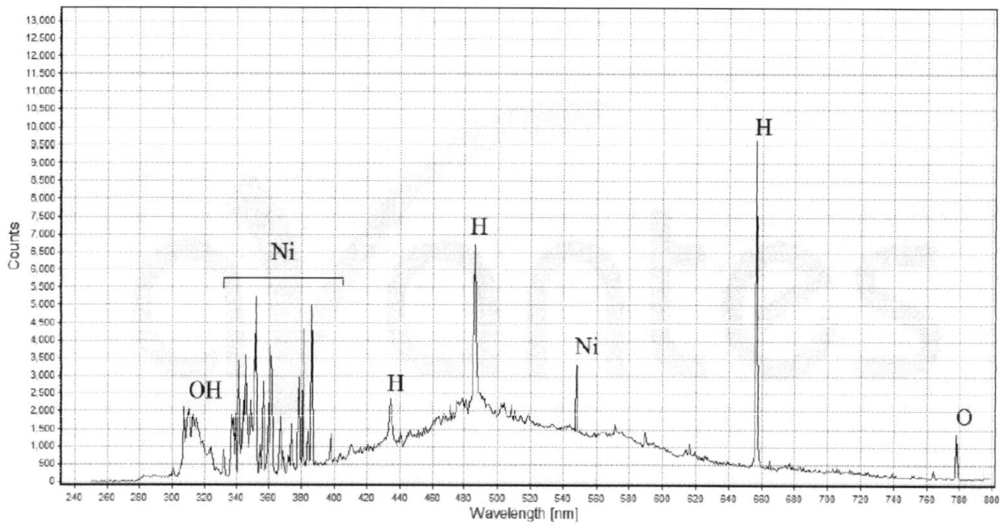

Figure 5. Optical spectrum in heterogeneous plasmoid in PVR. Starting stage of plasmoid creation. Argon mass flow rate 3 g/s, water steam mass flow rate 3 g/s .

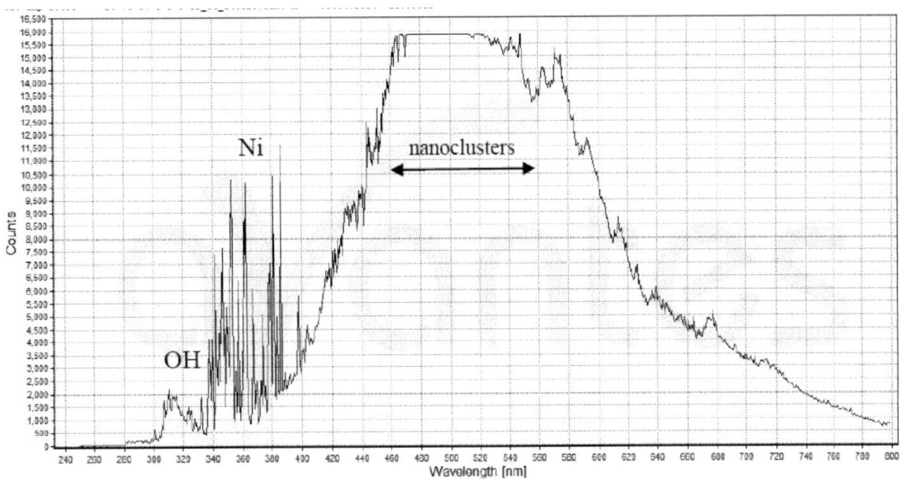

Figure 6. Optical spectrum in heterogeneous plasmoid in PVR. Final stage of plasmoid creation. Argon mass flow rate 3 g/s, water steam mass flow rate 3 g/s.

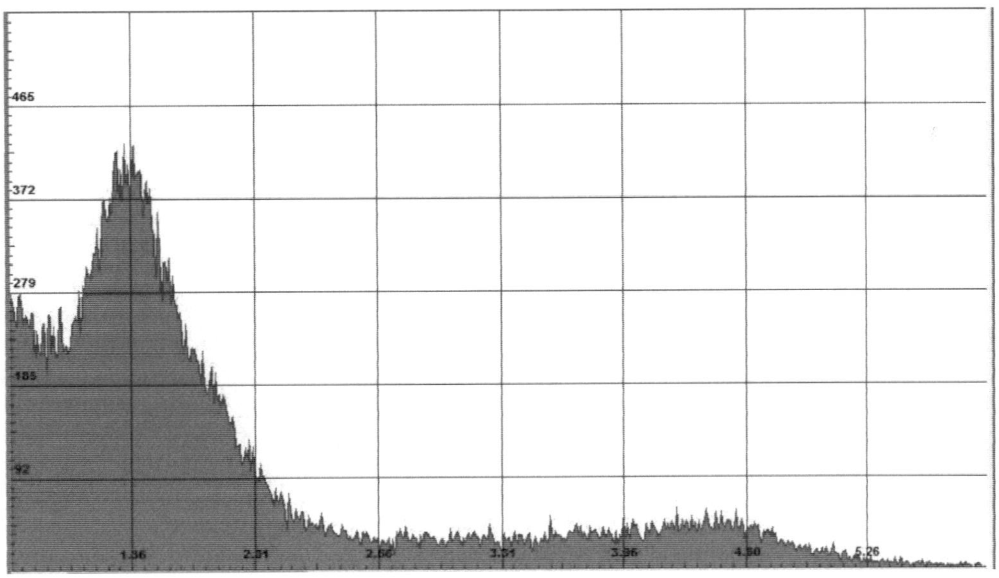

Figure 7. Soft X-ray radiation from heterogeneous plasmoid in PVR. Combined discharge (HF+DC). Power input in plasma 500 W. Mixture $Ar+H_2O = 1:1$.

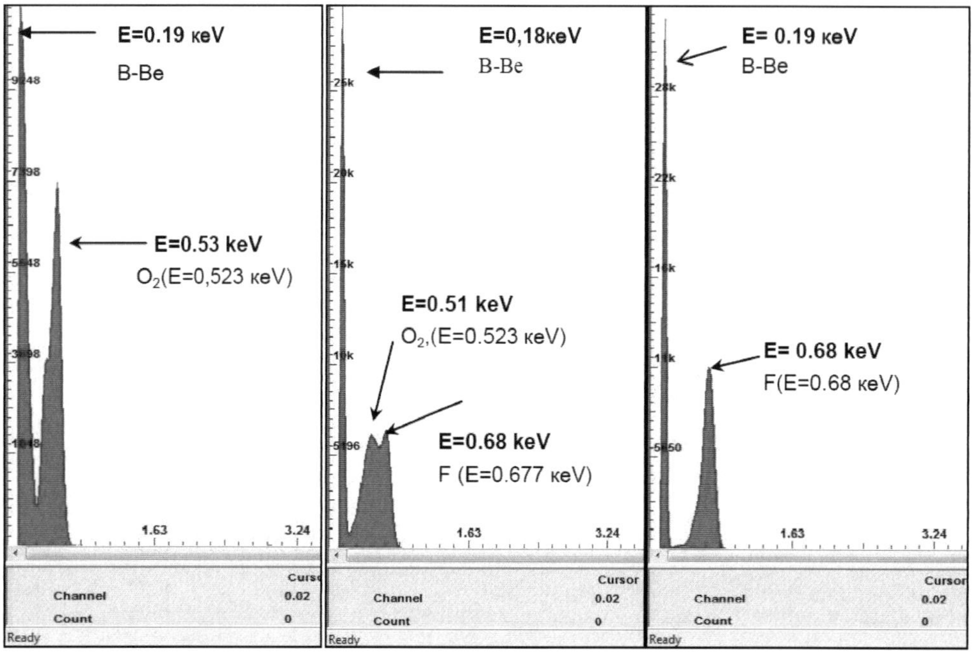

Figure 8. Soft X-ray radiation from heterogeneous plasmoid behind PVR's nozzle. X-intencity:-rel.units, quantum energy (x-axis –keV) Combined discharge (HF+DC). Power input in plasma 500 W. Mixture Ar + H_2O = 1:1. L_b = 75 mm- left, L_b = 100 mm- middle, L_b = 120 mm- right.

plausible hypothesis is that there is also hydrogen atom absorption by this stable heterogeneous plasmoid.

Calorimetric measurements of output gas flow behind the PVR nozzle in this plasmoid's regime are obtained. A high value COP = 6–10 is measured in this regime.

3. Soft X-ray radiation Spectra Created from Heterogeneous Plasmoid

A spectrometer (model X-123SDD) is used to record soft X-ray radiation (0.1–30 keV) from heterogeneous plasmoid. The X-ray receiver is arranged at different cross sections in PVR and behind its nozzle at L_b =1–100 cm from it. Small windows (holes) 6 mm diameters are cut in the Quartz testing section to measure X-radiation from heterogeneous plasmoids. These holes are covered by beryllium foil 40 μm thick.

Typical X-ray spectra obtained in heterogeneous plasmoid are shown in Figs. 6 and 7. One can see several peaks (maximums) in these spectra. The main peak is located near $E_1 \sim 1.3$ keV. Note that this value is close to quantum energy of $K_{\alpha 1}$ aluminum line $E_{\rm Al} = 1.487$ keV and $E_{\alpha 1}$ magnesium line $E_{\rm Mg} = 1.254$ keV. These chemical elements are consisting components of aluminum alloy which are used for test section manufacture. Additional peaks are located at the range E =3–5 keV. The peak E_2 = 4–4.6 keV corresponds *to the sum of resonant Ti, V, Cr lines.* These elements are consisting components of electrode material. It is impossible to explain of appearance of these lines by electron acceleration in discharge region and their interaction with metal clusters and electrodes. The measured discharge voltage is about U_d =1–2 kV only in this experiment. We determined that X-ray spectra are recorded in

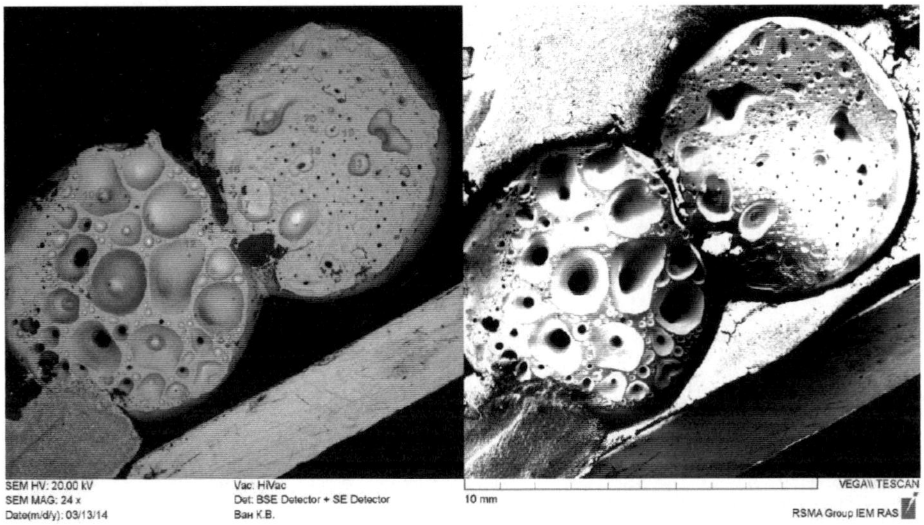

Figure 9. Melted cathode. Small holes and channels 0.1–1 mm diameter are shown in this image.

heterogeneous plasmoid, to be exact (a regime of intensive metal cluster creation in plasma). So, one can conclude that the charged and excited metal clusters can emit soft X-ray radiation.

An important result was obtained in calorimetric experiments with the PVR. We measured the maximum COP value at maximal X-radiation in this reactor. So, it is very simple to find the optimal PVR operating regime with an X-ray spectrometer.

The typical X-ray spectrum obtained behind the PVR's nozzle is shown in Fig. 7. One can see the main peaks in the range $E_1 = 0.1$–2 keV. Decay dependence of this X-ray radiation intensity on distance behind reactor's nozzle is about 30–50% at $L_b \leq 1000$ mm. So, the decrement of this radiation decay is very small.

Note that the results of analysis of consequent X-ray spectra at the different cross sections L_b behind the reactor's nozzle are very interesting:

- Nitrogen line is recorded at the cross section $L_b = 50$ mm,
- Oxygen line ($E_1 = 0.523$ keV) is recorded at the $L_b = 75$ mm,
- F- line ($E_1 = 0.68$ keV) is recorded at the $L_b = 120$ mm,
- Na, Mg, Al lines are recorded at the $L_b = 150$–200 mm.

This time evolution of the X-ray spectra at the different distance L_b is clear from possible chemical element transmutation mechanism. Fluorine is absent from the surrounding air near plasma-gas jet behind the PVR nozzle. One can see splitting of the X-ray line at the $L_b = 100$ mm. The first maximum is close to the oxygen line and the second one is close to the fluorine line. Then oxygen line intensity is decreased and fluorine line intensity is increased. One can see a single fluorine line at the L_b =120–150 mm. In our opinion this element evolution may be connected with the interaction of neutron-like particles with the atmospheric chemical elements. This key result is very important for theoretical LENR model creation via PVR operation. Remember that traditional X-ray radiation is connected with electron-metal electrode interaction in strong electric fields. But the electric field is absent in regions behind the PVR's nozzle. So, "combustion" (active reactions) of "nano-metal cluster-hydrogen fuel" continued behind the PVR nozzle. Measurement of gas flow temperature in this zone proves this suggestion. The gas temperature is increased up to 2 times at the $L_b = 1000$ mm in our experiment comparing the one at the $L_b = 0$.

4. Chemical Element Transmutation

Chemical analysis of exposed electrode, erosion droplets and dusty particles shows that there is chemical element transmutation connected with LENR in a heterogeneous plasmoid. A typical image of an exposed melted electrode's surface is shown in Fig. 8. One can see many thin holes (craters) on the electrode's surface. Note that crater diameter is close to the droplet's diameter. The main change of chemical composition of the electrode is located in these crater surfaces, to be exact.

Typical chemical element transmutation results are shown in Figs 9 and 10, and Table 1. These results are obtained by ion mass spectroscopy. One can see considerable change of surface electrode composition in Fig.10. There is concentration increase of the following chemical elements: sodium, aluminum, copper, chromium, manganese, magnesium, silicon and iron on the crater surfaces.

One can see that dusty particle composition is the following:- Ni ~15%, Si ~50%, Fe – 9%, Cu – 5% (see Fig. 9 and Table 1). Compare: - Ni-99.99% in the initial Ni-electrode.

The physical mechanism of chemical element transmutation was studied in detail.

Table 1. Dusty particle composition. Ion Mass spectroscopy analysis. Ni-electrodes, Gas mixture Ar: H_2O

Concentrations	Si	Ni	Fe	C	Al	Cu	Co	K	Mg
Atoms (%)	50	14,7	8,9	8,8	6,2	5,1	1,8	1,3	0,4
Molar mass (%)	31	19.2	10.9	2.3	4.7	7.2	2.4	1.1	0.2

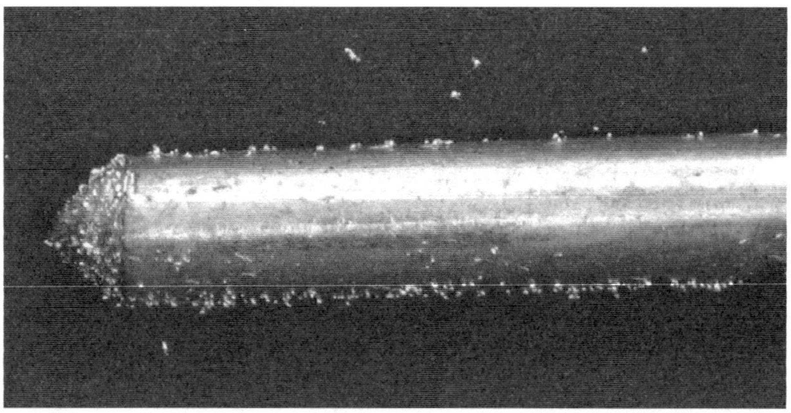

Figure 10. Anode surface is covered by tiny metal droplets injected from cathode. Cathode diameter is 5 mm .

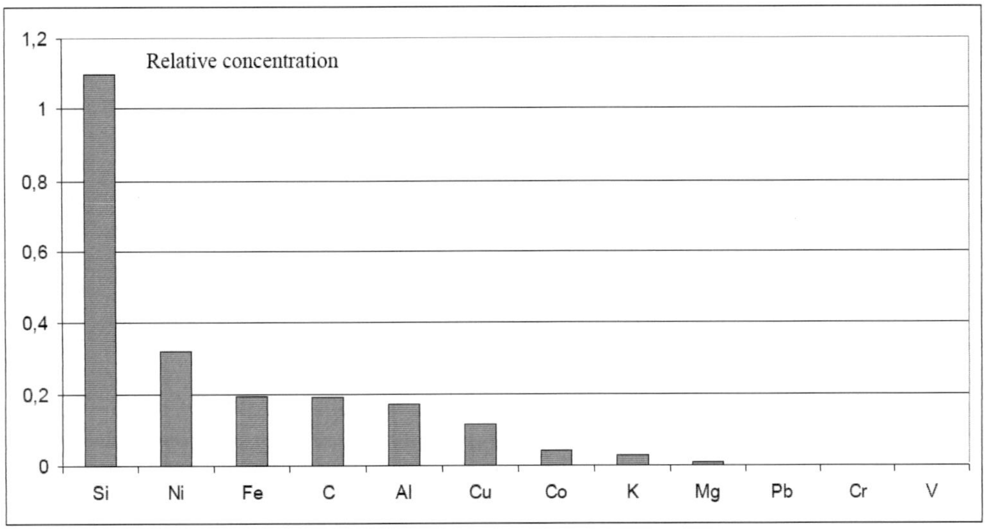

Figure 11. Relative concentrations of different chemical elements of erosive dusty particles from PVR's cathode.

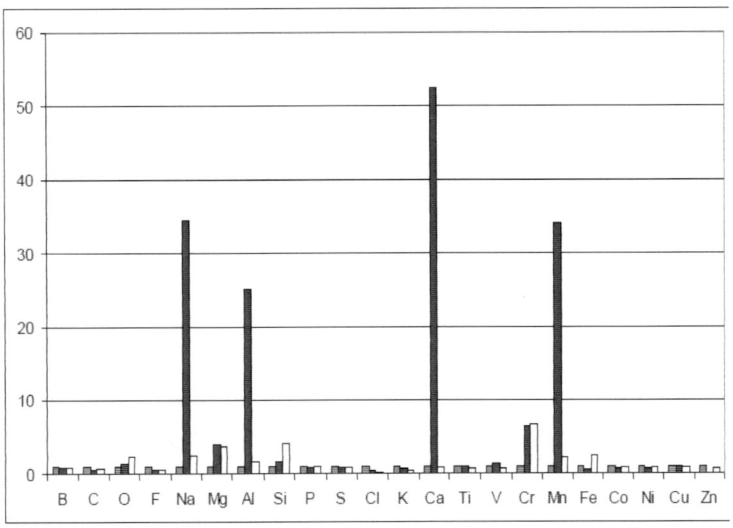

Figure 12. Relative concentrations of different chemical elements cathode's surface. All concentrations of impurities of exposed Ni-electrode are normalized on chemical element concentrations in initial Ni–electrode. Blue– initial electrode, brown– tested cathode (top side), yellow– tested cathode (lateral side).

5. Conclusions

(1) Parameters of non-equilibrium heterogeneous plasmoids in swirl flow have been measured. Electron temperature in this plasmoid is about of T_e ~6000–7000 K. Electron concentration is about $N_e \sim 10^{14} - 10^{15}$ cm^{-3}. Metal cluster temperature is about of $T_b \sim 2000$ K. Rotation and gas temperatures in plasmoid less than $T_R \sim 2300$ K. So, non-equilibrium heterogeneous plasmoids are created in the plasma vortex reactor PVR: - $T_e \gg T_b, T_R$ [2–9].

(2) We determined for the first time that heterogeneous plasmoid is intensive soft X-ray radiation generator (source). The typical value of quantum energy is about of E ~1–10 keV. We determined that the dusty plasma-gas flow behind reactor's nozzle is radioactive at long distance from it.

(3) There is chemical element transmutation in heterogeneous plasmoid. Results obtained by optical spectroscopy, ion mass spectroscopy, X-ray spectroscopy and α-ray spectroscopy prove our suggestion about transmutation.

(4) COP = 2–10 is measured in experimental setup PVR by the calorimetric method at a power release range of 1–10 kW. Estimation of specific energy release q in PVR is about 1 keV/atom. In our opinion LENR in a metal nano-cluster plasmoid may be responsible for the energy release in the PVR.

(5) We continued the Kapitsa's work [1] on stable heterogeneous plasmoid in swirl flow and proved his hypothesis about artificial ball lightning creation.

Acknowledgements

This work is supported by New Inflow, LLC. We thank Mr. V. Avdeychik, Mr. O. Grebenkin, Prof. F. Zaitsev, Prof. N. Magnitskii, for interest to this work and fruitful discussions.

References

[1] P. Kapitsa, Free plasma filament in high frequency field at high pressure, *Zhur. Exp. Teoret. Fiz.* **57**(6) (1969) 1801–1866.

[2] A. Klimov, Vortex plasmoids created by high-frequency discharges, *Atmosphere and Ionosphere: Dynamics, Processes, Monitoring,* Springer, Berlin, 2012, pp. 251–273.

[3] A. Klimov, N. Evstigneev, I. Moralev et al., Vortex control by combined electric discharge plasma, AIAA Paper 2013-1046, *51th AIAA Aerospace Sci ences Meeting*, Dallas, Texas, 2013, p.15.

[4] A. Klimov, Calorimetric measurements in vortex, *Proc. 12 Conf. CNTCE*, M. 2005, C. 246.

[5] A. Klimov, Calorimetric measurements in plasma vortex, *Proc. 14th Conf. on Cold Nuclear Transmutation and Ball Lightning*, Dagomys, 2006, pp. 246–253.

[6] A. Klimov, I. Moralev et al., Longitudinal vortex plasmoid created by capacity HF discharge, AIAA Paper 2008-956, *Proc. 46th AIAA Conf. Reno NV*, 2008, pp. 1–11.

[7] A. Klimov, V. Bityurin, I. Moralev et al., Study of a longitudinal plasmoid created by capacity coupled HF discharge in vortex airflow, AIAA Paper 2009-1046, *47th AIAA Aerospace Sciences Meeting,* Orlando, Florida, 2009, pp. 1–12.

[8] A. Klimov, Creation of new chemical elements in water by HF plasmoid, *Proc. Intern. Conf. RCCNT&BL-08*, Sochi- Dagomys, Russia, 2008, pp.15–25.

[9] A. Klimov, A. Grigorenko, A. Efimov et al., Vortex control by non-equilibrium plasma, AIAA Paper 2014, *Proc. 52nd AIAA Conf.*, 2014, pp.1–16.

J. Condensed Matter Nucl. Sci. 19 (2016) 155–163

Research Article

Energy Release and Transmutation of Chemical Elements in Cold Heterogeneous Plasmoids

A. Klimov*, †

Joint Institute of High Temperature RAS, 125412, Izhorskaya 13 bld.2, 127415 Moscow, Russian Federation

Abstract

Our main experimental results from excess energy release (COP >1) in cold heterogeneous (nano-cluster) plasmoid are considered and analyzed in this work. These experimental results were obtained by our team during the last 20 years. The main results may be divided into four groups.

(1) Excess energy release behind shock wave in a non-equilibrium heterogeneous plasma and plasma precursor creation before its front was revealed in our experiment. This measured energy is much higher than the electric energy consumed by the plasma, COP$\sim$ 4–10. This excess energy is estimated by measured gas density and pressure behind shock wave, its propagation velocity in plasma, and by using gas dynamics conservation laws.
(2) A systematic study of new chemical element creation in water by cold heterogeneous plasmoids was performed. The plasmoids are created by RF discharge over water surface. The plasmoid's chemical composition is determined by optical spectroscopy. The chemical compositions of water and its sediment are studied by independent diagnostic methods: IR spectroscopy, atomic absorption spectroscopy, ion mass spectroscopy, chemical analysis and micro-X-ray spectroscopy. We observed optical lines of the H, Li, K, Ca, OH-molecular bands, CaO-molecular bands and *unknown molecular bands* inside red plasmoids. The Li concentration and the Ca concentration are increased in a water sample by factor $10^2 - 10^3$. It is very important that the isotopic composition of the lithium is dramatically changed in this experiment. To be precise, it is possible these new elements are created by RF plasmoid in water.
(3) The energy release inside heterogeneous plasmoid in swirl gas flow was measured. The measured COP is about 2–4 in this experiment.

Keywords: Plasma aerodynamics, Plasma vortex reactor, Plasmoid, Excess energy release

*E-mail: klimov.anatoly@gmail.com
†Also at: Limited Liability Company – New Inflow, 119415, Leninsky Prospect, 96 Moscow, Russian Federation

© 2016 ISCMNS. All rights reserved. ISSN 2227-3123

1. Plasma Aerodynamics and Additional Energy Release in Weakly Ionized Non-equilibrium Plasma

A new branch of science called plasma aerodynamics (PA) was created 30 years ago thanks to a scientific discovery made by Russian scientists [1]. Flow control around flying vehicles and aerodynamic force control by weakly ionized non-equilibrium plasma formations (WINP) are now possible Scientists also observed shock wave dispersion and dissipation in WINP.

A dramatic change of a bow shock wave near spherical projectile in WINP is shown in the shadow picture, Fig. 1. One can see a considerable increase in the shock wave stand off and its dissipation in this picture. It is possible to create a regime of total shock wave dissipation at specific parameters of plasma formation and a specific Mach number. This is important because cold plasma formation with a gas temperature less than $T_g < 1500$ K is used in these plasma aerodynamic (PA) experiments. Note that plasma with a gas temperature $T_g > 8000$ K should be used in this experiment to explain of the experimental results obtained.

Excess energy release q behind a shock wave in pulse discharge plasma was measured in PA experiments and shock wave-plasma experiments many years ago [1,2]. A typical shadow picture of shock wave structure in WINP is shown in Fig. 2. One can see that there is shock wave splitting (a twofront structure) in this picture. There is considerable acceleration of shock waves in WINP also. Pulse discharge plasma created by preliminary charged capacity is used in this PA experiment. So, it is very simple to measure of maximum energy consumption in plasma by measuring of electric energy of a charged capacity. Specific energy consumption q in plasma is small in this experiment and determined by the following equation:

$$q = \int_0^{t_{\text{rop}}} j_p E_p \mathrm{d}t \leq 0, \quad 3c_p T_g \equiv 0.3h_0, \tag{1}$$

where j_p is the current density, E_p the electric field strength, t_{rop} the pulse time duration, C_p the heat capacity, T_g the gas temperature, and h_0 is the initial gas enthalpy. The value $E_p = U_d * /L$ is mean electric field in pulsed glow discharge, where $U_d^* = U_d - \delta U_c$ is mean voltage in discharge gap (δU_c is the cathode voltage gap), L is the discharge gap length. These values are measures by electric probes and voltage probes in this experiment. Formula (**??**) is simplified in a pulsed discharge. Charged capacity C is used for the glow discharge creation in this experiment. One can estimate the electric energy of this charged capacity $W = \mathrm{CU}^2/2$ accurately. The plasma volume V is estimated by high-speed frame very simple. So, maximum value q is estimated by the following formula $q = W/V$. The value W is so small that the value q is less than $0.3h_0$ in our experiment.

The measured energy release q behind the shock front was much higher than electric energy consumption q_0 (**??**) in the plasma. A typical COP of 4–10 was obtained in these experiments. Excess energy may be estimated reliably and carefully by measuring of gas density ratio ρ_2/ρ_1; pressure ratio P_2/P_1 behind the shock wave and its propagation velocity in plasma V_p. Using gas dynamics conservation laws the value q may be estimated by the well-known formula [7]:

$$q = c_p T \frac{\gamma - 1}{2\gamma} \left[\frac{p_2}{p_1} \left\{ \frac{\gamma + 1}{\gamma - 1} \frac{\rho_1}{\rho_2} - 1 \right\} - \frac{\gamma + 1}{\gamma - 1} \frac{\rho_2}{\rho_1} \right], \tag{2}$$

where the adiabatic index $\gamma = C_P/C_V = 1.4$ for air and WINP with ionization degree about $\alpha = N_e/N_a = 10^{-6}$. Note that air composition is not changed and chemical reactions are absent in this experiment. So, the adiabatic index γ_1 before shock front is equal adiabatic index γ_2 behind the shock front. Energy release q is connected with charge separation and Joule heating on the shock front only in this experiment.

The value $q = 4.2h_0 = 1.5$ kJ/g is estimated by this formula (2) and measured values (ρ_2/ρ_1 and P_2/P_1). So, this value q is higher than the initial value $q_0 = 0.3h_0$ by a factor of 10. So, there is additional energy release behind shock wave in WINP.

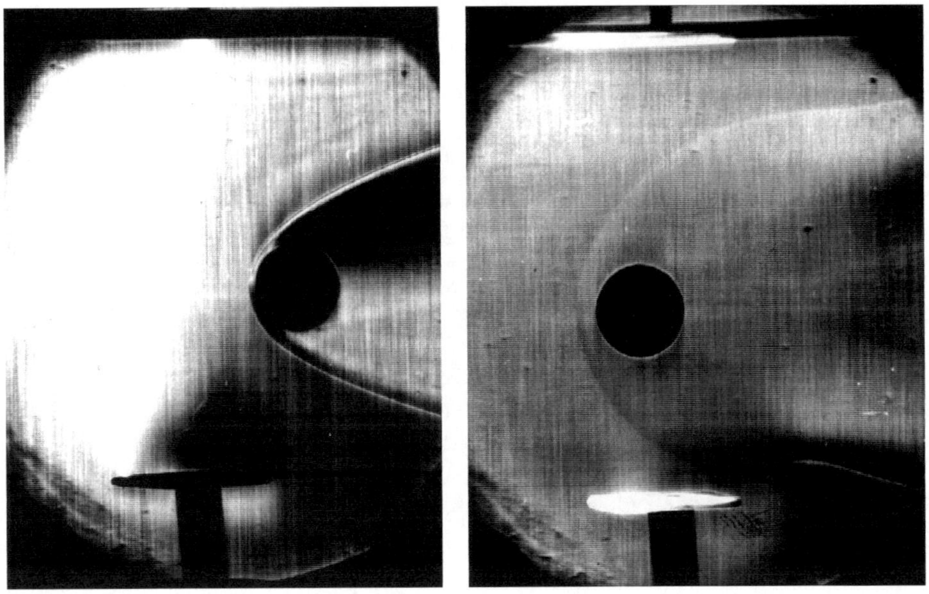

Figure 1. Shadow pictures flow around spherical projectile in WINP. Static pressure, $P_{st} \sim 40$ Torr, model's diameter, 20 mm, initial model's velocity, $V_0 \sim 2000$ m/s, and Pulse glow discharge, $j_d \sim 40$ mA/cm^2.

It is important to note that maximum change of shock wave parameters were obtained near the cathode region where *erosive nano-cluster creation takes place*, to be precise.

High value q (close to (2)) was obtained in plasma-ballistic experiment also, Fig. 1 [1].

2. Transmutation of Chemical Elements in Water Cluster Plasmoid

In the present work, a systematic study of new chemical element creation in *water cluster plasmoid* was performed [5]. This plasmoid is created by capacity type RF discharge (torch RF discharge) over the water surface. The experimental set up used in this experiment is shown in Fig. 3. This setup consists of a reactor (**??**), Tesla's HF generator (4,5) and diagnostic instrumentation (high-speed camera, optical spectrometer and ion mass spectrometer). Reactor is manufactured from Quartz glass. Its inner diameter is varied from 20 up to 80 mm. The reactor height H is about of 50–200 mm. Quartz tube thickness is about of 3–5 mm. Distillated water fills this reactor. Maximum water depth equals to 1/3 H. Tungsten electrode (2) is arranged under the water surface (10–20 mm below the surface). Tested metal target (7) from different material (aluminum, copper, nickel and others) is arranged in top region of this reactor in some experiment. The compositions of the metal sample, water, Quartz glass, and the lower electrode are analyzed by different diagnostic methods, before and after the experiment. The reactor is installed in a cylindrical HF electrode (6) connected with a ball electrode (6).

The experimental conditions are as follows: HF frequency, 0.5 MHz, HF power, 0.5–5 kW, output voltage, 40–60 kV, operation regime:-continuous and modulated, operation time <10 min.

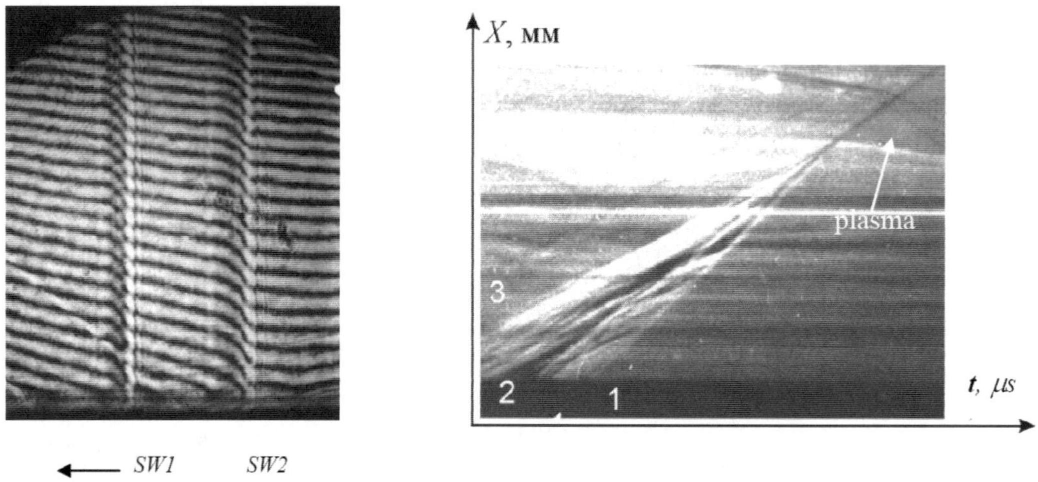

Figure 2. Shock wave structure in WINP (*left*, interferometer picture) and its scan shadow picture obtained by streak camera, $V_0 \sim 900$ m/s, static pressure, $P_{st,0} \sim 20$ Torr, Air, pulse glow discharge, $j_d \sim 30$ mA/cm^2.

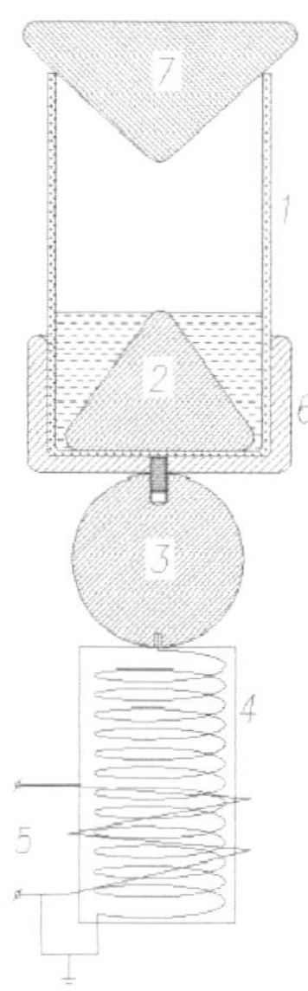

Figure 3. Schematic of experimental set up. 1- Quartz tube, 2- conical RF electrode, 3- ball RF electrode, 4, 5- HF Tesla's generator, 6- RF cylindrical electrode, and 7- metal tested sample.

Figure 4. Water cluster plasmoid created by HF torch discharge. Non-stable regime (*left*). Stable regime of plasmoid's creation (*right*). Reactor diameter, 60 mm. .

2.1. Experimental results

A red water cluster plasmoid created by HF torch discharge over water surface is shown in Fig. 4. The typical plasmoid's diameter is equal to 2–6 cm. This charged plasmoid has negative electric potential about – (4–8) kV. This potential is measured by electric probes. One can see that there is a streamer corona discharge around this charged plasmoid, Fig. 4 (*right*). It is shown that a red plasmoid is created inside blue torch HF discharge (the blue color is connected with OH-radical radiation).

2.2. Optical spectra in heterogeneous plasmoid

The plasmoid's chemical composition was studied by optical spectroscopy. The typical optical spectrum from cluster plasmoid is shown in Fig. 5. One can see the OH- radical molecular band, the Na-optical atomic lines, the Li -optical atomic lines and non-determined molecular bands X1 and X2. The optical lines of CaO-molecule, Ca I, Ca II, K, and H atoms are recorded in plasmoid also at different operation time T_p of HF generator. It is revealed that these optical lines appear and disappear at definite operation time intervals ΔT_p. The intensive H-lines and the OH-lines are recorded in the beginning of the experiment at $T_p < 10$ s. Then the Na-lines and the K-lines are recorded at the time delay $T_p > 3 - 5$ s after the discharge begins. The optical Li-line ($\lambda = 671$HM) and optical molecular bands X1, X2 appeared in the optical spectra at the time delay $T_p > 5 - 10$ s. It is remarkable that the Na-line and the K-line disappeared at the same time period $T_p > 5 - 10$ s simultaneously. Then the Li-line disappeared and bright Ca I-line and Ca II-line were recorded. The typical time evolution of the Li-line intensity is shown in Fig. 6. One can see that the Li-line dissipated finally at the time delay $T_p > 60 - 100$ s. Vibration temperature $T_V \sim 4000$ K and rotation temperature $T_R \sim 2000 - 3000$ K are determined by processing of the recorded OH-optical spectrum. Electron concentration is about $N_e \sim 10^{13} - 10^{14}\,\mathrm{cm}^{-3}$ estimated by processing of the H_β, H_α-lines. So, a hot charged cluster plasmoid is created in this experiment.

2.3. Chemical analysis of metal sample and water sample activated by plasmoid

The chemical compositions of the water and its sediment were studied by various independent diagnostic methods (IR spectroscopy, atomic absorption spectroscopy, ion mass spectroscopy, chemical analysis, and micro-X-ray spectroscopy). It was revealed that the Li-concentration and the Ca-concentration are increased *n* the water sample by factors of 10^2–10^3 (see Table 1). It is important to note that the Li-concentration of activated water is very high

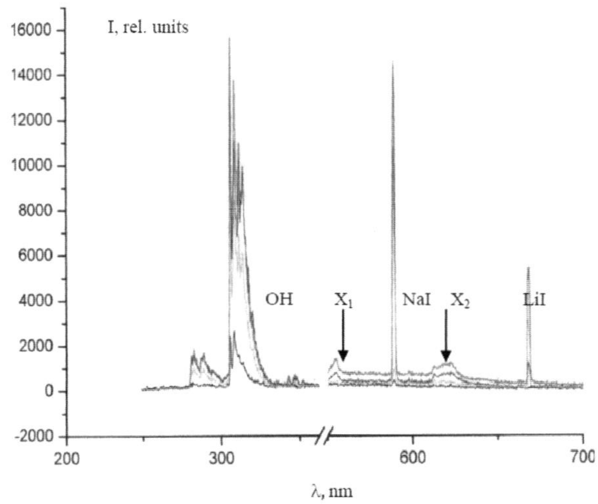

Figure 5. Optical spectrum of heterogeneous plasmoid. Operation time 10 s .

$\sim$ 1 mg/l at $T_p < 60$ s and very small < 0.01 mg/l at $T_p > 60 - 100$ s. So, one can suppose that the lithium is an intermediate element of chemical transmutation in the water reactor.

It is important to note that the isotopic composition of the lithium dramatically changed in this experiment. The initial ratio $Li_7/Li_6 = 13.192 \pm 0.004$ (initial sample) is increased up to the ratio $Li_7/Li_6 = 18.996 \pm 0.012$ (activated sample). Note that the lithium atoms and the calcium atoms are not measured in quartz tube wall of plasma-chemical reactor. So, it is possible these new elements are created by RF plasmoid in water. We observed bright red flashes (luminescence) inside water at a stable plasmoid creation regime. The typical frequency of these flashes is about of

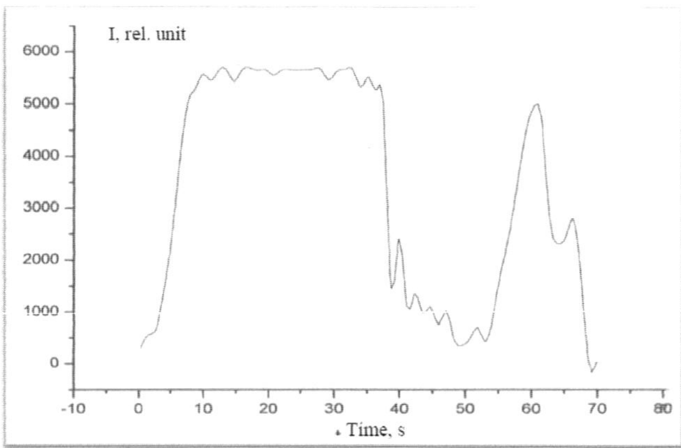

Figure 6. Time evolution of the Li-optical line $\lambda = 671$ nm in heterogeneous plasmoid.

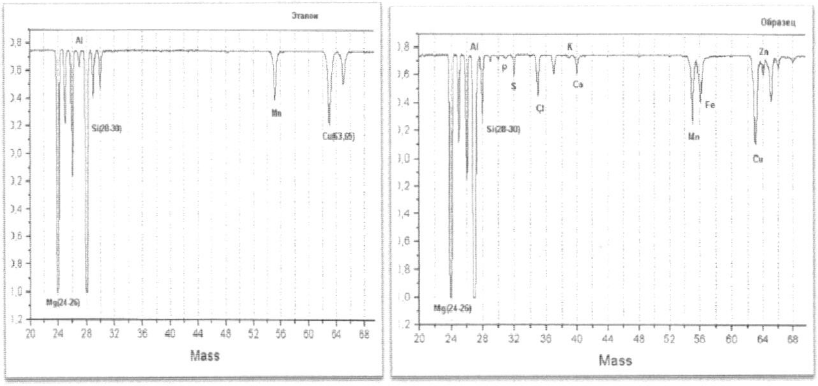

Figure 7. Ion mass spectrometer spectrum of aluminum alloy before experiment (*left*) and after (*right*). Time activation 10 min .

Table 1. Atomic absorption analysis of water sample

	Concentration (mg/ l)	
	Li	Ca
Initial water, non-activated by plasmoid	< 0.01	2.2
Tested water, activated by HF plasmoid	0.8	64.7

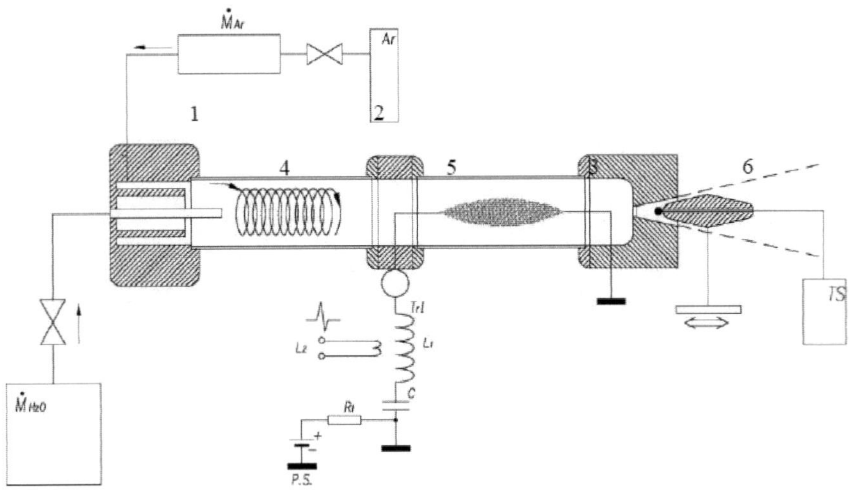

Figure 8. Schematic of experimental set up PVR. 1- swirl generator, 2- argon high-pressure tank, 3- nozzle-cathode, 4- Quartz glass tube, 5- anode, 6(TS)- valve with thermocouple, M_{H_2O}- water steam generator, and M_{ar}- mass argon flow meter.

1–10 Hz (Fig. 4, *left*). The composition of a metal sample activated by heterogeneous HF plasmoid was analyzed by ion mass spectrometer. The typical ion mass spectra of aluminum alloy sample before its activation by plasmoid and after it are shown in Fig. 7. One can see creation of number of new chemical elements (such as phosphorus, sulfur, chlorine, calcium, iron, and zinc) in this mass spectrum. These elements are absent in the initial metal sample, initial Quartz tube and water.

So, *new chemical elements and changes in their isotopic compositions by cluster plasmoid* are created in this experiment. Many of new elements (lithium, potassium, calcium, ...) are intermediate ones. They are created by a plasmoid at definite time delay T_p and then these elements are transmuted to the new ones.

3. Energy Release Inside Heterogeneous Plasmoid in Swirl Gas Flow

A new step of our research on heterogeneous plasmoid physics has been made possible by support from the limited liability company "New Inflow". A new plasma vortex reactor (PVR) was designed and then improved on the basis of experimental background described above. *Heterogeneous metal cluster plasmoid* is created by combined discharge (HF+DC) in various gas flows. High value COP $= 2-10$ is obtained in this PVR, to be precise [6]. Detail experimental results obtained in this PVR are considered in this work.

A schematic of the PVR is shown in Fig. 8. Experimental conditions are as follows: argon–water steam mixture $= 1:1$, mass gas flow rate <10 g/s, gas flow Mach number $M < 0.2$, static pressure $P_{st} < 2$ bar, mean electric power consumption <1 kW, mean thermal output power <10 kW, and the erosion electrode mass <1 mg/s.

A heterogeneous metal cluster plasmoid (HP) created by combined discharge in swirl flow is shown in Fig. 9. It is revealed that this plasmoid consists of many metal charged micro-droplets and metal nano-cluster halos around them (created by cathode electrode's erosion). The typical negative electric potential of these particles is about –(2–4 kV). Its value is close to the cathode electric potential.

We determined that these nano-clusters absorb hydrogen atoms very effectively. An analysis of optical spectra proves this conclusion. The measured gas temperature of this plasmoid is about 2000 K.

It is important to note that a high value COP of 2–10 is obtained in swirl flow in this PVR. The A COP of 0.5–0.7< 1 is obtained in straight gas flow (non-swirl flow). So, swirl flow plays a very important role in excess energy release.

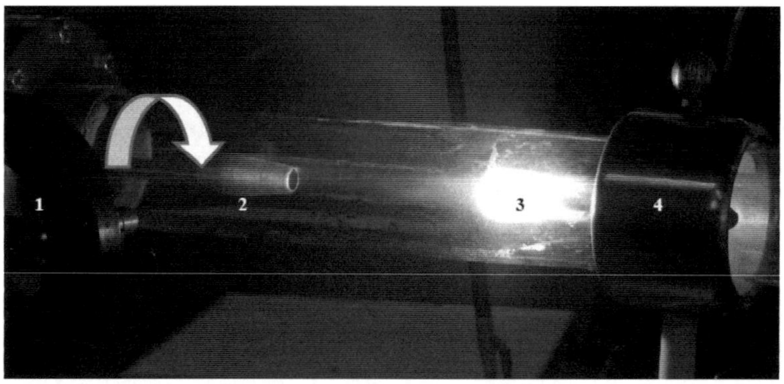

Figure 9. Heterogeneous plasmoid in PVR. Mixture: argon - water steam= 1:1. 1- swirl generator, 2- water vapor injector-anode, 3 – plasmoid, and 4- cathode.

We observed chemical transmutations in this cluster plasmoid. The transmutations may be connected with LENR. Optical spectra, ion mass spectra, soft X-spectra, stimulated α-radioactivity of activated Bi-electrodes prove this conclusion [6].

Acknowledgements

This work is supported by the limited liability company "New Inflow". We thank Mr. V. Avdeychik, Mr. O. Grebenkin, Prof. F. Zaitsev and Prof. N. Magnitskii for their interest to this work and their fruitful discussions.

References

[1] R. Avramenko, A. Klimov et al., Diploma on scientific discovery #007, Moscow, State Committee on discoveries and patents, USSR, 1985.

[2] Ball Lightning in Laboratory, R. Avrameko, A. Klimov et al. (Eds), *Chemistry*, Moscow, 1994, pp.1–256 (in Russian).

[3] A. Klimov, Vortex plasmoids created by high-frequency discharges, *Atmosphere and Ionosphere: Dynamics, Processes, Monitoring,* Springer, Berlin, 2012, pp. 251–273.

[4] A. Paschina and A.Klimov, Structure of long-lived high-energetic plasmoid and its interaction with incoming gas flow, *Chem. Phys.* **33**(2) (2014) 78–86.

[5] A. Klimov, Creation of new chemical elements in water by HF plasmoid, *Proc. Int. Conf. RCCNT&BL-08*, Sochi- Dagomys, Russia, 2008, pp 15–25.

[6] A. Klimov, A. Grigorenko, N. Evstigneev et al., High-energetic nano-cluster plasmoid and its soft X-radiation, *Proc. ICCF-19,* to be published.

[7] F. Barteleme, *Combustion Gas Dynamics*, M. Energoizdat, 1981, pp.1–351 (in Russian).

J. Condensed Matter Nucl. Sci. 19 (2016) 164–172

Research Article

Lithium – An Important Additive in Condensed Matter Nuclear Science

Chang L. Liang, Zhan M. Dong, Yun P. Fu and Xing Z. Li*

Department of Physics Tsinghua University, Beijing 100084, People's Republic of China

Abstract

A ($p+^6Li$) low energy resonance state is found using the 3-parameter formula for fusion cross-section based on the selective resonant tunneling model. The electron capture is the possible weak interaction to make use of this low energy resonance. When the resonance energy level is close to zero, the width of the resonance peak in the fusion cross-section is much greater than the width of the resonance energy level; therefore, the absorption through this low energy resonance level is no longer a problem for resonant tunneling of Coulomb barrier at low energy. Both hot fusion data and the CMNS experiment data support this resonant tunneling concept. As a result, lithium turns out to be an important additive in both CMNS and Hot Plasma fusion research.

Keywords: Resonance and resonant tunneling, $p+^6Li$ low energy resonance, Resonance accompanied with weak interaction, The width of resonance at low energy, 3-Parameter formula for fusion cross-section

1. Introduction

A formula for fusion cross-section has been presented in ICCF-14 (Washington D.C. 2008). It was based on selective resonant tunneling model [1], and has three parameters only [2–8]; however, it fits with a set of five major fusion cross-section experimental data much better than the 5-parameter old formula in the famous NRL Plasma Formulary [9,10] does. This 3-parameter formula has been further applied to almost all the comparable light nucleus fusion cross-section data (14 sets of data including p+p, p+D, $p+^6Li$, $p+^7Li$, $p+^7Be$, $p+^9Be$, $p+^{10}B$, d+d, d+T, $d+^3He$, $d+^6Li$, $d+^{11}B$, t+t and $t+^3He$), and the good agreement with those hot fusion data strongly confirmed the validity of this 3-parameter formula and its physics – selective resonant tunneling model.

Using this 3-parameter formula, we are going to discuss the following three issues:

(1) What is the resonance and resonant tunneling in CMNS phenomena? Although resonance has been mentioned almost everywhere in CMNS literature, nevertheless, what is the definition of a resonance in CMNS? Whenever one needs to overcome Coulomb barrier, a resonance is invoked as a Panacea, however, why does a resonance

*Corresponding author. E-mail: lxz-dmp@tsinghua.edu.cn

overcome the Coulomb barrier? How can a resonance make a resonant tunneling? What are the necessary conditions to have this important resonant tunneling? Is there really such a low energy nuclear resonance in the nature? How can we identify such a resonance while there is no experimental data for a proton projectile of thermal energy? This 3-parameter formula is able to answer these questions, and shows clearly that (p+^{6}Li) has just this resonance.

(2) Lithium has been used as an additive widely in CMNS research; however, it was based on the consideration of electrochemistry ("conductivity" for electrolyte), or based on chemical engineering ("catalyzer" in Ni–H systems), or based on thermodynamics ("long range coherence in liquid lithium" – Chemonuclear reaction). Since the year of 2002, we attempted to find an explanation for this lithium issue based on nuclear physics. This presentation is an effort to pursue this explanation using the 3-parameter formula for fusion cross-section.

(3) The width of a resonance peak in *cross-section* and the width of a resonance *energy level* are different concepts; however, there are some misunderstanding which would be clarified using the selective resonant tunneling model

2. 3-Parameter New Formula for Fusion Cross-section

For low energy projectile the cross-section, $\sigma(E)$ is used to be expressed by a phase-shift of S-partial wave function, δ_0, provided that S-partial wave is dominant,

$$\sigma(E) = \frac{\pi}{k^2}\left(1 - \left|e^{i2\delta_0}\right|^2\right). \quad (1)$$

Here, k is the wave number in the centre of mass system. This expression does not show clearly where Gamow factor for charged particle interaction is, and how the resonance would overcome the Coulomb barrier. Therefore, we derived another expression which is identically equal to Eq. (1) [2]

$$\sigma(E) = \frac{\pi}{k^2}\frac{(-4W_\mathrm{i})}{W_\mathrm{r}^2 + (W_\mathrm{i} - 1)^2}. \quad (2)$$

Here $W \equiv \cot\delta_0 \equiv W_\mathrm{r} + iW_\mathrm{i}$ is introduced to replace δ_0. The imaginary part, W_i, describes the absorption in the nuclear potential well. This new formula clearly shows the physical meaning of a resonance and resonant tunneling: resonance corresponds to an energy which makes $W_\mathrm{r} = 0$, and resonant tunneling corresponds to both $Wr = 0$ and $Wi = -1$. Indeed, W is the coefficient of a linear composition of two independent solutions of Schrodinger equation. In case of charged nuclei collision, $\varphi(\mathrm{r}) = W \cdot F_0 + G_0$, here $\varphi(\mathrm{r})$ is the reduced radial wave function in the Coulomb field, F_0 and G_0 are the regular and irregular Coulomb wave function, respectively. Here r is the radial distance from the centre of nuclear potential well. When the resonant tunneling happens,

$$\varphi(r) = (W_\mathrm{r} + iW_\mathrm{i}) \cdot F_0 + G_0 \underset{r\to\infty}{\longrightarrow} (W_\mathrm{r} + iW_\mathrm{i})\sin(kr) + \cos(kr) \propto \frac{1 + W_\mathrm{i} - iW_\mathrm{r}}{1 - W_\mathrm{i} + iW_\mathrm{r}}e^{ikr} + e^{-ikr} \to e^{-ikr}. \quad (3)$$

The resonant tunneling implies an incoming spherical wave without any reflection, or a *perfect absorption* of incoming wave by a nuclear potential boundary. This is the physical meaning of a resonant tunneling. Then, where is the Gamow factor? It is embedded in W. Based on the continuity of wave function at the interface between nuclear potential well and Coulomb barrier, we may further explore the meaning of $W_\mathrm{r} = 0$ as follows:

$$W = -\left(\frac{G_0}{F_0}\right)_a \left[\frac{D_\mathrm{L} - \frac{k}{G_0}\frac{\partial G_0}{\partial \rho}\Big|_a}{D_\mathrm{L} - \frac{k}{F_0}\frac{\partial F_0}{\partial \rho}\Big|_a}\right] = -\left(\frac{G_0}{F_0}\right)_a \left[\frac{D_\mathrm{L} - \frac{k}{G_0}\frac{\partial G_0}{\partial \rho}\Big|_a}{\left(D_\mathrm{L} - \frac{k}{G_0}\frac{\partial G_0}{\partial \rho}\Big|_a\right) - \frac{k}{F_0 G_0}\Big|_a}\right], \quad (4)$$

where D_L is the logarithmic derivative of inner wave function at the interface between Coulomb field and nuclear interaction region, k is the wave number in the Coulomb field in the centre of mass system. $\rho = k \cdot r$, a is the radius of the nuclear interaction region. When the real part of numerator in Eq. (4) approaches to 0 (which implies that the inner wave function is mainly connected to G_0 at the interface), we have $W_r = 0$. This is the first requirement to have a resonant tunneling.

In order to clearly show where the Gamow factor is, we may separate W into two factors in Eq. (4): the fast varying factor in round brackets and the slow one in square brackets. Since G_0 is exponentially rising and F_0 is exponentially decreasing when r is approaching nuclear boundary, a, the ratio of

$$\left(\frac{G_0}{F_0}\right)_a \propto \left(\frac{e^{2\pi/(ka_c)} - 1}{2\pi}\right) \equiv \theta^2$$

is a fast varying factor at low energy, and is a very large number.

$$a_c \equiv \frac{4\pi\varepsilon_0\hbar^2}{z_a z_b \mu e^2}.$$

ε_0 is the vacuum dielectric constant, $\hbar$ the Planck constant divided by2π, e the charge of proton, μ reduced mass, z_a and z_b are the charge numbers of the colliding nuclei, respectively. On the contrary, both

$$\left(D_L - \frac{k}{G_0}\frac{\partial G_0}{\partial \rho}\bigg|_a\right) \quad \text{and} \quad \frac{k}{F_0 G_0}\bigg|_a$$

are slowly varying factors when $k \to 0$; therefore, we may assume:

$$W \equiv \theta^2(w_r + iw_i) \approx \theta^2(C_1 + C_2 E_{lab} + iw_i) \tag{5}$$

and it leads to the $\sigma(E)$ expression with Gamow factor $(\frac{1}{\theta^2})$,

$$\sigma(E) = \frac{\pi}{k^2}\frac{1}{\theta^2}\frac{(-4w_i)}{w_r^2 + (w_i - \frac{1}{\theta^2})^2} \approx \frac{\pi}{k^2}\frac{1}{\theta^2}\frac{(-4w_i)}{(C_1 + C_2 E_{lab})^2 + (w_i - \frac{1}{\theta^2})^2}, \tag{6}$$

$$\theta^2 \equiv \left(\frac{e^{2\pi/(ka_c)} - 1}{2\pi}\right). \tag{7}$$

This new formula of fusion cross-section clearly shows that the Gamow factor would disappear only if $w_r = 0$ and $w_i \approx -\frac{1}{\theta^2}$. Indeed w_i is proportional to the imaginary part of D_L, i.e. the probability current density across the interface between the nuclear interaction region and the Coulomb field. Hence, $w_i \approx -\frac{1}{\theta^2}$ implies a very weak absorption rate inside the nuclear interaction region. This is the second necessary condition for a resonance to overcome the Coulomb barrier.

This assumption (5) is strongly supported by the experimental data for eight fusion cross-sections: d+T, d+^{3}He, d+D, t+T, t+^{3}He, p+D, p+^{6}Li and p+^{7}Li in Fig. 1. In the eight plots of Fig. 1, the solid lines are the fitting curves using Eq. (6) with three parameters: C_1 C_2 and w_i; circles are experimental data points from National Nuclear Data Center in Brookhaven National Lab. [11]. Using the least-squares method, we may find three parameters for each reaction as shown in Table 1. Logarithmic scales are used to show the good fit at very low energy region, and the famous resonance peak for d+T is shown in Fig. 1 in linear scale as well.

Recently this new formula has been further justified by 14 sets of light nuclei fusion data. However, we discuss first the fact that in Table 1, among the eight fusion data, only p+^{6}Li cross-section data might be fit by three parameters with diminishing C_1 and C_2. This is the first hint of low energy resonance in p+^{6}Li fusion, because it means that $w_r \to 0$ when $E_{lab} \to 0$, and the width of the resonance would be determined by the term of $(w_i - \frac{1}{\theta^2})^2$ in the denominator of Eq. (6).

Table 1. Three parameters in formula of cross-section for eight reactions

Reaction	C_1	C_2 (1/keV)	w_i	Norm(barn)/data point number	$(\sigma(E))_{max}$ (barn)	$(E)_{max}$ (keV) in Lab.
d+T	0.544	−0.00558	−0.390	0.227/24	5.0	280
d+^{3}He	1.13	−0.00304	−0.670	0.0520/800	0.8	1034
d+D	4.78	−0.00226	−0.186	0.00567/39	0.177	1045
t+T	36.8	−0.00928	−24.6	0.0129/757	0.115	4300
t+^{3}He	2.79	0.000959	−1.04	0.00331/225	0.0214	1000
p+D	8.04×10^7	-1.80×10^6	-5.31×10^7	3.35×10^{-8}/74	1.45×10^{-7}	48.1
p+^{6}Li	**0**	-5.02×10^{-9}	−6.62	0.00125/15	0.0197	190
p+^{7}Li	30.9	−0.00367	−4.18	0.000310/42	0.00633	998

3. The Width of this Resonance Energy Level near Zero Energy

The width of the resonance peak in the cross-section is very important after knowing the existence of this p+^{6}Li resonance near the zero energy, because it will determine the absorption integral of the injected proton in the CMNS experiments.

The diminishing C_1 and C_2, and the large value of $(-w_i)$ for p + ^{6}Li → ^{3}He + ^{4}He cross-section data makes an unexpected behaviour of the resonance peak. It does not show a sharp peak near the zero energy, instead it keeps climbing up the in the range of 50–190 keV without any peak. Indeed, in the range of 50–190 keV, we have

$$(C_1 + C_2 E_{\text{lab}})^2 \ll \left(w_i - \frac{1}{\theta^2}\right)^2 \tag{8}$$

because in this range, θ^2 varies from 97 150 to 148, hence, $\frac{1}{\theta^2}$ is much smaller than $(-w_i)$ such that $(w_i - \frac{1}{\theta^2})^2 \approx (-6.62)^2$, and $(C_1 + C_2 E_{lab})^2 < (0 - 5.02 \times 10^{-9} \times 190)^2 \ll (w_i - \frac{1}{\theta^2})^2$.

Thus we may write the expression of cross-section as

$$\sigma(E) \approx \frac{\pi}{k^2}\frac{1}{\theta^2}\frac{(-4w_i)}{(w_i - \frac{1}{\theta^2})^2} = \frac{\pi}{k^2}\frac{(-4\theta^2 w_i)}{(\theta^2 w_i - 1)^2} = \frac{\pi}{k^2}\frac{(4x)}{(x+1)^2}. \tag{9}$$

Here, $x \equiv -\theta^2 w_i$. It should be noticed that if $E \to 0$, $x \to \infty$ then, $\frac{(4x)}{(x+1)^2} \to 0$ and if $E \to \infty$, $x \to 0$; then, $\frac{(4x)}{(x+1)^2} \to 0$ as well. Only if $x = 1$, then $\frac{(4x)}{(x+1)^2} \to 1$ reaches its peak value. Its peak position is determined by

$$x \equiv -\theta^2 w_i = 1. \tag{10}$$

For p + ^{6}Li → ^{3}He + ^{4}He, $w_i = -6.62$, in the range of 50–190 keV,

$$x \equiv -\theta^2 w_i \sim 10^5 \text{ to } 10^2 \gg 1. \tag{11}$$

Therefore, $\sigma(E)$ would not show any peak until $x \equiv -\theta^2 w_i = 1$. In order to have a peak in this energy region it is necessary to reduce the value of $(-w_i)$. In Fig. 2, four curves with different values of w_i are drawn using Eq. (6) for p+^{6}Li reaction. The thin solid (red) line (the curve for $w_i = -6.62$) is for the real p + ^{6}Li → ^{3}He + ^{4}He data, the curve is almost overlapping with the abscissa without any peak before 800 keV. However, if there is any other weaker interaction which would make $(-w_i)$ much smaller; then, there might be a resonance peak before 800 keV (dotted line, dashed line, and the thick dash–dotted line). It is noticed that: the width of the resonance peak in cross-section is decreasing while $(-w_i)$ is getting smaller and smaller. Then one may ask if the width of resonance becomes extremely small when the peak position is approaching zero energy?

Fortunately, the resonance peak position of the cross-section and its width approaches zero slowly in the scale of $\log[-w_i]$. It is very slow; hence, the absorption integral near this resonance energy level would approach to zero very

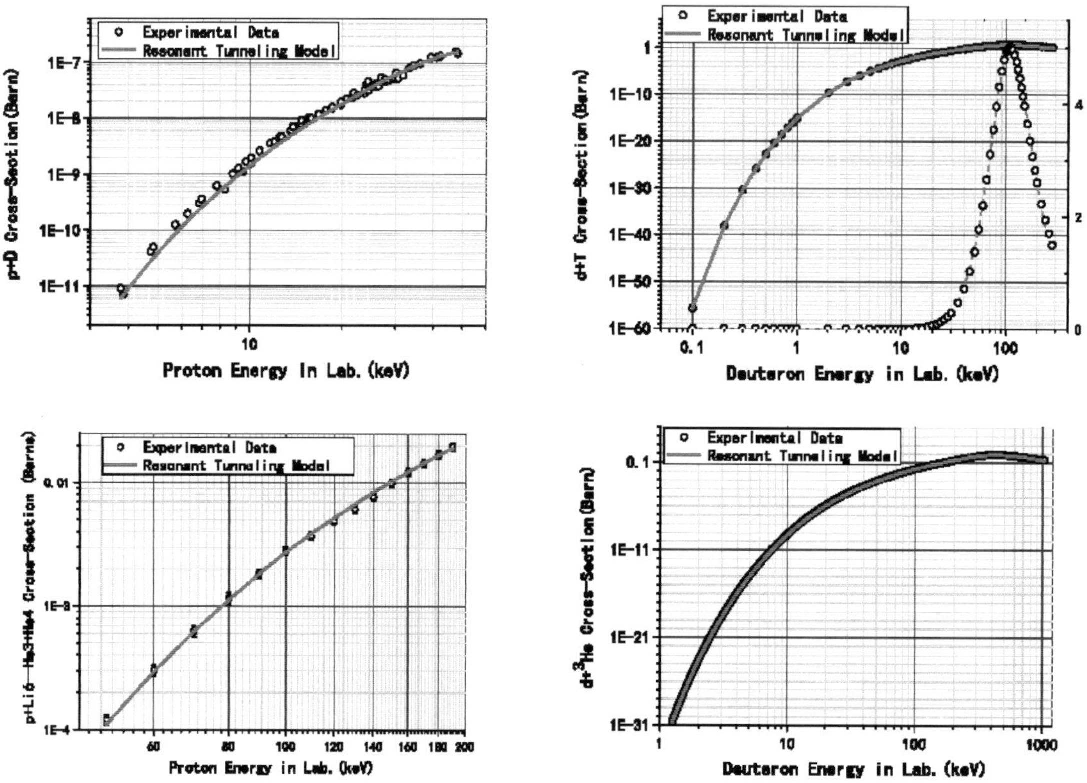

Figure 1. 3-Parameter formula, Eq. (6) is justified by eight sets of fusion cross-section data.

slow as well. Figure 3 shows that while the peak position of $\sigma(E)$ in Eq (9) is determined by $x \equiv -\theta^2 w_{\mathrm{i}} = 1$, the width of this peak is determined by $x = 3 + 2\sqrt{2}$ and $x = 3 - 2\sqrt{2}$. Therefore, when $-w_{\mathrm{i}} \to 0$, the corresponding energy of resonance peak and width are approaching zero in logarithmic scale as $\log[E]$ due to Eq. (7)

$$\left(\theta^2 \approx \left(\frac{\mathrm{e}^{2\pi/(\sqrt{E}a_c)}}{2\pi}\right)\right).$$

Indeed, what affects the absorption of injected beam is the integration under the curve between $x = 3 - 2\sqrt{2}$ and $x = 3 + 2\sqrt{2}$ (Fig. 3). The absorption integral – the shaded area under the cross-section curve in the Full Width Half Peak (FWHP) decreases with $(-w_{\mathrm{i}})$ slowly in a logarithmic scale (Fig. 4) as well. Consequently, we do not worry about the width of the resonance peak of the cross-section for this low energy resonance induced by a weak interaction.

There have been two misunderstanding about the width of this low energy resonance: (1) The life-time of the resonance would be very long due to the weak interaction; therefore, the energy level would be extremely narrow because of the energy uncertainty relation. As a result, it would be very difficult for the injected beam to be collimated

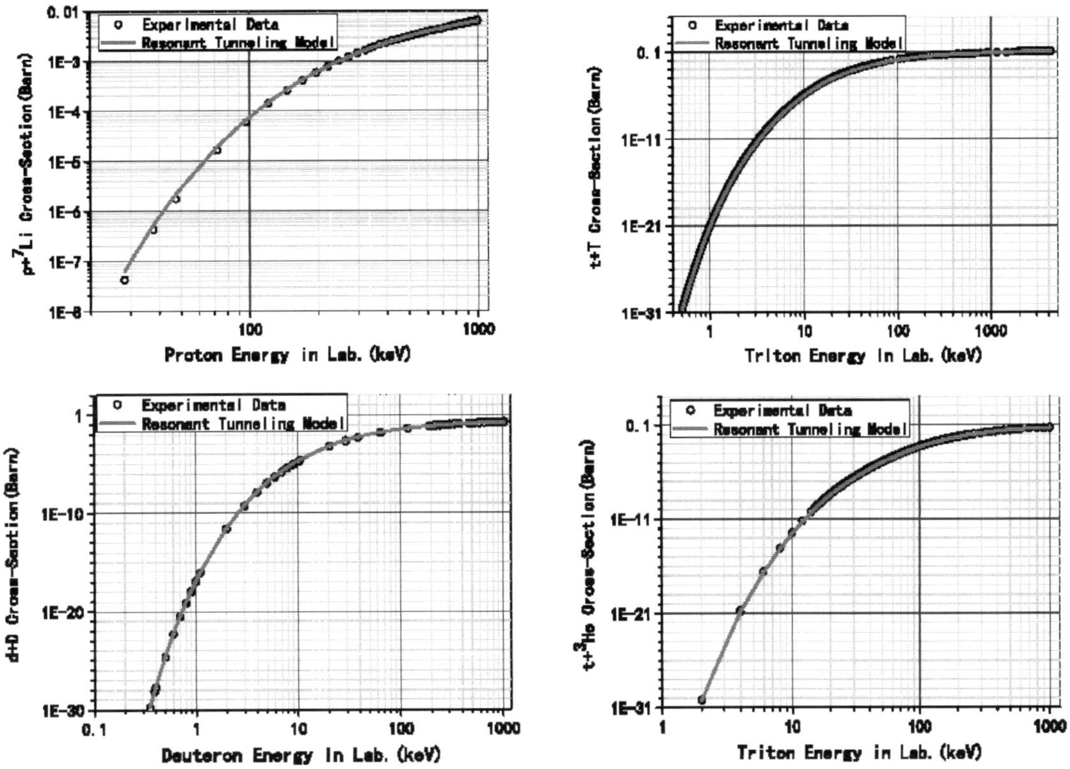

Figure 1 (continued).

in such a narrow energy width. (2) In Eq. (4), the difference between two logarithmic derivatives,

$$\left(D_{\mathrm{L}} - \frac{k}{G_0} \frac{\partial G_0}{\partial \rho}\bigg|_a \right) = 0,$$

might be valid only at a single value of energy, $E = E_0$. Any deviation from E_0 would introduce a non-zero value $(w_{\mathrm{r}})^2$ which would be enlarged by the big factor, θ^2; therefore, the width of resonance in E would be extremely narrow, such that there would be no chance for any resonance to happen in a real world.

To answer the first question, we should notice that the *width of the resonance energy level* is not necessary equal to the *width of the resonance peak of the cross-section*. Only in the case of single energy level far away from the zero energy and far away from the other energy levels, we might use the relation,

$$\mathrm{e}^{-i(E_0 - i\Gamma)t/\hbar} \propto \int_{-\infty}^{\infty} \frac{\Gamma/2}{(E - E_0) + i(\Gamma/2)} \mathrm{e}^{-iEt/\hbar} \mathrm{d}E;$$

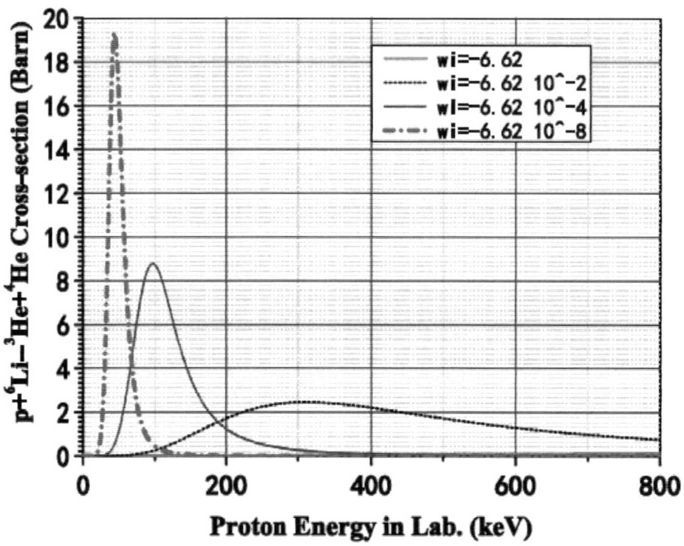

Figure 2. The low energy resonance peak in the cross-section would appear only if $(-w_{\rm i})$ is small enough.

$$\sigma(E) = \frac{4\pi}{k^2}\frac{(\Gamma/2)^2}{(E-E_0)^2 + (\Gamma/2)^2}. \tag{12}$$

to derive that the width of the resonance peak in the cross-section would be equal to the width of the energy level, Γ while the energy dependence of Γ is ignored. For the case of tunneling Coulomb barrier near zero energy, only $(-w_{\rm i})$ is related to the ratio of the life-time to bouncing time inside the nuclear interaction region, but the width of the cross-section peak is proportional to the $\log[-w_{\rm i}]$, when the energy dependence of Γ is considered explicitly.

To answer the second question, we should notice that when the resonance energy approaches zero, the cross-section is no longer determined by $w_{\rm r}^2 \approx (C_1 + C_2 E)^2$ term, instead, it is determined by $(w_{\rm i} - \frac{1}{\theta^2})^2$ in the denominator of Eq. (6), because

$$w_{\rm r}^2 \ll \left(w_{\rm i} - \frac{1}{\theta^2}\right)^2.$$

θ^2 affects only on the larger one among

$$w_{\rm r}^2 \quad \text{and} \quad \left(w_{\rm i} - \frac{1}{\theta^2}\right)^2$$

.

Consequently, we do not worry about the absorption near this low energy resonance level, but we have to identify a weak interaction channel which makes use of this low energy resonance level.

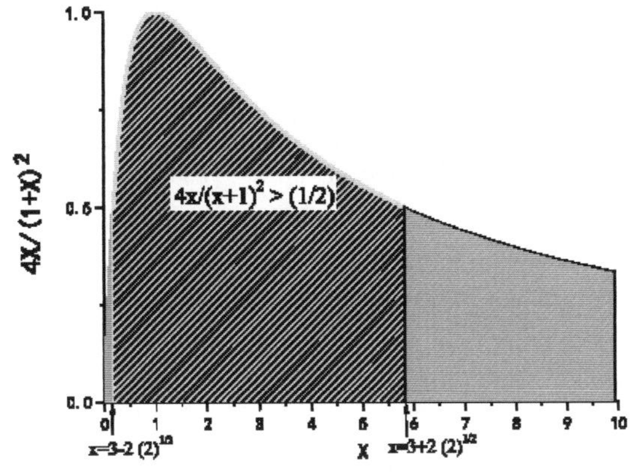

Figure 3. The width of the resonance peak in cross-section is $4(2)^{1/2}$.

4. A Possible Weak Interaction Channel accompanied with Low Energy Resonance for p+^{6}Li Fusion

In order to see the resonance peak in cross-section, it is necessary to identify a weak reaction channel which satisfies the second resonance condition: $-\theta^2 w_{\rm i} = 1$. Indeed there is an orbital electron capture (weak interaction) which may be accompanied with the resonant tunneling.

$$\begin{aligned} \mathrm{p} + {}^6\mathrm{Li} + e &\to {}^7\mathrm{Li}^* + \text{neutrino} \\ {}^7\mathrm{Li}^* &\to T + {}^4\mathrm{He} + (\sim 4\,\mathrm{MeV}). \end{aligned} \tag{13}$$

This reaction is very slow in comparison with the fast fusion reaction accompanied with the strong interaction:

$$\mathrm{p} + {}^6\mathrm{Li} \to {}^3\mathrm{He} + {}^4\mathrm{He} + 4.001\,\mathrm{MeV}. \tag{14}$$

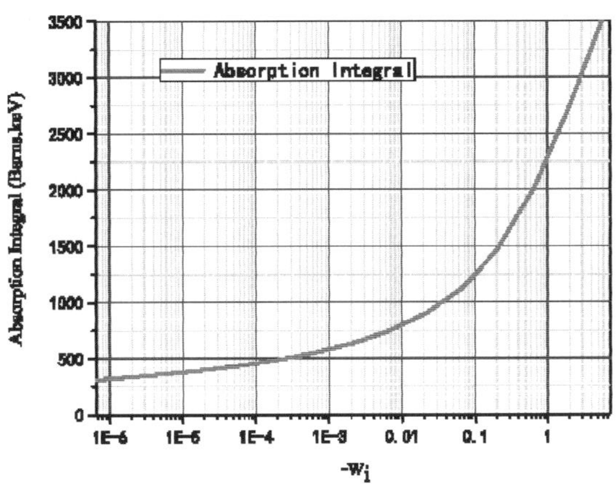

Figure 4. The absorption integral decreases with the $(-w_{\rm i})$ slowly in a logarithmic scale.

Based on the life-time of ^{7}Be due to electron capture, 53.22 days, it is possible to make a rough estimate of the $w_{\mathrm{i}} = -6.62 \times 10^{-29}$. Using Eqs. (7) and (10), we may show that the resonance peak of the cross-section would be close to 2 keV, and the width of this peak in cross-section would be 0.2 keV. As we pointed out in ICCF-17 [12], this excited state ^{7}Li* keeps most of the reaction energy, and this neutrino emission carry away only a small part of the reaction energy. Eventually the final fusion products, T + ^{4}He, which are all charged particles, would deposit the most of fusion energy in the crystal lattice and form the "excess heat" in CMNS experiments.

5. Conclusion Remarks

The hot fusion data for p+^{6}Li have shown the existence of a low energy resonance using the new 3-parameter formula for the fusion cross-section at low energy. It explains not only the importance of lithium additives in the series of CMNS experiments, but also consistent with the large cross-section of ^{6}Li for absorbing thermal neutron. Indeed, the solar energy is from a fusion reaction accompanied by a weak interaction as well [13]. Recently, in the hot fusion experiments, the hydrogen plasma pressure and confinement were greatly enhanced at the edge of the plasma when a little lithium was added into plasma [14]. This might be another evidence to show the existence of the p+^{6}Li low energy resonance.

Acknowledgments

This work is supported by The Ministry of Education (#20091770437), The Ministry of Science and Technology (Fundamental Research Division, #2009CB226113), Natural Science Foundation of China (#10475045 & #21153003) and Tsinghua University (Basic Research Fund (985-III)).

References

[1] X.Z. Li, A new approach towards fusion energy with no strong nuclear radiation, *Nucl. Fusion and Plasma Phys.* **16** (2) (1996) 1–8 (in Chinese), (see also *J. New Energy* **1**(4) (1996) 44−54 in English).

[2] X.Z. Li, J. Tian, M. Y. Mei and C.X. Li, Sub-barrier fusion and selective resonant tunneling, *Phys. Rev.* C **61** (2000) 024610.

[3] X.Z. Li, Nuclear physics for nuclear fusion, *Fusion Sci. Technol.* **41** (2002) 63.

[4] X.Z. Li, B. Liu, S. Chen, Q.M. Wei and H. Hora, Fusion cross sections for inertial fusion energy, *Laser Part. Beams* **22** (2004) 469.

[5] X.Z. Li, Q. M. Wei and B. Liu, A new simple formula for fusion cross-sections of light nuclei, *Nucl. Fusion* **48** (2008) 125003.

[6] M. Kikuchi, *Frontiers in Fusion Research – Physics and Fusion*, Springer, London, 2011, p. 31.

[7] X.Z. Li, Z.M. Dong and C.L. Liang, Studies on p+^{6}Li fusion reaction using 3-parameter model, *J. Fusion Energy* **31** (2012) 432.

[8] C.L. Liang, Z.M. Dong and X.Z. Li, Selective resonant tunneling – turning hydrogen-storage material into energetic material *Current Sci* **108**(4) (2015) 519.

[9] B.H. Duane, Fusion cross section theory, in Annual Report on CTR Technology 1972, W.C. Wolkenhauer (Ed.), Rep. BNWL-1685, Battelle Pacific Northwest Laboratory, Richland, WA, 1972.

[10] D.L. Book, NRL Plasma Formulary, Publ. 0084-4040, Rev., Naval Research Laboratory, Washington, D.C., 1987.

[11] C.L. Dunford, Data retrieved from the Cross Section Information Storage and Retrieval System (CSISRS) data base (Feb. 27,1996); available on Internet (http://www.nndc.bnl.gov) (EXFORC00 23001); plot produced using the code BNL 325. National Nuclear Data Center, Brookhaven National Laboratory.

[12] X.Z. Li, Z. M. Dong and C.L. Liang, Excess heat in Ni–H systems and selective resonant tunneling, *J. Condensed Matter Nucl. Sci.* **13** (2014) 299–310.

[13] H.A. Bethe and C.L. Critchfield, The formation of deuterons by proton combination, *Phys. Rev.* **54** (1938) 248.

[14] S.M. Kaye et al., An overview of recent physics results from NSTX, *Nucl. Fusion* **55**(10) (2015) 104002.

J. Condensed Matter Nucl. Sci. 19 (2016) 173–182

Research Article

LENR Anomalies in Pd–H_2 Systems Submitted to Laser Stimulation

Ubaldo Mastromatteo*

A.R.G.A.L. Via S. Stefano, 27 – 20010 Bareggio (MI), Italy

Abstract

In a continuation of the research started in the late 1990s at the University of Lecce, and then repeated in 2004, recently in the second half of 2014 in the A.R.G.A.L. lab., it was decided to thoroughly investigate the LENR anomalies of Pd–H_2 system (in gaseous environment), having as reference the preparation techniques and surface analysis used in the microelectronics industry, to demonstrate convincingly that material contamination was not affecting the transmutation findings. As previously, the activation of the anomalies occurred during irradiation by low power lasers (633 nm, 1 mW, or 403 nm, 3 mW) on 250 nm thick palladium layers deposited by high vacuum e-beam evaporation on silicon oxide substrates, and with 30 nm of Cr as adhesion layer. In summary we did the following: in a stainless steel chamber filled with hydrogen, through a glass viewport, the sample surface was irradiated for two weeks and then accurately inspected in a SEM equipped with an EDX microprobe. A reference sample was also accurately analyzed before irradiation.

Keywords: Laser, LENR, Palladium, Transmutations

1. Introduction

In order to better understand the results of the present experiments on the interactions between transition metals and hydrogen, assisted by with laser light excitation, it is a good idea to examine some data from the scientific literature that have strong similarities with said type of interactions.

The first reference is [1]. This paper describes the behavior of a pure nickel rod in a hydrogen environment heated from room temperature to over 400○C, and the subsequent examination under the electron microscope and contextual EDX analysis of the surface of said sample, after this has been shown to react with the generation of excess heat in a quantity not attributable to reactions of the chemical type. For our purposes, what is interesting is the EDX result shown in Fig. 1.

A second reference comes from a private communication to me by Dr. Luca Gamberale, referring to experiments he conducted at the Pirelli advanced research laboratories in the late 1990s. In this communication there are a couple of findings concerning transmutation useful for a deeper comprehension of these peculiar aspects of LENR anomalies. The first one comes from the analysis of a palladium cathode observed by electronic microscope and analyzed by

*E-mail: ubaldo.mastromatteo@libero.it

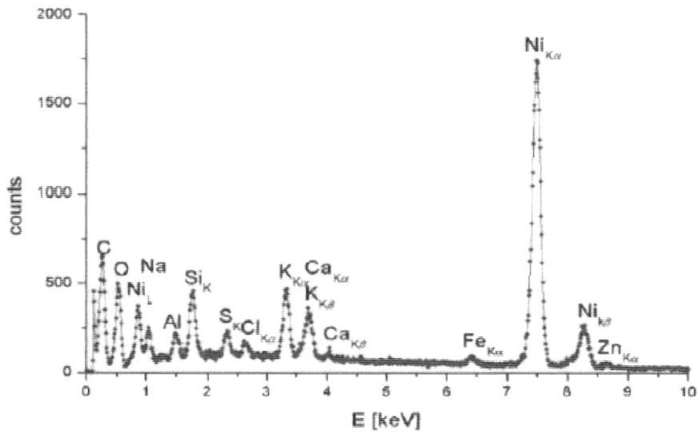

Figure 1. The list of elements from the EDX excluding nickel include: C, O, Na, Al, Si, S, Cl, K, Ca, Fe, and Zn.

EDX micro probe after an experiment of heavy water electrolysis that had shown excess heat: Fig. 2 shows the EDX spectrum.

A second interesting transmutation finding from Gamberale's experiments concerns a copper electrode submitted to glow discharge in a hydrogen environment. The EDX analysis of the electrode after the experiment showed a spectrum with many elements extraneous to the starting material, all lighter than copper (Fig. 3).

Another background reference about transmutation is [2]. In this paper the material submitted to ultrasound stimulation is stainless steel. The shape of the sample was a small rod on which, after one hour of ultrasound stimulation (16 W), several spots appeared on the surface. The results of an EDX analysis of the spots are reported in Fig. 4, Table 1.

Another reference with interesting findings on transmutation concerns experiments on constantan wires is [3]. Several experiments highlighted small excess heat effects and, even though the post experiment SEM analysis was not easy, several spots were detected on the wire surface where an EDX analysis was possible: Fig. 5 shows a typical spectrum.

An interesting quite recent paper about a palladium/deuterium system in an electrochemical cell describes the presence of anomalous contaminations on the cathode surface, the origin of which was not clearly identified [4].

The circumstances of the findings described in the paper led us to think that there is some analogy with the mentioned experiments showing transmutation of palladium to lighter elements. Some common circumstances are as

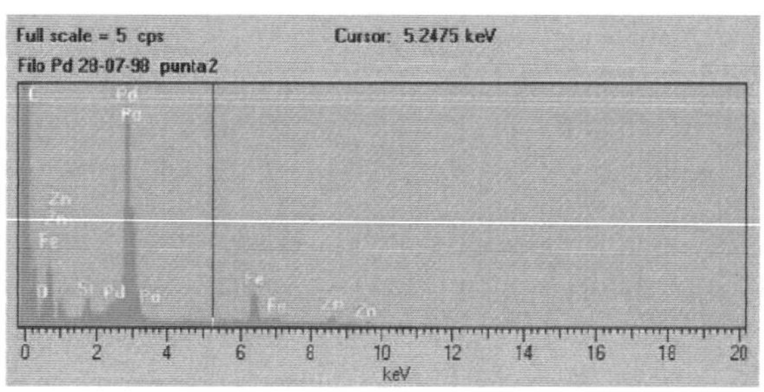

Figure 2. Excluding palladium the new elements are: C, O, Si, Fe, and Zn.

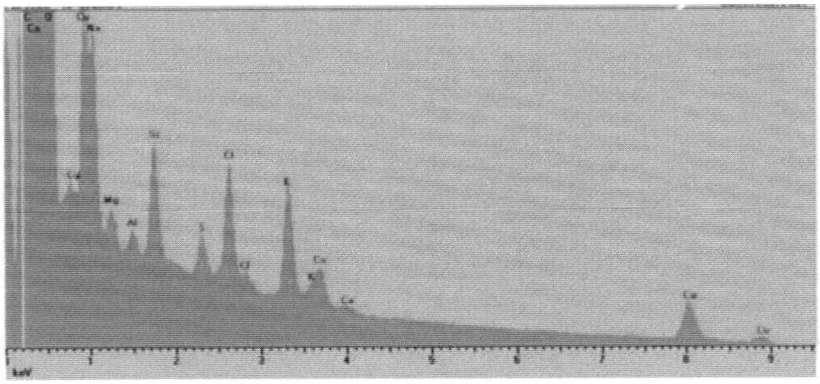

Figure 3. EDX after glow discharge experiment, elements in the spectrum: C, O, Na, Mg, Al, Si, S, Cl, K, and Ca.

Table 1. The results of an EDX analysis.

Element		Weight
C	Carbon	19.80
O	Oxygen	29.27
Na	Sodium	1.20
Mg	Magnesium	0.19
Al	Aluminium	0.53
Si	**Silicon**	**0.49**
S	Sulfur	0.27
Cl	Chlorine	1.61
K	Potassium	0.54
Ca	Calcium	0.68
Mn	Manganese	0.47

follows:

- RF was detected at the cathode during the test,
- anomalous heat was produced,
- then, EDX on three suspected contaminated areas of the palladium cathode after electrolysis highlighted the following elements: C, O, F, Mg, Al, Si, Ca, Cr, Fe, and Ni.

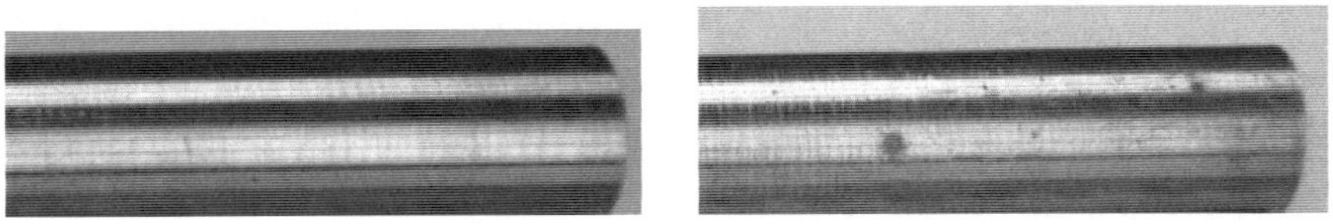

Figure 4. Pictures of the rod before and after the treatment. Table 1 shows in non-bold characters the new elements detected in the spots and the increase of C and O, plus a decrease of the amount of iron.

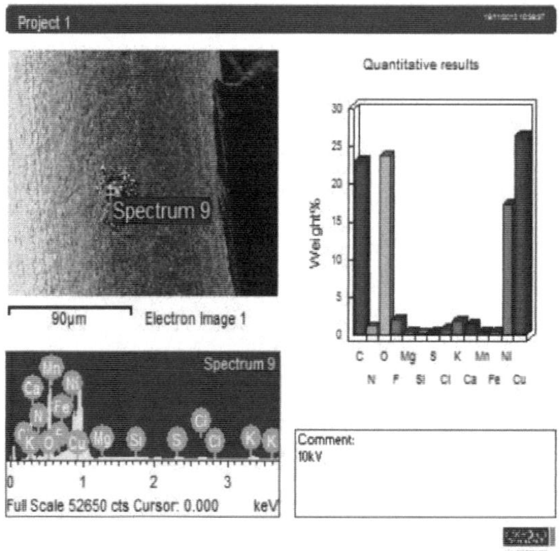

Figure 5. EDX on one of several hot spots detected on the constantan wire. Elements not in the wire composition are: C, N, O, F, Mg, S, K, Cl, and Ca .

The most important background reference about transmutation findings related to our experiments on palladium/hydrogen systems and laser stimulation is Ref. [5].

The experiments reported in this paper, done in 2004 at the University of Lecce, show the effects of laser stimulation of palladium thin layers in hydrogen gas. After several weeks of low-energy irradiation, several hot spots were observed on the surface of the palladium. Careful observation of the spots clearly showed the local melting of the material and

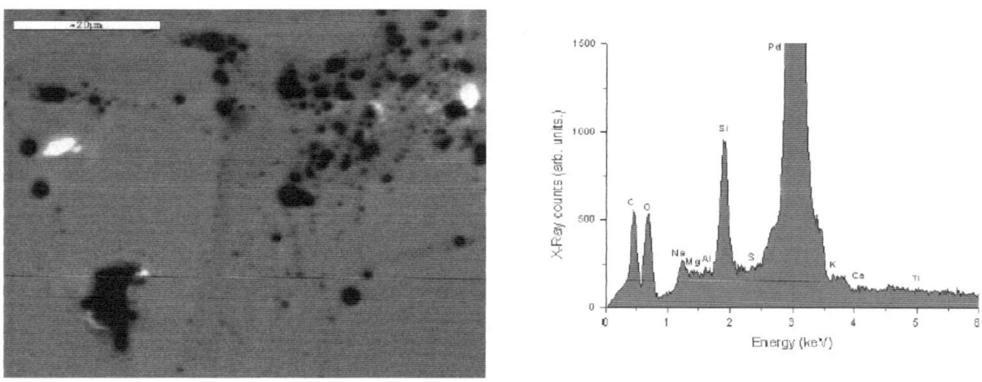

Figure 6. EDX on hot spots after laser stimulation of Pd thin film in H_2 gas. The list of the elements excluding Pd, Si, and Ti (present in the virgin sample) is the following: C, O, Na, Mg, Al, S, K, and Ca.

Figure 7. Photos of the reactor and the bottom part where the sample with the palladium film will be placed.

the EDX analysis highlighted the presence of elements foreign to the composition of the material before the experiment, or the material outside the hot spots.

In the above-mentioned paper, we also reported on experiments with identical palladium thin film samples in deuterium gas; in this case also the reference sample that not irradiated, placed together the irradiated one in the reaction chamber, was affected by hot spots on which EDX analysis highlighted possible transmutation elements. Figure 6 shows a typical picture of a hot spot and the EDX spectrum.

2. New Experiments on Palladium Thin Films and Laser Stimulation

The purpose of the experiments described in what follows was mainly to confirm the activation of LENR "anomalies" in the interaction between palladium and hydrogen (or deuterium) in the presence of coherent electromagnetic radiation (laser or other), in order to understand the detailed mechanism that activates this phenomenon. The choice of this type of embodiment is suggested by the simplicity of the experimental conditions and by the ease of results analysis on small size samples and therefore suitable to a careful examination by SEM microscopy.

Moreover, operating at ambient temperature and without energy input from the outside in addition to that extremely low powered laser, allows easy portability of the entire experimental setup.

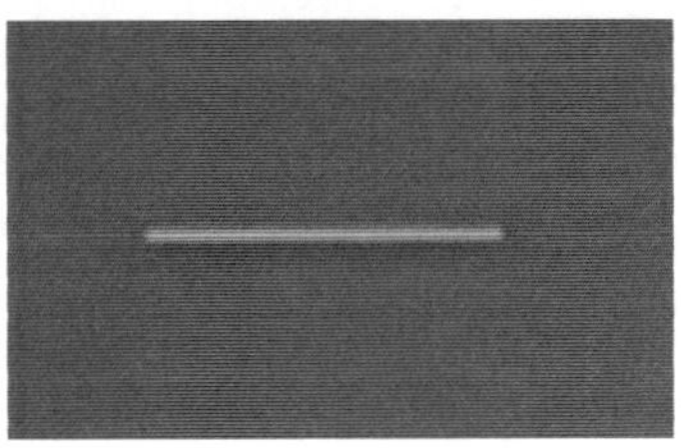

Figure 8. Photo of the sample.

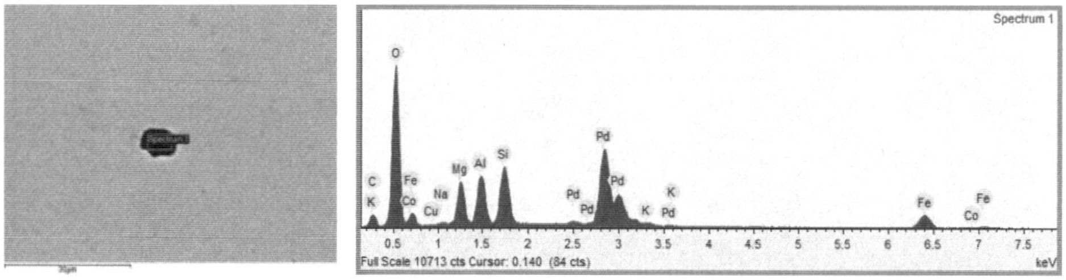

Figure 9. EDX inside the cavity, spectrum element list: C, O, Na, Mg, Al, Si, K, Fe, Co, and Cu.

3. He–Ne Laser Stimulation Experiments

Figure 7 shows the reactor. The sample was an oxidized silicon chip on which it was deposited a layer of palladium 250 nm thick. The size of the chip is: width 1 mm, length 28 mm. See Fig. 8.

4. Test Description

The sample shown in Fig. 8 was inside a reactor filled with hydrogen at 1.5 bar for 2 weeks and irradiated through a glass viewport with a low-power He–Ne laser (633 nm, 0.9 mW) at ambient temperature. An optical system was used to enlarge the laser spot up to 1 cm^2 size. The palladium surface was then accurately explored with a SEM to

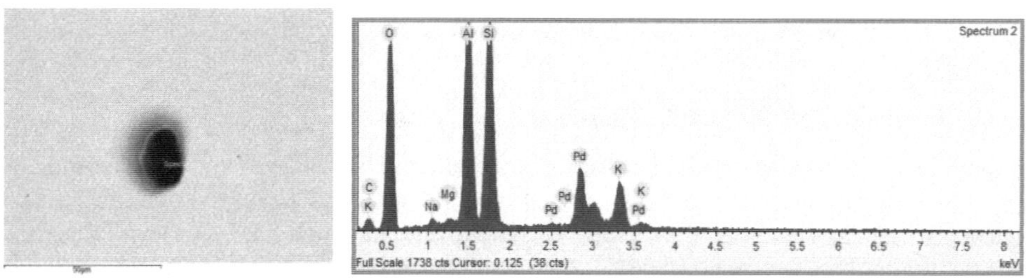

Figure 10. EDX inside the cavity, spectrum element list: C, O, Na, Mg, Al, Si, and K .

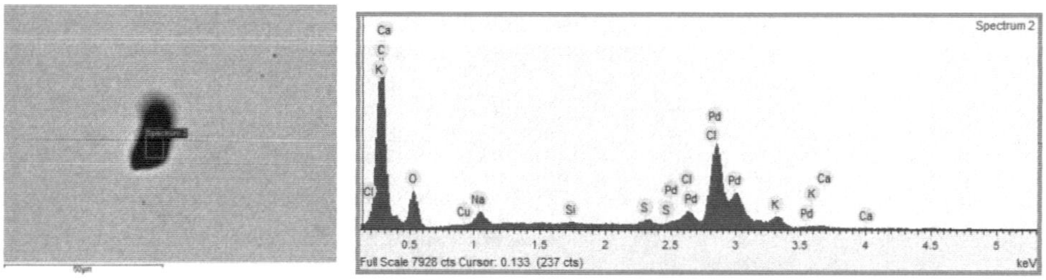

Figure 11. EDX inside the cavity, spectrum element list: C, O, Na, Si, S, Cl, K, Ca, and Cu.

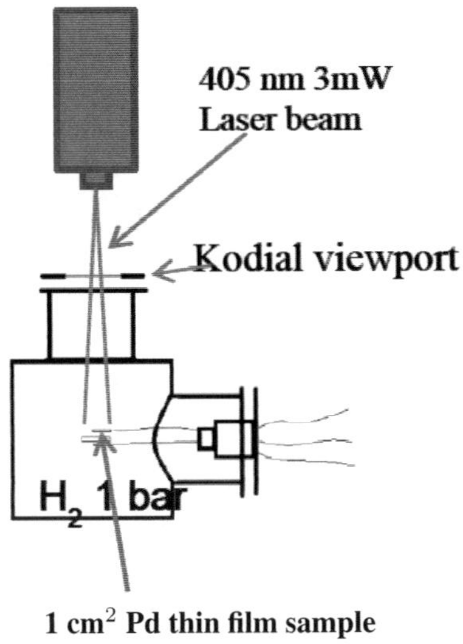

Figure 12. Sketch of the experimental set up.

find possible traces of morphological changes: several cavities were found, looking like similar findings of previous experiments. Images and EDX analysis follow.

5. Irradiated Palladium Film Cavities EDX Analysis

In contrast to the experiments conducted at the University of Lecce in 2004, in these experiments we did not observe hot spots with evident traces of melting of the material, but a certain number of cavities with dimensions around tens of micrometers, inside of which the EDX analysis showed the presence of elements not present in the thin palladium layer as deposited. In reference to this, it is important to consider that the samples were manufactured inside a very clean environment of a microelectronics facility (STMicroelectronics), with maximum care to avoid contamination issues.

Figures 9–11 are the pictures and the spectra of some EDX analysis of cavities.

Cavity pictures and EDX analysis spectra

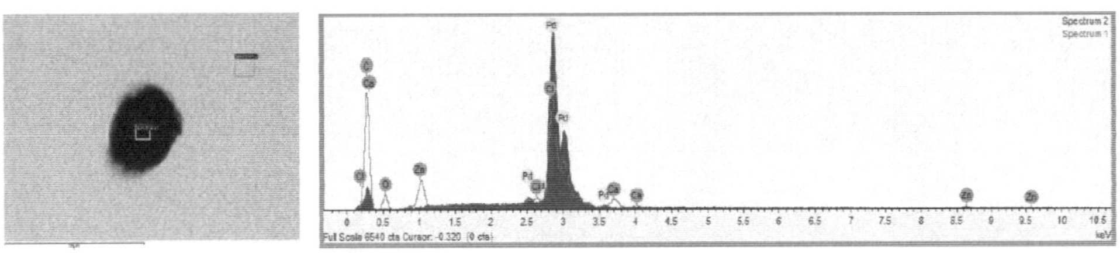

Figure 13. EDX spectrum element list: C, O, Cl, Ca, and Zn (blue spectrum outside the cavity) .

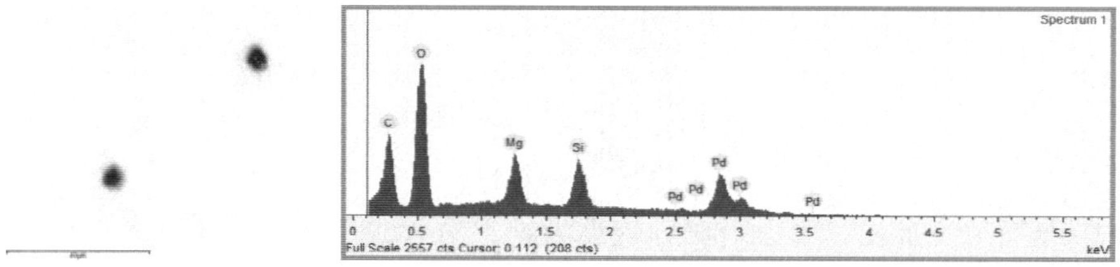

Figure 14. EDX spectrum element list: C, O, Mg, and Si.

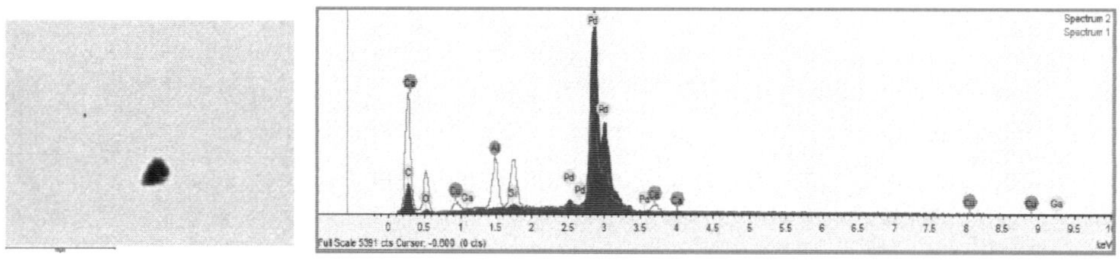

Figure 15. EDX spectrum element list: C, O, Al, Si, Ca, and Cu (blue spectrum outside the cavity) .

6. 405 nm Wavelength Laser Experiments

Another experiment was conducted on a square silicon chip (1 cm^2) covered with the same thin palladium film (250 nm) as the previous tests, but now irradiated with a solid state laser, 405 nm, 3 mW, widening the spot of the beam to about 1 cm^2. After two weeks of irradiation, the sample was extracted and examined by SEM and EDX analysis using a Hitachi TM 3030 system. Also in this case, carefully exploring the palladium surface revealed several cavities on which EDX analysis was conducted.

With the reactor used in these experiments, it is possible to heat the sample up to 450○C. In spite of that, the experiments were conducted at ambient temperature, the same as the previous ones. Some heating was used, but only to out-gas the sample before the test started in order to best clean at the reactor inside.

Another feature of this reactor, due to the presence of several electrical feedthroughs, allowed us to measure the

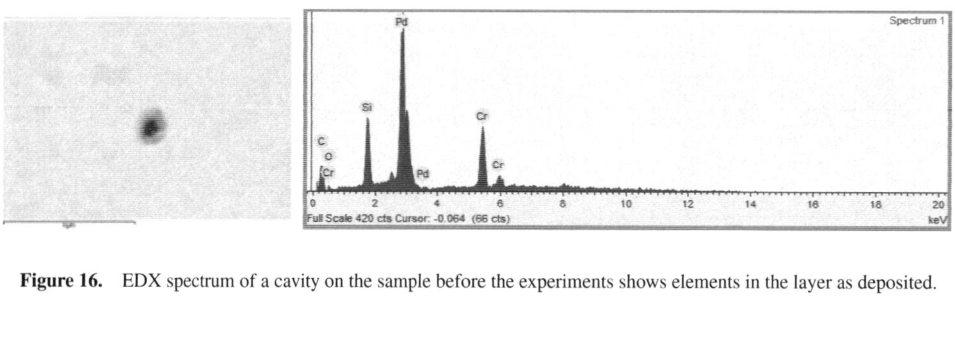

Figure 16. EDX spectrum of a cavity on the sample before the experiments shows elements in the layer as deposited.

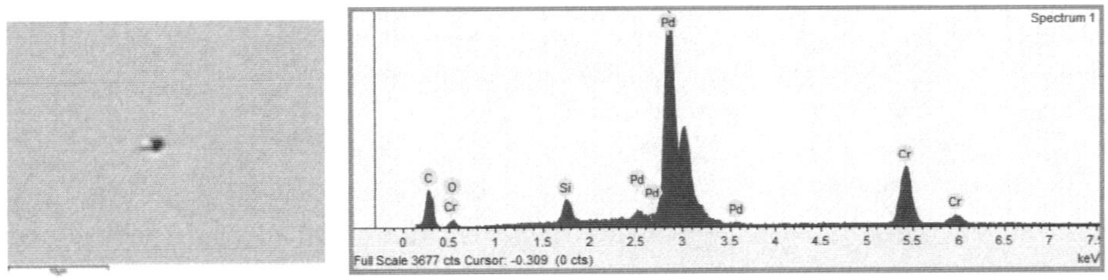

Figure 17. EDX spectrum of a cavity on the sample before the experiments shows elements in the layer as deposited.

electrical resistance of the layer during hydrogen filling. It was observed that the absorption of the hydrogen in the thin film is extremely rapid and reaches equilibrium before the pressure is close to one bar. So, when the experiment starts the environment is at thermodynamic equilibrium.

Cavity pictures and EDX analysis spectra are shown in Figs. 13–15.

7. Reference Sample Analysis

In order to verify that the cavities found on the irradiated palladium surface of the tested samples were due to some kind of new nuclear reactions (LENR) in a laser stimulated H–Pd interaction, the surface of an untested sample has been very accurately explored using the SEM system (Hitachi TM 3030).

What was revealed by the analysis is the almost complete absence of holes, apart from some very small palladium layer imperfections which likely occurred during the layer e-beam evaporation (Pd and Cr adhesion layer on oxidized Si).

Imperfections of the palladium film on the reference sample analysis are shown in Figs. 16–18. These are the related pictures and EDX spectra.

8. Final Remarks

- The experiments shown confirm previous findings, although to a lesser extent (because the duration of the tests was two weeks instead of 10 weeks).
- The 405 nm solid state laser appears to be less effective than the 633 nm He–Ne one.
- In spite of in our continuous monitoring for neutrons and gamma emission, our instruments never registered deviations from the background spectrum, which is the same negative result we saw with previous experiments

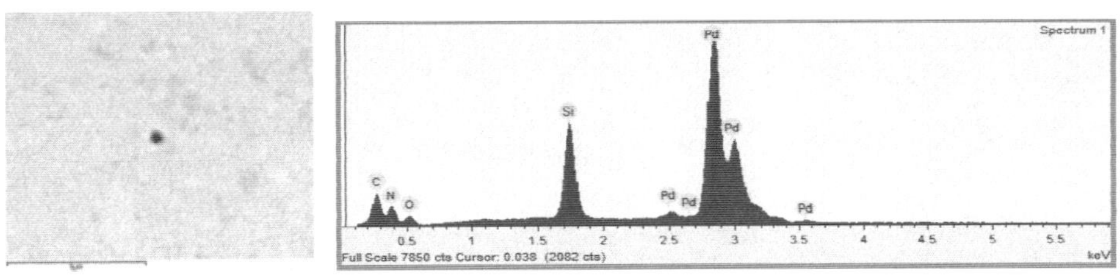

Figure 18. Figure 18. EDX spectrum of a cavity on the sample before the experiments shows elements in the layer as deposited.

done at Lecce University with the Pd/H_2 system. We did observe some neutron emission from the Pd/D_2 system. Moreover, as described above in the introduction, neutron bursts were also detected in the experiments with stainless steel rod and ultrasound [2]. This suggests that different evidences of nuclear activity are strictly connected with the materials used in the experimental set up.

- An accurate analysis of samples not irradiated did not show the presence of cavities with the elements found in the irradiated samples.
- All the elements attributable to transmutation are lighter than palladium.

9. Conclusions

- The experiments cited as background indicate the presence of transmutations similar to those shown in these replication experiments described above. Often the transmutation findings were associated with other LENR anomalies, such as emission of neutrons and excess heat generation.
- Induced oscillations in the material, by ultrasound, radiofrequency, or laser radiation, sometimes appears essential for the occurrence of the anomalies.
- Since some elements attributable to transmutations can easily come from environmental contamination (Na, Si, Al, Mg, Ca), we plan to replicate controls on the samples with accurate SEM analysis before the laser treatment in the reactor.
- Finally, because almost all the experimental evidence reported, new and old, points to transmutation of the reacting material to lighter elements, a deeper investigation of possible fission reactions of the material submitted to the described tests is strongly suggested.

References

[1] S. Focardi, V. Gabbani, V. Montalbano, F. Piantelli and S. Veronesi, On the Ni–H System, *SIF Conference Proceedings*, Vol. 64, 1997, pp. 35–47.

[2] F. Cardone, G. Cherubini and A. Petrucci, Piezonuclear neutrons, *Physics Letters* A **373** (2009) 862–866.

[3] U. Mastromatteo, A. Bertelè and F. Celani, Hydrogen absorption and excess heat in a constantan wire with nanostructured surface, *J. Condensed Matter Nucl. Sci.* **15** (2015) 240.

[4] V. Violante, E. Castagna, S. Lecci, G. Pagano, M. Sansovini and F. Sarto, RF detection and anomalous heat production during electrochemical loading of deuterium in palladium, *Energia, Ambiente e Innovazione* **2-3** (2014) 63–67.

[5] V. Nassisi et al., Modification of $Pd–H_2$ and $Pd–D_2$ thin films processed by He–Ne laser, *J. Condensed Matter Nucl. Sci.* **5** (2011) 1–6.

J. Condensed Matter Nucl. Sci. 19 (2016) 183–191

Research Article

Cold Fusion – CMNS – LENR; Past, Present and Projected Future Status

Michael C.H. McKubre*

SRI International, Menlo Park, CA, USA

Abstract

A brief overview of the SRI effort over 26 years is provided as a precursor to suggestions on how we might best proceed to validate the vision of Martin Fleischmann and Stanley Pons, and proceed towards its logical conclusion.

Keywords: Cold fusion, Excess heat, Helium, LENR, Reproducibility

1. Introduction

On March 23, 1989, Martin Fleischmann and Stanley Pons announced their observation of a non-chemical heat source in the D_2O/Pd system. In their initial publication they speculated on the possibility that the effect they had observed may derive from nuclear processes and included the interrogative "Fusion?" in the preliminary title of their submission [1]. While they presented no direct evidence for nuclear or fusion processes, the evidence of a heat effect considerably larger than chemical or lattice storage effects was strong. In their initial Patent submission [2] Fleischmann and Pons reported results from the electrochemical cell and calorimeter shown in Fig. 1 that would be difficult or impossible to explain by conventional chemistry or electrochemistry. Operating at constant current in a thermodynamically open system, the cell was topped up with D_2O approximately every 12 h to compensate for D_2 and O_2 losses via electrolysis (the downward excursions in Fig. 2). As the cell resistance rises at constant current due to loss of electrolyte conductivity and increased anodic and cathodic overvoltages, the power in the cell also rises and so does the cell temperature as shown in Fig. 2. With addition of D_2O at $\sim 3.1\times 10^6$ s (after $\sim$36 days of electrolysis) the cell temperature in the example shown abruptly and inexplicably rose $\sim$ 20°C, and the temperature vs, time slope changed, then changed again and the cell temperature rose to boiling, all at constant current and sensibly constant input power. Given the power and energy densities of the heat effect claimed by Fleischmann and Pons, only one of two rational possibilities existed in 1989:

- Fleischmann and Pons were wrong in their excess heat determinations.

*E-mail: michael.mckubre@sri.com

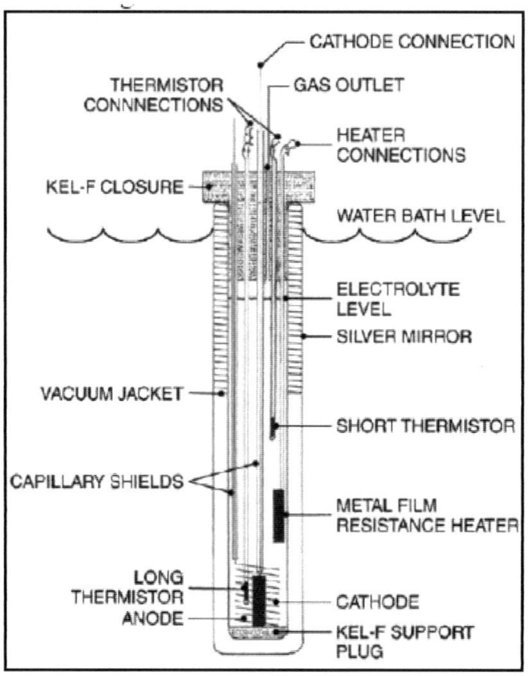

Figure 1. Fleischmann Pons calorimeter.

- Nuclear reactions occur in metallic lattices by mechanisms and with product distributions different from similar reactions in free space.

This opened considerable speculation about the two following questions. Is there such a thing as Condensed Matter Nuclear Science (CMNS)? Does our free-space view of nuclear physics need to be extended in potentially interesting directions?

2. Experimental effort at SRI

Scientists at SRI have engaged in approximately 70 person-years of study stimulated by the original Fleischmann and Pons announcement [3–5]. The resulting conclusion is that nuclear level thermal energy density can be produced from the electrochemical loading of deuterium into palladium under difficult-to-achieve but well-defined conditions.

Known critical conditions include:

(1) Maintain high *average* D/Pd ratio (*Loading*)
(2) For times > 20–50 times the deuterium diffusional time constant $\tau_{D/D}$ (*Initiation*)
(3) At electrolytic current densities[a] i >50–500 mA cm^{-2} (*Activation*)

[a]This current density threshold is reasonably well specified (and large) for wire cathodes with concentric anode geometry but is less well defined (and generally lower) for parallel plate geometries that exhibit far more heterogeneous current distributions.

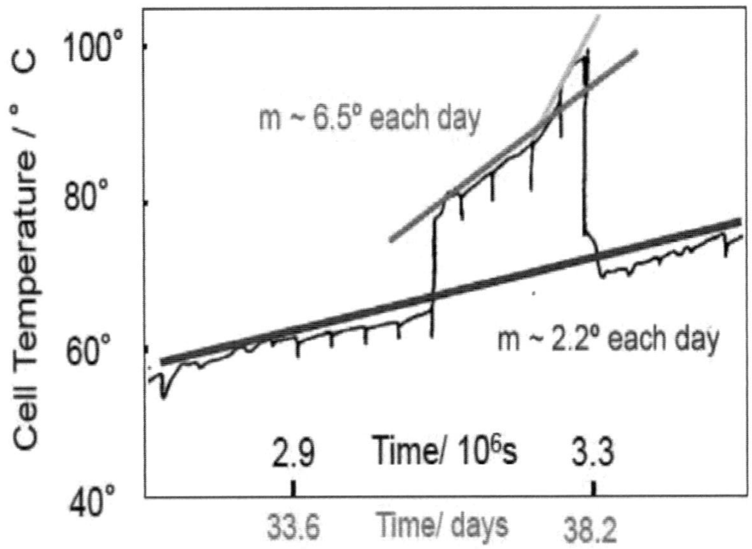

Figure 2. Early reported excess heat excursion [2].

(4) With an imposed or adventitious deuterium flux *(Disequilibrium)*

An empirical "prediction function" was developed for excess power (P_{xs}) as a result of these observations:

$$P_{xs} = M(x - x^{\circ})^2(i - i^{\circ})|i_D|,$$

where x = D/Pd, x°~0.875, i = 50–400 mA cm^{-2}, i_D = 5–20 mA cm^{-2}, t >20 $\tau_{D/D}$.

In addition to helping to define the conditions under which the Fleischmann Pons Heat Effect (FPHE) can be seen and explored, another major thrust of the SRI team was to attempt replication of experiments of other scientists and inventors in the field. A brief itemization follows:

(1) *1989–1991 Fleischmann Pons heat effect.* Initial criticism of the FPHE was that "*it could not be reproduced*". Obviously, given consistent input conditions in non-stochastic processes everything reproduces (particularly systematic error), so this rather unsophisticated criticism can be interpreted to mean one of two things: Fleischmann and Pons did not observe an anomalous heat effect (even as a result on systematic error); the input conditions needed to produce their observation were not reproduced in subsequent attempts. For those who knew Fleischmann even if only by repute, the first possibility was not highly plausible so efforts were undertaken at SRI and elsewhere to understand under what conditions the FPHE could be observed. Once understood we successfully reproduce the FPHE on approximately a hundred occasions with electrodes of various forms.
(2) *1992 – Kevin Wolf gamma activation.* Attempting to reproduce SRI excess heat observations in electrolytic cells with addition of Si and B, Kevin Wolf observed bursts of both neutrons and gamma rays from three out of three cells running under real-time nuclear monitoring. At the end of the experiment these cathodes all were gamma active [6]. Under Wolf's guidance SRI reconstructed his neutron and gamma spectrometers but were not able to recreate the observations of prompt and activated emission in about 2 years of focused effort.

(3) *1993 Ni – Natural water heat effect.* Considerable interest (and confusion) was introduced into the cold fusion field in 1993 with multiple reports of excess heat production from large area nickel cathodes exercised in light water (typically in carbonate) electrolytes. One highly credible report of such observations was by M. Srinivasan from Bhabha Atomic Research Center (BARC) in India. In 1993 we invited Dr. Srinivasan to demonstrate his experiment during a sabbatical at SRI. He was able to reproduce the effect but we were able to identify an interesting (but mundane) source of the thermal anomaly.

(4) *1993–1996 Mel Miles heat-^{4}He correlation.* In one of the more important papers in cold fusion history Mel Miles presented evidence at ICCF2 in Lake Como [7] that excess heat from Pd/D_2O electrolysis, and ^{4}He, were statistically well correlated nearly quantitatively with a Q value of $\sim$24 MeV/^{4}He atom. One criticism of Miles' technique (later shown to be unfounded) was his use borosilicate glass that is permeable to ^{4}He. To guard against the possibility of error due to ^{4}He inleakage from ambient, SRI used an all metal sealed cell and calorimeter to successfully reproduce the Miles Bush heat-helium results with the experimental assistance of one of Miles' co-authors (Ben Bush).

(5) *1995 Patterson "Light" water excess heat.* At ICCF5 in Monaco in 1995 Dennis Cravens presented an impressive demonstration of a novel excess heat generating device invented by Jim Patterson. The device featured a flow-through, packed-bed of polystyrene latex spheres coated with multiple thin layers on metals including Cu, Ni and Pd. Simple calorimetry was performed by measuring the temperature difference in the electrolyte as it transited a packed bed of coated spheres that was polarized cathodically with respect to a downstream anode. The electrolyte was said to be "water", the demonstration appeared to show energy gains of 3–5 even ignoring known parasitic losses. We attempted to replicate this impressive result with considerable assistance from Dennis Cravens under more carefully controlled conditions at SRI. Although we were able to observe potentially anomalous ∂T's these were nowhere near the magnitude of the effects demonstrated in Monaco. Moreover, whatever excess heat was present if any in the Patterson cell at SRI it was insufficient to demonstrate convincing evidence in a purpose-built mass-flow calorimeter configured to surround the Patterson cell. Only later was it revealed that Patterson employed $\sim$10% D_2O in his "water" electrolyte in earlier apparently successful measurements (including ICCF5). By the time this was known Paterson's original batch of coated spheres had been exhausted; subsequent batches evidenced little of no heat effect.

(6) *1996–1998 Les case heat and ^{4}He.* One of the most startling pronouncements ever made in the cold fusion world was by Les Case at ICCF7 in Vancouver in 1996. After similar pre-treatment in H_2 gas Case introduced D_2 at a pressure of a few atmospheres to a bed of chemical hydrogenation catalyst consisting of "coconut shell charcoal" infiltrated with approximately 0.5% platinum group metals. This was confined in a 1.7-liter stainless steel vessel. In a very narrow temperature range, 150–250°C, and with a few selected catalysts Les was able to produce a 5–35°C temperature rise in D_2 compared with H_2 gas. He also claimed anecdotally to have had a post-test D_2 sample analyzed by mass spectrometry at Oak Ridge National Labs (ORNL) where they observed $\sim$100 ppm of helium-4. Since the concentration of ^{4}He in room air is 5.22 ppm, if verifiable this finding was stunning.

Funded by DARPA SRI mounted a major campaign to replicate Case's reported findings. This effort initially failed repeatedly as we were not able to evidence significant temperature differences between catalysts in D_2 and H_2 or any indication of ^{4}He increase. It was not until we stood beside Les Case and watched his detailed handling of catalyst and gas that we were able to make progress. The differences between the "Case process" and our understanding of the Case process were subtle, but significant, and we were rewarded by an ultimately successful replication of the Case Heat Effect accompanied by some of the largest ^{4}He concentrations so far observed (up to 11.8 ppm) in cold fusion experiments.

(1) *1997–1998 Arata–Zhang "DS" cathodes heat and ^{3}He.* Arata contributed one of the more interesting and

reportedly replicable innovations with his novel "double-structured" (DS) cathode. The Arata claim was that every case in which his DS cathodes were exercised in heavy water electrolytes resulted in excess heat and both ^{3}He and ^{4}He in the sealed cathode void space that contained nano-Pd. Blank cells employing H_2O produced null results. Using DARPA funding in a replication effort and trajectory similar to that for the Case effect, SRI with help from the Violante Group at ENEA Frascati, was unable to replicate the Arata DS heat effect without direct assistance from Professors Arata and Zhang. Using their cell and calorimeter design, experimental protocols and a pair of DS cathodes fabricated under their supervision, SRI was able to replicate important aspects of the Arata–Zhang claims. Two identical cells were operated simultaneously, side-by-side, with the same current protocols. The cell and cathode exercised in D_2O produced significant excess power (a maximum of $\sim$10% of P_{In}) and energy (a total of $\sim$100 MJ) while the sister cell and cathode operated in H_2O showed no evidence for excess heat. Considerable effort had been put into planning and preparing for gas phase analysis of the contents of the sealed cathode voids to verify the Arata reported finding that ^{3}He and ^{4}He build up concentration in these voids to very significant levels, possibly commensurate with the excess heat. Due to a "gas handling lapse" by the selected measurement laboratory this particular claim could not be verified by SRI. We were, however, able to prove unambiguously that tritium was created in the sealed void space of the D_2O cathode at some time during the cathodic electrolysis. The initial evidence of ^{3}H creation was a very considerable buildup of the decay product ^{3}He. No ^{3}H, ^{3}He or excess heat were observed for the H_2O cell.

(2) *2003–2011 Energetics "SuperWave" excess heat and ^3H.* At ICCF10 in Cambridge in 2003 Arik El-Boher representing Energetics Technologies Incorporated (ETI) surprised the audience with a report of substantial excess power generation and energy gain at useful working temperatures in both gas discharge cells and electrochemical cells, encountering the heat-after-death phenomenon and the production of tritium. In less than two years of effort Energetics had progressed from uninvolved to contribute what I reported to Martin Fleischmann to be "the best paper of the conference". The secret of Energetics' success appeared to be what their progenitor Irving Dardik called a "SuperWave" – a complex multi-frequency, multi-amplitude driving function designed to stimulate simultaneously a wide range of resonant processes. Of immediate significance to the SRI team was that the "SuperWave" generating function appeared to be capable of engendering simultaneously both high deuterium loading and high interfacial deuterium fluxes in electrochemical systems, both then known to be correlated with excess heat.

With funding from DARPA and significant assistance from the Violante Group at ENEA Frascati, SRI mounted a formal replication effort to demonstrate the efficacy of the SuperWave protocol in creating the conditions necessary for excess heat production. This effort resulted in the most successful campaign of replication so far performed at SRI with one set of Violante-produced Pd foil cathodes and Energetics-SuperWaves resulting in significant excess heat in $\sim$73% of the experiments run. Furthermore, using the criteria of "critical conditions" enumerated above we were able to explain our failures to produce excess heat as a failure to meet one or more critical conditions, in every case except one electrode that produced no measurable heat but considerable tritium [8].

(1) *2012–2015 Brillouin excess heat.* SRI is presently engaged with Brillouin Energy Corporation (BEC) to help them develop and scale up their triggered excess heat effect into parameter space that supports practically engineered heat and/or electricity production devices. The end objective of BEC is (as was ETI) practical technology. The primary experimental driver and foundational IP claim of BEC resides in their use of very fast, closely tailored electrical pulses (that they call a Q-pulse) that is supplied axially along carefully structured cathodes (initially just Ni or Pd wires). This pulse has been shown to trigger excess heat production providing Brillouin with a level of technological control that has not previously been evidenced in the cold fusion/LENR/CMNS world. SRI's engagement with Brillouin is ongoing.

3. Present Status

The present situation of the field offers both good news and bad. In the former category it seems apparent that Fleischmann and Pons were correct in more ways than they anticipated The accumulated evidence strongly supports the conclusion that nuclear effects take place in condensed matter by pathways, at rates and with products different from those of simple, isolated, pair-wise nuclear reactions in free space (i.e.two-body interactions). The group of influential physicists self-assigned to prevent or delay the full elucidation of the effect that Fleischmann and Pons reported on March 23, 1989, have largely disappeared. Although there has been terrible attrition from "team CMNS" our critics have suffered much worse and have not been replaced. The "evidence" presented at the May first Meeting of the APS in Baltimore is seen (retrospectively) as essentially irrelevant due to:

- The failure of the group lead by Lewis [9] to anticipate now known threshold conditions: Loading; Current Density; Flux; long Initiation Times.
- The practical irrelevance of the Koonin calculation [10] due to the lack of anticipation of or attention to the massive electron screening effects seen on PdD (and other condensed matter structures [11]), and disregard for the fact that hot fusion products (and thus processes) are clearly not seen and that the D–D spacing in PdD is greater than that in D_2O which does not undergo spontaneous fusion at measurable rates.

As a result of these critical failures it is now very difficult to find a place or person to approach for up-to-date objective criticism. This is mostly not good news, however, as progress depends on constructive criticism.

Other news is not good and there remain a number of critical challenges. Despite considerable effort there is a lack of clear data supporting the existence of an effect and an understanding of what drives the results. In particular there exists no fully documented experiment with raw data set available, and no agreed to theory (despite a plethora of offerings). The confidence that many feel about the reality of the FPHE is largely based on the accumulation and synthesis of the outcome of many experiments (see [12]) rather than the existence of a single clear, simple, unambiguous experimental result. This need could be avoided if there were clear demonstration of technology based on the FPHE but, although many claim to be developing this, none has been brought to the market. In the future, open or communal experimentation may serve as a useful pathway to proof.

The academic stigma is maintained mostly by an inability or barrier to publish in mainstream Journals. The editors of these Journals all are of sufficiently advanced age that they remember the furor and claimed "slam dunk" rejection that attended the May 1989 APS. Because of the paucity of clear results and the experimental complexity, these individuals have not been adequately exposed to the work following May 1998 that largely invalidates the thinking behind that rejection and the global significance of the null results reported at that time. This timidity and unwillingness to re-investigate the data of the FPHE and related effects by editors of main stream Journals, effectively ostracizes the CMNS community from the main body of science and may be the single most effective barricade to progress. Without the ability to publish in recognized journals talented young people with academic aspirations are discouraged from entering the field. Fortunately as positive results persist and improve, and with the passage of time from 1989, this problem is beginning to fade. Innovation, by and large, is for the young, a group essentially unaware and unaffected by the almost hysterical condemnation of Fleischmann and Pons in 1989.

Help presented itself to the CMNS community in a most unlikely manner in the person of Andrea Rossi. Not a recognized scientist, having not presented at or even attended any major meeting in the field, and having published nothing on the topic in a recognized Journal of any type Rossi created what might be called "The Rossi Effect". Using combined showmanship and demonstrated operational scale[b] Rossi brought to the attention of a new generation of innovators the possibility that "cold fusion" might not only be real (on which point the world was largely apathetic)

[b]Rossi's largest claimed result was 475 kW thermal [13].

but also practical. Rossi has inspired a significant number of experimenters to pursue the demonstration of heat on moderate scale (hundreds to thousands of Watts) with significant power and energy gain (2–10) at significantly elevated temperatures (up to and above 1200°C). All of these systems employ Ni (usually in finely divided form, often claimed to be nano-metric), thus obviating the intrinsic issues of cost and scarcity associated with Pd. They also employ natural hydrogen (rather than deuterium) sourced in various forms often from the thermal decomposition of inorganic hydrides. The possibility exists that lithium also may be crucial. To this point the proof of performance remains "tantalizing" but it would be a mistake not to recognize the positive impetus and stimulus that "The Rossi Effect" has had on the CMNS community. Without Rossi (who was not even present) it is unlikely that ICCF19 would have been the largest cold fusion conference to date.

4. What is the Future?

The years 2015 and 2016 will not be "business as usual" for the CMNS field – we do not want it to be. Things have changed fundamentally although the extent and direction of change is not obvious to all. We are watching the CMNS field convert from "resource limited" to "talent limited". The capable experimentalists who have brought the field to the point that it is – always in short supply – have diminished with age and attrition and have not been replaced at anything like a steady state rate. Partly inspired by "The Rossi Effect" but also by the weight of positive CMNS results, the time of study (26 years and counting), and the increased awareness that cold fusion may have the possibility to obviate malignant effects of conventional nuclear (fission and hot fusion) and fossil fuel combustion as a primary energy source, significant commercial and governmental interests are directing attention to the questions: "*Is cold fusion/LENR/CMNS real?*" and "*Can it be scaled up to make a difference for the planet?*"

How we proceed will depend on how we convert this "once in a quarter century" opportunity. It is time for us to "*get real*"! Take this opportunity and proceed with due deliberation towards a goal that is not just good for us but may offer real potential for mankind. We have all been drawn to this field at different times and with different motivations but I want to offer a challenge to this community that "business as usual" has not taken us as far as we needed to go. We have contributed mostly as individuals or as individual organizations. We get together every 18 months or so – have been doing that for 26 years – and yet the whole had not exceeded the sum of its parts. Are we too comfortable with this situation? We complain about the lack of respect and resources, how do we really feel about the entry of new people and new money into the field? We would have a lot more competition, for starters. This is coming, which forces the question: *what do we want the future of this field to look like in 5 years*? Do we want to see this field expand and make the world a better place? Think about it. The reflex answer is "*Yes, of course*", but too many (possibly all of us) want to do this "*my way*" using the medium of advancement most familiar to us (business, academic, etc.). How do we escape the myopic trap in which we seem to be ensnared?

We have a tremendous opportunity to come together as a community and expand the field, by making it easier for more resources to participate. Let us draw consensus around our best experiments and have them widely replicated. That would improve the quality of the dataset and improve the science. It would make it harder for critics to deny the verity of LENR, and it would make it easier to encourage students to study it. The goal of any community is to make itself stronger. Those of us who have worked in this field for decades now have a duty to teach. It's time to give back, so that we may all move forward. An understanding of LENR will shock the foundations of science and will leave a legacy for generations. After 26 years and approaching retirement I *really* want to see this. Let's focus on proving to the world we are right. Do we really want to compete amongst ourselves? Do we want to keep keeping the same secrets from each other? I have watched this happen now for nearly all the 26 years I have been engaged in this pursuit. It is time to stop hiding information and start helping each other. If any of us "wins" we all do, and so potentially does the broader community from which we must draw our resources and support.

I personally believe that LENR is too big for any one person or company, or country, to control. Everyone who has contributed to the development of the field is a pioneer. We are well-positioned for glory and recognition, and possibly wealth if it comes. The notion of competition – other than friendly competition among like-minded individuals – is absurd in this field. A monopoly is not possible or productive. If LENR is what I believe it may be it has the potential to change society in the way that Newcomen's steam engine did in England in 1712 – or before that the discovery of fire. It would be ludicrous and anti-societal to attempt to assert exclusive or proprietary "ownership". Even the argument sometimes made that the free enterprise system is the best way to advance such a radical change fails to appreciate the scope and importance of the potential change. Having studied cold fusion/LENR/CMNS now for 26 years (and before that a further decade studying the D/Pd system) I still have discovered no intrinsic reason why the effect cannot be utilized to create a primary energy source that is: essentially unlimited (whether H or D are employed as fuels); accessible to all mankind irrespective of nation or social class; considerably cheaper than existing power sources (especially if Pd can be employed only as a catalyst or avoided altogether); environmentally benign; operable on the scale of single humans or small communities (kW – tens of kW) rather than massive centralized operation with the attendant disbenefits of large capital cost, institutional or governmental control, and the needs and weaknesses of distribution and reticulation.

I know this sounds too good to be true – but I think it may be true and would like to prove it. What single entity should or could control such a boon ethically? Which one of us or which of our institutions? I sincerely believe that a monopoly is not possible or productive. We are the leaders of this field, so let us take that responsibility seriously. The faster and more broadly that LENR has a positive impact, the better off we all are. Think about the commercialization of the transistor. Bell Labs invented most of the enabling technology, and then licensed it broadly and relatively cheaply creating the communication and computing environment we all enjoy today. It would be hard to estimate the number of industries that derive from that decision and the rapid advancement in quality of life that has resulted. We can contrast the licensing of the transistor with the development of power by nuclear fission. In the latter case the technology is controlled by a few large companies and is terminally distrusted by the public. If we want to optimize for impact, the former path is preferred.

I suggest the following as a practical pathway of the joint solution to the very welcome problem of making LENR practical. Let us work together to make each other and the field stronger. I encourage the preparation of more publications with author lists from multiple institutions. Let us reinforce and reference each other's best work and work as a community. Far too many of our publications in our field cite primarily the author's own work. Whether from ignorance or arrogance this needs to stop. Let us make a commitment to learn from each other. From direct observation I know that huge resource has been squandered by failing to take advantage of knowledge earlier hard won in our tiny community. "*Not invented here*" needs to stop and our literature (fortunately fairly well codified in the ICCF proceedings volumes and now the *Journal of Condensed Matter Nuclear Science*) needs to be read, understood and respected, prior to serious experimental effort.

Most importantly we need to invite and include a new generation of researchers hoping they will be able to see farther than we have or can. The future is owned by the young. In our community this is even more true as only those with established reputation in 1989 (and thus seniority) could afford to enter the field. We are all, those who have survived, now 26 years older. Young scientists, unaffected by the academic trauma of the birth of cold fusion are interested and enthusiastic to enter a field with so many open questions and so much potential. It is our duty to help: to teach, to train, to mentor, to lead – if needed – but I expect it will not be. I would like us all to come together now and put our best ideas forward. The opportunity? – obligation? – destiny? – of this community is to birth a new science!

References

[1] M. Fleischmann, S. Pons and M. Hawkins, Electrochemically induced nuclear fusion of deuterium, *J. Electroanal. Chem.* **261**: 301 and errata in **263** (1989).

[2] S. Pons, M. Fleischmann, C. Walling and J. Simpson, International Patent Publication No. 90/10935 (1990).

[3] M. M.C.H. *et al.*, Excess power observations in electrochemical studies of the D/Pd system; the influence of loading, in *Proc. ICCF8, Frontiers of Cold Fusion*, Nagoya, Japan, *Universal Academy Press*, Tokyo, Japan, 1992.

[4] M.C.H. McKubre, F.L. Tanzella, P. Tripodi and P.L. Hagelstein, The emergence of a coherent explanation for anomalies observed in D/Pd and H/Pd systems: evidence for ^{4}He and ^{3}He production, in *Proc. ICCF8*, Italy, 2000.

[5] M.C.H. McKubre, J. Bao and F.L. Tanzella, Calorimetric studies of the destructive stimulation of palladium and nickel fine wires, in *Proc. ICCF17*, South Korea, 2012.

[6] T.O. Passell, Charting the way forward in the EPRI research program on deuterated metals, in *Proc. ICCF5*, Monte-Carlo, Monaco, IMRA Europe, Sophia Antipolis Cedex, France, 1995.

[7] M. Miles, B.F. Bush, G.S. Ostrom and J.J. Lagowski, Heat and helium measurements in cold fusion experiments, in *Proc. ICCF2*, Italy, 1991, p. 363.

[8] M.C.H. McKubre *et al.*, in *Low-Energy Nuclear Reactions Sourcebook*, Jan Marwan, Steven B. Krivit, (Eds.), Vol. 998, American Chemical Society, 2008.

[9] N.S. Lewis et al., Searches for low-temperature nuclear fusion of deuterium in palladium, *Nature* **340** (1989) 525–530.

[10] S.E. Koonin and M. Nauenberg, Calculated fusion rates in isotopic hydrogen molecules, *Nature* **339** (1989) 690.

[11] J. Kasagi, Screening potential for nuclear reactions in condensed matter, in *Proc. ICCF14*, Washington, DC, 2008.

[12] E. Storms, *The Science of Low Energy Nuclear Reactions,* World Scientific, Singapore, 2007.

[13] M. Lewan, *An Impossible Invention*, M. Lewan, Sweden, ISBN/EAN13 , 2014.

J. Condensed Matter Nucl. Sci. 19 (2016) 192–201

Research Article

Nature of the Deep-Dirac Levels

Andrew Meulenberg*
Science for Humanity Trust Inc., USA

Jean-Luc Paillet
Aix-Marseille University, France

Abstract

Maly and Va'vra (M&V) in 1993 and 1995 presented a computational evaluation of the Dirac equations that included the 'anomalous' solution. The regular solutions of these equations are the basis for modern quantum mechanical predictions for comparison with the experimental values of atomic-electron orbital energies. The other solution, discussed in the literature for over 55 years, is relativistic and considered anomalous because its predicted levels are very deep (up to 511 keV) and have never been observed. Nevertheless, the existence of these deep levels provides a ready explanation of the mechanism for penetration of the Coulomb barrier and the means of D–D fusion below the ^{4}He fragmentation levels. Since these levels also provide the basis for all of the other cold fusion observations (both PdD and NiH systems), it is important that arguments for and against the Dirac model be examined. The theoretical support for this anomalous solution is provided in a companion paper in this conference. This presentation seeks: to update the deep-orbit information provided in a poster at ICCF-17, to describe the nature of these deep–Dirac levels (DDLs), to report on additional, but unpublished, results presented by Va'vra in 1998, and to correct some interpretations of the model that Va'vra has provided in 2013. There are some unusual properties of the DDLs relative to those of the known atomic orbitals. Interpretation of the DDL properties, based on the non-relativistic solutions, leads to misunderstandings and further rejection of the concept of the deep levels. We hope to clarify this situation and indicate the importance of the calculations for cold fusion models. Cold fusion results provide a basis for understanding the DDLs and the proposed new fields of femto-physics and femto-chemistry.

Keywords: Anomalous solution, Dirac equations, Electron deep levels, Hydrogen atom, Relativity

1. Atomic Orbitals VS. the Relativistic Schrodinger Electron Deep Levels (EDLS) and the Deep Dirac Levels (DDLS)

Atomic-electron orbitals are very well known and understood. Until Maly and Va'vra (M&V, [1,2]) did so in the early 1990s, the multiplicity of relativistic electron deep levels (EDLs) had not been explored in the literature. The basic

*E-mail: mules333@gmail.com

premise of *any* deep levels has been improperly challenged for over 45 years (see companion paper [3]); therefore no one has 'bothered' to look beyond the 'ground-state' of these predicted orbitals.

With the advent of powerful computers, the task of solving the relevant equations became much easier and M&V did so in the context of cold fusion (CF) [1,2]. In another context, their results might have been acceptable. Even within the CF community, the results and implications were misinterpreted and ignored for over 15 years. Their works are still not accepted, with few exceptions, despite the obvious answers they provide for the theory of cold fusion [4,5].

What are the properties of the DDLs and why are they important to CF? First, why do we refer to the electron deep levels as Dirac Deep levels when they are also predicted by the Klein–Gordon (K–G) equation that is equivalent to the relativistic Schrodinger equation? In many of the early arguments against the K–G prediction of deep-orbits, the fact that it was valid for spinless particles was interpreted to mean that it was invalid for spin $1/2$ particles. These arguments ignored the fact that the Schrodinger equation (also valid for spinless particles) had been used for decades to describe electron orbitals. M&V buried this argument by showing that the relativistic Schrodinger equation also predicted the electron deep levels (EDLs). Furthermore, the effect of spin on the hydrogen atom calculation for the deep levels had an effect of only 2 keV out of over 500 keV. Thus:

Relativity introduces new electron binding energies at ~ 500 keV.

Spin affects these deep levels by only ~ 2 keV.

Second, a proton with a DDL electron is a neutral femto-atom. In matter, it acts like a 'fat' neutron.

2. What are Some Similarities between Atomic Orbitals and EDLs?

(1) Both sets of orbits are solutions of the same equations – they have identical starting assumptions and potentials.

(2) The solutions are a sequence that,

   (a) in the atomic case, depends on the electron orbital radius, energy, and angular momentum, along with the momentum, mass, and velocity (i.e., the deBroglie wavelength, $\lambda_c = h/mv$)

   (b) in the EDL cases, they also depend in a major way on relativity (in some form ?)

   (c) thus, while relativity could be involved in both atomic and DDL orbits, it is just not as noticeable in the low-energy atomic orbits.

(3) The sequences are related to the cyclical nature of something that relates to a wave, which periodically returns to identical conditions.

   (a) This depends on the conservation laws of energy and momentum (linear and angular).

   (b) Mathematically, integration about a closed path gives a zero result. Thus, no change occurs and a stable point is established. (See Appendix A.)

   (c) Closure depends on conditions being identical in all 'dimensions'. Orientation of a spin axis is one of those dimensions.

   (d) In an orbit, if the angular momentum vector remained 'fixed', it could not be a source of the periodicity.

      i. However, if the vector were to precess, then periodic motion, beyond that of the orbit, would be introduced.

      ii. It is assumed that the deBroglie wavelength, fundamental to atomic orbitals, is also based on some form of closed path. Is it based on vector precession? Precession of a spin vector would be a logical assumption for such paths, even for a linear trajectory.

      iii. The relativistic-Schrodinger equation predicts the discrete deep levels without explicitly resorting to spin. Nevertheless, the assumption of quantum mechanics is an inherent wave motion that, along with the deBroglie wavelength, could be related to a particle's spin.

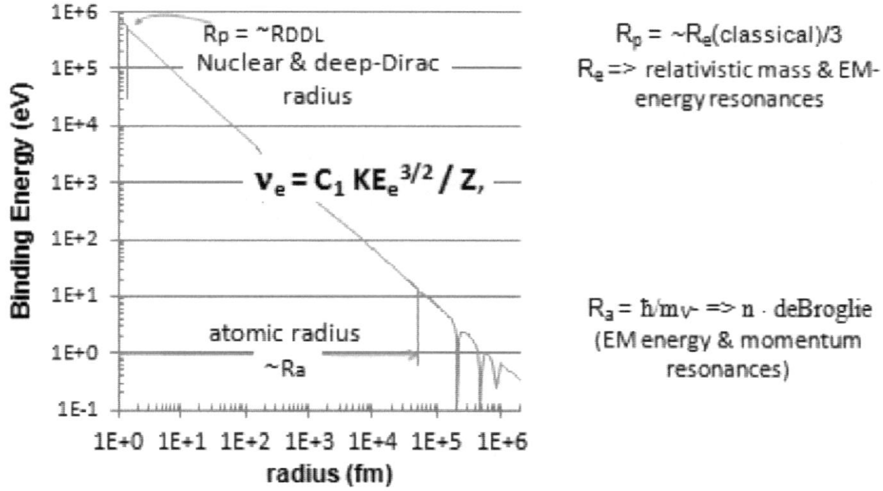

Figure 1. Binding energy vs. average electron-orbital radius. Both atomic-electron and deep-level electron orbital radii and energies are displayed as 'dips' in the Coulomb potential.

iv. Does the dependence on the deBroglie wavelength for atomic orbitals provide an equivalent to precession of a spin vector? (see Appendix A.) If so, can this concept be extended further to provide the relativity-induced deep orbits?

(e) A consistent model results if precession of different angular momentum vectors (e.g., spin and rotational) is considered for electrons in linear and orbital motion. We assume that relativity-induced forces (torques on a body with angular momentum) are the source of such precession.

i. Relativity and deBroglie waves (based on a spin vector?)
ii. Relativity and angular velocity vectors.

(4) Relativistic effects, strong enough to significantly increase effective electron mass, would be the cause of the deep orbits in both the Schrodinger and Dirac models.

3. What are Some Differences between Atomic Orbitals and EDLs?

Comparisons are based mainly on the Maly and Va'vra results from the tables in Appendix B.

(1) The obvious differences are expressed in Fig. 1 as the binding energy and average orbital radii. The frequencies (dependent on kinetic energy) have a difference of nearly eight decades for the difference in binding energy of five decades.
(2) The EDL orbits are so close to the nucleus (see Appendix C) that the $1/r$ Coulomb potential, valid at the atomic-orbit levels (see Appendix D), is no longer a valid assumption. The KE = $|PE|/2$ relation predicted by the virial theorem for a $1/r$ potential no longer holds here. A deep-orbit electron is very energetic and the relativistic virial theorem gives a different relationship between the kinetic and potential energies. Furthermore, because of proximity, the nucleus is so large relative to the electron's deep-orbit radius that only near-circular electron orbits stay in the $1/r$ potential region.

(3) For near-circular orbits (with a high angular momentum for its orbit), an electron only sees the $1/r$ potential, not the repulsive centrifugal core. This requirement is similar to that of the super-large Rydberg atomic orbitals that also avoid perturbations of the central regions by having high angular momenta and circular orbits. In the deep-electron orbitals, the central region must be avoided [6]. This is not because of the cloud of bound electrons in the center of a Rydberg orbit but because of the large centrifugal barrier and the non-$1/r$ potential inside the nucleus (e.g., $r_n =\sim 1$ fm), which is large relative to the electron deep orbit ($r_{\rm EDL} =\sim 2$ fm).

(4) The angular momentum of atomic-electron orbitals is sufficient to produce and receive photons as the energy-exchange medium. This is not the case for the EDLs where the angular momentum is on the order of $h/100$.[a] This is where Va'vra's EDL story goes wrong [7]. It is not his solutions of the Schrodinger or Dirac equations that are in error; it is in their misinterpretation (based on Va'vra's false assumption of the Heisenberg Uncertainty Principle being applicable to this region, see the paragraph beneath his Fig. 3) and then its extrapolation (his Figs. 5 and 6).

(5) The relativistic models predict many new and different energy levels:
   (a) Negative energy levels (positrons).
   (b) Electron deep levels that are not matched by positron deep levels.
   (c) Probably a symmetry-breaking effect.

(6) The relativistic mass increase, as a DDL electron gets closer to the nucleus and as its velocity approaches that of the speed of light, means that the stable orbits for increasing angular momentum must be closer to the nucleus than those of the low-l electrons (unlike Va'vra's predictions in his Figs. 5 and 6 of [7]), but more circular. This allows the electrons to move deeper into the Coulomb potential well and still avoid the centrifugal barrier in the center.

(7) Relativistic effects break the degeneracy in many of the atomic-electron levels. They are fundamental to all of the EDLs.

(8) The kinetic energy and mass (and our suggested angular-momentum increase in orbits nearer to the nucleus) are consistent with decreasing radii of more-nearly circular DDL orbits. We note that the low-angular momentum orbitals of high-n deep levels cannot exist since such highly elliptic paths would penetrate into the nucleus and encounter the altered Coulomb potential.
   (a) The selection rules differ for the EDLs and DDLs and the atomic-electron orbitals.
   (b) The atomic-electron levels exist for all positive integer n values.
   (c) The relativistic levels exist only for alternating integer n values.
      i. The relativistic Schrodinger levels exist only for odd integers.
      ii. The relativistic Dirac levels exist only for even integers.
   (d) The alternating levels is a relativistic effect.
   (e) The odd/even selection rules result from the spin effect.
   (f) The *multiple* levels predicted for the deep orbits are similar to those for the atomic orbits in some respects, not in others.
      i. They exist. There is not just a single deep level as predicted in some developments.
      ii. There is an integer change between energy levels in the solutions of the wave equations for both deep and atomic orbitals.

[a] In the solutions for the relativistic equations, the angular momentum quantum number of the electron is l. This value must work in both sets of solutions (the atomic and deep orbits), so it would appear that l should have the same value in both. This is not the case because the term that includes l has a multiplicative 'constant', which has different values for the two solutions. Consequently, when the binding energy changes, the angular momentum value does also, even though the meaning of the quantum number l does not.

- The difference between atomic levels is associated with h (the deBroglie wavelength).
- The difference in deep levels cannot be directly associated with h. (The angular momentum of any state below the atomic ground state is less than h.)

iii. Nevertheless, since the relativistic equations all include h and correctly predict the atomic-orbital angular momenta, there must be a relationship between the discrete deep levels and the discrete atomic-electron levels.

iv. **There must be an additional source of periodicity.**

v. We believe that the binding energy increases with angular momentum for the deep orbits (as indicated in Appendix B, but not in Fig. 6 of [7]–see Fig. C-1). This binding energy decreases with higher angular momentum in atomic orbits, but not in the deep orbits).

vi. The source of periodicity in the deep levels must be related to something beyond the deBroglie wavelength that dominates at atomic levels. There is insufficient electron angular momentum at these deep levels for any integer h change.

4. Comparisons between, and Questions about, Electron Deep Levels (EDLs) Predicted by the Relativistic Schrodinger Equation and the Deep Dirac Levels (DDLs)

(1) The EDLs and DDLs are not symmetric between the electrons and positrons.

(2) There are the same number of positron states and electron states within the Schrodinger results.

(3) There are more positron states than electron states in the Dirac results. Is this a result of spin? If so, why?

(4) Relativity has broken symmetry in both cases. Is it a valid conclusion, a real effect, or an artifact of the calculation? Would these same results appear if the calculation were instead carried out for positrons and anti-protons?

(5) Spin has little effect on the atomic-electron orbitals (beyond that of the Pauli Exclusion Principle).

(6) Spin has a much larger effect (keV) on the deep-levels.

(7) Spin introduces a new quantum number and, in the deep region, a reordering of the old.

(8) *The new quantum number (not found in the relativistic Schrodinger equation results) is related to spin, not to relativity.*

(9) *The source of the deep orbits is related to relativity, not to spin.*

(10) If the deep orbits are separated by even or odd integers, *what do the integers represent physically?*

5. Conclusions

Our analysis of the first papers [1,2] to both provide a non-singular potential for the relativistic Schrodinger and Dirac equations *and* then to find the solutions for the known and anomalous bound-electron levels has led to a number of surprises. These computational solutions include a coherent structure for an unusual family of electron deep levels (EDLs) with binding energy of their higher angular momentum orbitals asymptotically approaching that of the electron mass. While the papers have provided an immense amount of information, they have also left some unanswered questions.

Relativity has brought a solution to the central-body problem out of the center of the potential. We suggest that this and a definite, but low, angular momentum (i.e., within the limitations of $l = 0$ for the 'standard' centrifugal barrier of the atomic orbitals) provides an electron orbital close to the proton by using the new lower values of angular momentum (for deep orbits with its quantum numbers l).[b] The Coulomb potential is modified by the 'reduced' angular

[b]The possible confusion caused by the use of the same quantum numbers for different series of levels will have to be reduced in the future unless

momentum barrier, which introduces a series of deep levels available to electrons, as they, in turn, are modified by their increasing mass with kinetic energy. The deep orbits are 'quantized' with respect to an unknown source, apparently *not* one related to the deBroglie wavelength or the Heisenberg Uncertainty Relation and thus the Planck constant. Increased angular momentum within the deep-orbit framework (and at much less than h, therefore still with $l = 0$ in the atomic-electron framework) circularizes the deep-electron orbits and allows them to move closer to the proton.

While it would be nice to have experimental evidence directly supporting these deep electron levels, the low numbers and short lifetime of electrons in these levels prevent this. (This problem is suggested by the ease with which the resulting neutral femto-atoms can penetrate through an electron cloud and into atomic nuclei, thus producing the observed transmutation in cold fusion experiments.) If cold fusion, which depends on one or more electrons spending more time between protons or deuterons than possible in accepted physics, can provide enough used reactor material from its activity, then these levels might be indirectly, or even directly, observed. It is probable that sufficient used 'reactor' fuel material already exists (from 2013 or 2014 tests) for such additional analysis, if it were to be made available for that purpose. However, only a few facilities in the world may provide adequate analysis capabilities. Access to such facilities requires convincing their controllers and funding agencies that both cold fusion and the electron deep-level model are real and that such material with a long-enough life time for measurement could be produced in sufficient quantity for the tests. It is improbable that such tests will be made this year. However, we could work for that goal in 2016.

Appendix A. Integration along a Closed Path – the deBroglie Wavelength and Electron Orbitals

Integration along a 'closed' path gives a zero result (unless the path encircles a singularity). Extending this concept to wave motion along a linear path, x, allows a constant periodicity (i.e., to be a wave equation with $\mathrm{e}^{-2\pi ni} = 0$, for integer values of n) with a repeat distance R of the deBroglie wavelength. ($R = nx/\lambda_{\mathrm{dB}}$, where $\lambda_{\mathrm{dB}} = h/p$ and p is the momentum of the object [8] and h being the Planck constant. Thus, we get the quantum mechanical wave equation of $\mathrm{e}^{-2\pi ip\cdot x/h)}$). What does this signify? That will depend on some assumptions. Our assumption is that relativity-induced precession of a particle's spin-vector (from its velocity) is the basis of the wave motion and that the closed path is one traced out by a full cycle of the pointing angle of this vector. At a full cycle, the pointing angle of the vector is the same as at the beginning of the cycle and the conservation of momentum and energy says that they are independent of the position along the linear path. Thus, the integral over each cycle is zero.

In a non-relativistic bound orbit, the similar conditions hold; but now, the closed path must include both the orbital circumference and the deBroglie wavelength. The resonance of these two frequencies establishes the atomic-electron orbitals. In the relativistic case, additional closed orbits are predicted classically [9,10]. In relativistic quantum mechanics, these new deep orbits are also predicted [3], but cannot be associated with the Planck constant (since $l = 0$). Therefore, to be compatible with the closed-orbit model, there must be another wave mechanism. The orbital motion may provide the additional mechanism, via the relativity-induced precession (or nutation) from the Coulomb force, for the higher-frequency resonances needed to create the deep orbits.

Appendix B. Excerpts from Maly and Vavra's First Paper on Deep Dirac Levels

Published in *Fusion Technology*, Vol. 24, November 1993 (excerpts with permission, Copyright 11/93 by the American Nuclear Society, La Grange Park, Illinois [1])

people get used to the different values of angular momentum for different series. This is not confusing for the different values of energy (from the principal quantum numbers) for the atomic and deep orbits because these energies are both still related to hν. However, the fixation with only the Planck constant being associated with angular momentum and the Heisenberg uncertainty relation makes 'suspect' the new, lower, angular momenta values of the deep orbits.

Electron Transitions on Deep Dirac Levels I

J.A. Maly and J. Va'vra

Abstract

The original solutions of the Schroedinger relativistic equation and the Dirac equation for hydrogen-like atoms were analyzed for the possible existence of some other electron levels, which were not originally derived. It was found that besides the known atomic levels, each atom should also have the Deep Dirac Levels (DDL). The electron transition on such DDL would produce large amounts of atomic energy (400–510 keV per transition depending on the Z of the atom).

4. Calculations of New Energy Levels

A computer program was written which calculates atomic energy levels for the Relativistic Schroedinger and Dirac levels. For comparison, the non-relativistic Schroedinger levels given by a simple Bohr formula are also shown.

The Schroedinger levels are calculated in Table 1 with the plus sign inside the $s\ldots$ in the E1S(+) column and with the minus sign inside the s in the E2S(-) column. The Dirac levels are calculated in Table 2 with the plus sign of $s\ldots$ in the E1D(+) column and with the minus sign of s in the E2D(-) column. In describing Schroedinger levels, the notations $n = N =$ main quantum number, $l = L =$ angular quantum number, $n' = M =$ radial quantum number will be used. The nln' notation used in the equations of this paper is defined as the NLM level in Table 1 ($l = 0, 1, 2, 3, 4, 5, \ldots$ are also called s,p,d,f,g, ... levels in the spectroscopic notations). Similarly, we will use the notation $n = N =$ main quantum number, $n' = M =$ radial quantum number, $k = K = a(j + 1/2) =$ Dirac k number, $l = L =$ angular quantum number to describe the Dirac energy levels ($a = +$ or $-$ sign at k).

Table 1. Relativistic Schroedinger levels for H ($Z = 1$) in eV

	$E(N, Z)$	N	M	L	E1S	E2S	
1s	−13.605826	1	0	0	−13.606597	−507171.937500	
2p	−3.501457	2	0	1	−3.401449	−13.605632	*
2s	−3.401457	2	1	0	−3.401570	−13.603699	
3d	−1.511759	3	0	2	−1.511747	−3.401425	*
3p	−1.511759	3	1	1	−1.511755	−509755.250000	
3s	−1.511759	3	2	0	−1.511790	−3.401207	
4f	−1.511764	4	0	3	−0.850357	−1.511744	*
4d	−0.850364	4	1	2	−0.850358	−13.605434	*
4p	−0.850364	4	2	1	−0.850361	−13.604666	
4s	−0.850364	4	3	0	−0.850376	−1.511683	
5g	−0.850364	5	0	4	−0.544228	−0.850356	*
5f	−0.544233	5	1	3	−0.544228	−3.401415	*
5d	−0.544233	5	2	2	−0.544229	−510264.468750	

*Negative energy states, not observable.

Appendix C. Excerpts from Vavra's Presentation at Siegen University, Germany, Nov. 25, 1998 (excerpts with permission of J. Va'vra)

On a possibility of existence of new atomic levels, which were neglected theoretically and not measured experimentally.

Table 2. Dirac levels of hydrogen-like atoms for H ($Z = 1$) in eV.

	$E(N, Z)$	N	M	K	L1	E1D	L2	E2D	
1s	−13.605826	1	0	1	0	−13.605873	1	−13.605873	*
2p	−3.401457	2	0	2	1	−3.401434	2	−3.401434	*
2s	−3.401457	2	1	1	0	−3.401479	1	−509133.375000	
3d	−1.511759	3	0	3	2	−1.511746	3	−1.511746	*
3p	−1.511759	3	1	2	1	−1.511750	2	−13.605512	*
3s	−0.850364	3	2	1	0	−1.711764	1	−13.604422	
4f	−0.850364	4	0	4	3	−0.850356	4	−0.850356	*
4d	−0.850364	4	1	3	2	−0.850357	3	−3.401419	*
4p	−0.850364	4	2	2	1	−0.850359	2	−510064.125000	
4s	−0.850364	4	3	1	0	−0.850365	1	−3.401298	
5g	−0.544233	5	0	5	4	−0.544228	5	−0.544228	*
5f	−0.544233	5	1	4	3	−0.544228	4	−1.511744	*
5d	−0.544233	5	2	3	2	−0.544229	3	−13.605389	*
5p	−0.544233	5	3	2	1	−0.544230	2	−13.604785	
5s	−0.544233	5	4	1	0	−0.544233	1	−1.511710	
6h	−0.377940	6	0	6	5	−0.377936	6	−0.377936	*
6g	−0.377940	6	1	5	4	−0.377936	5	−0.850356	*
6f	−0.377940	6	2	4	3	−0.377936	4	−3.401412	*
6d	−0.377940	6	3	3	2	−0.377937	3	−510381.343750	

*Negative energy states, not observable.

The electron density distribution is calculated as follows and as pictured in Fig. C-1.

$$\mathrm{Eld} = 4\pi r^2 R^2(r) = 4\pi r^2 \mathrm{e}^{-\rho} \rho^{2s} L^2(\rho).$$

The radial distributions for the deep-Dirac levels have different quantum numbers, but nearly identical shapes for

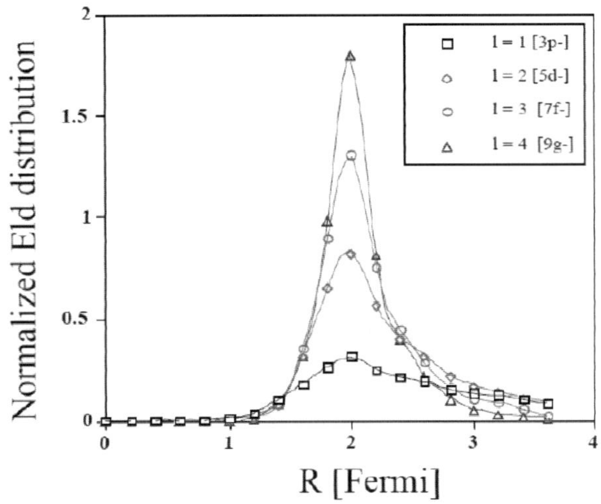

Figure C-1. Normalized radial distribution of the first four deep-electron levels based on the relativistic Schroedinger equation with Nix potential.

the different orbitals. Note that the first level is extended to higher and lower radii (i.e., it is composed of more elliptical orbits). Sequential orbitals (with higher angular momenta) become more confined into a spherical shell as the distribution settles deeper into the potential well.

Appendix D. Deviation from the $1/r$ Coulomb potential

The Hamiltonian in all of the models, non-relativistic and relativistic, used the $1/r$ Coulomb potential. Feynman states that this potential is valid, at least up to the nucleus. Nevertheless, he introduces the angular momentum, in the form of a centrifugal force as a pseudo-potential, into the solution for the hydrogen atom [11]. Thus, unless the angular momentum contribution is removed (leaving $l = 0$), the virial theorem must be determined for the $1/r$ Coulomb potential V_C as modified by the addition of significant angular momentum with quantization, e.g., $l = 1, 2, ..., n-1$):

$$|V_C| = |ee'|/r = e^2/r \Rightarrow |V_C'| = e^2/r - l(l+1)h^2/2mr^2(D-1). \qquad \text{(D.1)}$$

When these two potentials are equal in magnitude (and opposite in direction) the attractive potential of the nucleus for an electron becomes repulsive.

At the Bohr radius, $a_0 = 4\pi\varepsilon_0 h^2/me^2 = h/mc\alpha$, and the turnover potential is found at:

$$e^2/a_0 = l(l+1)h^2/2ma_0^2 = l(l+1)h^2/2m(h/\alpha * mc)^2 = \alpha^2 l(l+1)mc^2/2(D-2). \qquad \text{(D.2)}$$

Since the attractive Coulomb potential energy at a_0 is 27.2 eV (= $\alpha^2 mc^2$), this is equal to the repulsive energy at $l(l+1) = 2$. This allows $l = 0$ to be the only possibility for a ground state, since the $l = 1$ possibility would require an exactly circular (a classical) orbit with a nearly exact energy rather than the spread in orbital parameters as required by the Heisenberg relation. However, this quantization of angular momentum does not prevent all non-integer angular momenta (for $0 \leq l < 1$). The uncertainty relation permits some variation, despite the $l = 0$ restriction. Again, *any* angular momentum prevents the singularity at $r = 0$. However, for angular momentum approaching *h*, e.g., $l = 1$, the effective central potential is no longer well approximated by the $1/r$ Coulomb potential for the $n = 1$ orbital. For the $n = 2$ levels (or above) and $l = 1$, the approximation is still adequate. Thus, there is a physical basis for the normal selection rules (integer values $0 < l < n-1$).

At the classical electron radius, $r_c = \alpha * h/mc$ (and $e^2/r_c = e^2/\alpha * h/mc$, and the turning point is at:

$$e^2/r_c = l(l+1)h^2/2mr_c^2 = l(l+1)h^2/2m(\alpha * h/mc)^2 = l(l+1)mc^2/2\alpha^2(D-3). \qquad \text{(D.3)}$$

The point where Coulomb potential and centrifugal barrier are equal in magnitude gives $l(l+1) = 2\alpha^2 = 2/137^2$. It is clear that any orbits based on the centrifugal barrier, for angular momentum quantized in units of the Planck constant, is forbidden except for l very near to zero. This same restriction also applies to any such localization where the Heisenberg Uncertainty Relation holds. How can the Dirac anomalous solution predict orbits that oppose these restrictions? The condition for stable orbits in the sub-atomic regime must result from a momentum operator that is not based on the Planck constant. An understanding and description of this requirement is the basis of another paper. However, in the M&V papers, their analysis included a finite-sized nucleus with a non-singular potential for electrons. In the atomic-electron case, this has little effect since it is nearly 5 orders of magnitude smaller than the orbitals. In the deep-orbit electron case, the nucleus is nearly one-half the orbital radius and this severely restricts the possible orbits. At the femtometer orbits, the nuclear size, relativity, quantized angular momentum, and their effects on the electron must be added to the centrifugal potential that alters the virial theorem results for the Coulomb potential.

Acknowledgement

This work is supported in part by HiPi Consulting, New Market, MD, USA; by the Science for Humanity Trust, Bangalore, India; and by the Science for Humanity Trust Inc., Tucker, GA, USA.

References

[1] J.A. Maly and J. Vavra, Electron transitions on deep Dirac levels I, *Fusion Technol.* **24**(3) (1993) 307–318. http://www.ans.org/pubs/journals/fst/a_30206

[2] J. Maly and J. Va'vra, Electron transitions on deep Dirac levels II, *Fusion Sci. Technol.* **27**(1) (1995) 59–70. http://www.ans.org/pubs/journals/fst/a_30350

[3] J. L. Paillet, A. Meulenberg, "Basis for electron deep orbits of the hydrogen atom, this conference proceedings 19th Int. Conf. on Cond. Matter Nuclear Science, ICCF-19, Padua, Italy, 15/05/2015, to be published in *J. Cond. Matter Nucl. Sci.*, Vol. 19, 2016 . More detail in J.-L. Paillet and A. Meulenberg, Arguments for the Anomalous Solutions of the Dirac equations, http://viXra.org/abs/1506.0177 .

[4] A. Meulenberg and K.P. Sinha, Deep-electron orbits in Cold Fusion, *17th Int. Conf. on Condensed Matter Nuclear Science*, Daejeon, South Korea, 12–17 August, 2012, *J. Condensed Matter Nucl. Sci.* **13** (2014) 368–377.

[5] A. Meulenberg, Deep-orbit-electron radiation absorption and emission, *ICCF-18, 18th Int. Conf. on Cond. Matter Nucl. Sci.*, Columbia, Missouri, 25/07/2013, http://hdl.handle.net/10355/36501

[6] J. Va'vra, On a possibility of existence of new atomic levels, which were neglected theoretically and not measured experimentally, presented at Siegen University, Germany, November 25, 1998.

[7] J. Va'vra, A new way to explain the 511 keV signal from the center of the Galaxy and some dark matter experiments, *Astronomy & Astrophysics* (2013), http://arxiv.org/abs/1304.0833

[8] R.P. Feynman and R.B. Leighton, *The Feynman Lectures on Physics*, Vol. III, Section 3.1, Eq. (3-7), 2nd Edn., Addison–Wesley, 1964.

[9] T.H. Boyer, Unfamiliar trajectories for a relativistic particle in a Kepler or Coulomb potential, *Am. J. Phys.* **72** (2004) 992, http://dx.doi.org/10.1119/1.1737396.

[10] J. Chen, Trajectories for a relativistic particle in a Kepler potential, Coulomb potential, or a scalar field, Senior research thesis, (2005), College of William and Mary, go to: http://www.wm.edu/as/physics/undergrad/major/seniorhonors/previoustheses/2005/index.php (then click on Chen).

[11] R.P. Feynman and R.B. Leighton, *The Feynman Lectures on Physics*, Vol. III, Section 19-4, 2nd Edn., Addison–Wesley, 1964.

J. Condensed Matter Nucl. Sci. 19 (2016) 202–209

Research Article

Basis for Femto-molecules and -Ions Created from Femto-atoms

Andrew Meulenberg*
Science for Humanity Trust Inc., USA

Jean-Luc Paillet
Aix-Marseille University, France

Abstract

Starting with the assumption of validity of the Dirac equations (relativistic quantum mechanics), which are fundamental to much of atomic physics today, we also assume that the anomalous solutions to these equations are valid. If they are valid, then short-lived femto-atoms with electron orbitals in the low femto-meter range should exist. If femto-atoms exist, then the existence of femto-molecules could be expected. This paper addresses the possibility and nature of the femto-molecules and the nature of the forces creating them. The approach is that of Feynman's molecular-hydrogen ion derivation using the Yukawa potential. The result is a molecular ion with femto-meter order spacing between the nuclei and an attractive potential identical with a medium-range Yukawa potential for an exchange particle with the mass of an electron. There are significant implications for both cold fusion and for nuclear physics and chemistry.

Keywords: Deep-Dirac levels, Femto-hydrides, Halo nuclei, Isotopic anomalies, Medium-range nuclear binding

1. Introduction

Given the existence of femto-hydrogen atoms with electrons in Deep-Dirac Levels (DDLs) having binding energies at 509 keV [1], what support can be given for the existence of molecules composed of such atoms? Starting with Feynman's quantum mechanical, but non-relativistic, derivation of the molecular-hydrogen ion [2], we will extend it to the femto-meter and relativistic region – to the femto-molecular level.

2. Feynman's Derivation of H_2^+ Molecular-ions based on Electron-exchange Potentials

Feynman assumes a paired two-state system consisting of two protons and an electron. A primary potential E_0 is the electron-screened repulsive Coulomb potential between the protons. Secondary potentials, $\pm A$,exist because the electron can be exchanged between the two protons in two ways (with $A < 0$). Two resulting time-independent states

*E-mail: mules333@gmail.com

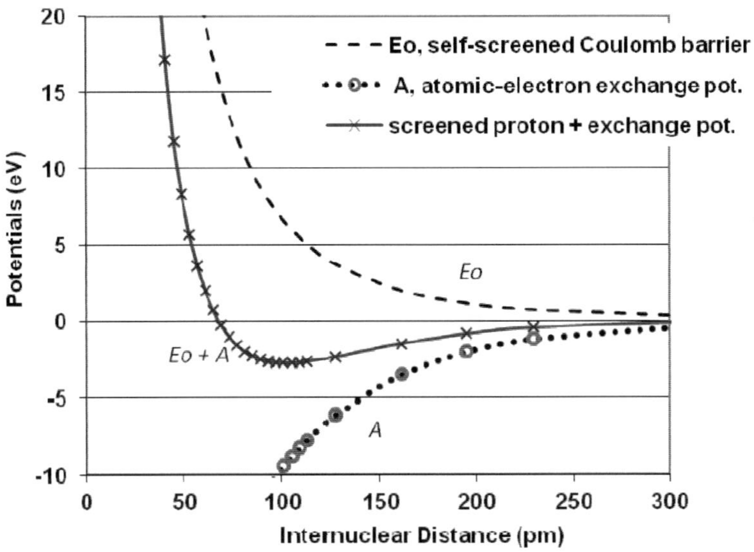

Figure 1. Potentials, as a function of internuclear separation, between two protons with a bound electron.

of the system (EI and EII) are composed of the primary potential, E_0, plus the secondary potential (A or $-A$). Since in one case, the summed potential is always repulsive, no bound state can exist. In the other case, the summed potential has a region of negative energy – the bound state. Figure 1 provides sample values of these potentials based on the H_2^+ ion.

Feynman does not go into the details of the screened potential, so neither shall we [3]. However, the critical issue for us is the nature of the 'exchange' potential, A. The unlabeled precursor to his equation (Eq. (10.10)) for this term is the amplitude for a free particle to get from one place to another a distance R away:

$$A \sim (1/R)\exp(ipR/\hbar), \tag{1}$$

where p is the momentum of the electron [4] and R is the distance between the protons. (Notice that p/h is just 2π divided by the deBroglie wavelength, λ_{dB}.) For a non-relativistic atomic electron with mass, m, and kinetic energy, KE, its momentum $p = (2m\,\mathrm{KE})^{1/2}$. However, when encountering a barrier and tunneling into or through it, the wave equation predicts an increase in deBroglie wavelength of the particle as its momentum goes toward zero at the barrier. At the barrier, the wavelength goes to infinity and the wave function turns into a reflection from, and an exponentially decaying function (an evanescent wave) inside, the barrier.

For a particle trapped in a well, where its potential (relative to outside the well) is less than zero, $V < 0$, the kinetic energy, $\mathrm{KE} = \mathrm{TE} - V$, is a positive value equal to the difference between the total energy, TE, and the potential energy ($V < \mathrm{TE} < 0$). However, if the particle tries to penetrate the wall of the well, where $V = 0$, then $\mathrm{KE} = \mathrm{TE} < 0$. As a result and since $\mathrm{KE} < 0$, the momentum p is imaginary:

$$p = (2m\mathrm{KE})^{1/2} = (2m(\mathrm{TE} - V))^{1/2} \quad \text{and} \quad \mathbf{p = i(2m|TE|)^{1/2}} \text{for } V = 0 \text{ and } \mathrm{TE} < 0. \tag{2}$$

The non-relativistic virial theorem for particles bound in a Coulomb $(1/r)$ potential states that $\mathrm{KE} = \mathrm{PE}/2$ for stable orbits. Since the difference between total energy and potential energy is the binding energy BE of the trapped particle, a positive energy equal to the absolute value of the binding energy, $|\mathrm{BE}| = |\mathrm{TE} - V|$, is necessary to lift the particle out of the well. For a bound particle, the potential outside of the well (amplitude for penetrating into the barrier), no longer with the imaginary exponent, is

$$A\sim (1/R)\ \exp((-(2m|\mathrm{BE}|)^{1/2}/\hbar)R). \tag{3}$$

In his 'Lectures' [2], Feynman used W_{H} for the binding energy of the hydrogen electron (W_{H} is the work required to take an electron from the H-atom ground state to infinity) but, outside of the well (where $V = 0$), he set BE equal to the negative of the kinetic energy in the well ($\mathrm{KE} = -\mathrm{BE}$). Thus, he was able to make p imaginary and arrived at the same operative expression for the exchange potential, Eq. (3).

3. H_2^+ Femto-molecular-ion Electron-exchange Potentials

The DDL-electron exchange effect produces a much shorter-range attractive potential than does the atomic-electron effect. In Fig. 2, the log-log version of Fig. 1, the primary potential E_0 is again the electron-screened repulsive Coulomb potential between the protons and, here, it only represents the effect of loosely bound (5.8 eV) molecular-orbit electrons. The absolute values of various attractive exchange potentials (A, A_2, and $A_2' < 0$) are displayed so that their magnitudes can be compared directly against the repulsive potentials.

The attractive secondary potentials for molecular- and DDL-electron orbits (A, A_2, and A_2') indicate the differences between the bound-electrons of different average kinetic energies (5.8 eV, 100 keV, and 1.5 MeV, respectively). The reason that they, unlike the E_0 curve, 'flatten out' (saturate) below the picometer- and lower-fermi-ranges respectively is related to the facts that E_0 is the electron-screened proton–proton Coulomb repulsion and that the A's are for the proton–electron interaction. The 'size' of the electron orbital distribution exceeds the spatial separation

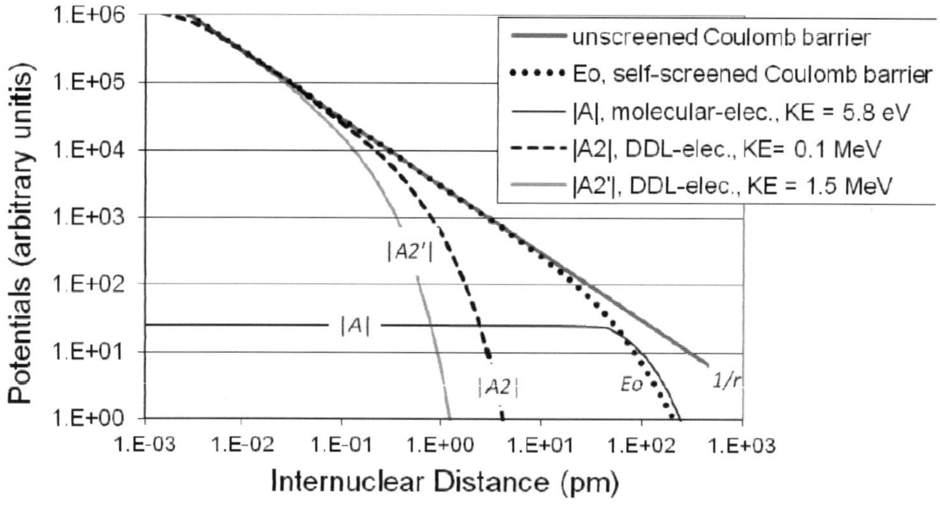

Figure 2. Representative potentials (some from Fig. 1 but now on a log-log plot) as a function of internuclear separation, between two protons with a bound electron.

between the protons within this range and therefore the screening is reduced as the protons come together and the electron spends less and less time between the protons. The DDL orbitals in the low femto-meter range means that the screening-saturation effect does not really begin until the protons get within the sub-picometer range of one another.

This exchange-particle range is based on the particle's momentum (as in the deBroglie wavelength and Eq. (1)). In the Yukawa nuclear potential, the relativistic momentum is related to the mass energy, $E^2 = (pc)^2 + (mc^2)^2$ and, with the assumption of total energy near zero, $E =\sim 0$, leading to $p =\sim imc$, the resulting potential has a range dependence on the particle's Compton wavelength, λ_C (Feynman's Eq. (10.14)).

$$A\sim (1/R)\exp(-mc/\hbar)R= (1/R)\exp(-2\pi R/\lambda_C) \tag{4}$$

This equation is the basis of both the short-range pion-exchange force and the long-range photon-exchange model. However, in Feynman's electron-exchange model of the molecular binding potential, the non-relativistic equation for momentum relates it to the square root of the kinetic energy, $p = (2m\,\mathrm{KE})^{1/2}$– and that to the binding energy (Eq. (3)).

The femto-molecule analysis is in-between the strongly relativistic and the non-relativistic regimes. The relativistic momentum is still strongly related to the electron mass energy, $E^2 = (pc)^2 + (mc^2)^2$, since E, at 2–4 keV (depending of whether the Dirac or Klein–Gordon anomalous solutions are used) [5] for the hydrogen atom, is much smaller than mc^2 (at 511 keV). The main differences between the atomic molecule and the femto-molecule are in the greatly increased screening provided by the deep electron orbits (reducing the range of E_0) and in the greatly enhanced exchange potential as seen with A_2 and A_2' in Fig. 2.

Our analyses, of both the molecular and the femto-molecular ion exchange energies, are representative rather than accurate. They assume that the screening field of the electron in the molecule is twice the Bohr radius, $a_0 =\sim 53$ pm; but, for the femto-molecule, it is reduced to the classical electron radius, $r_e =\sim 2.8$ fm. The fact that the classical electron radius is the fine structure constant times the Compton electron radius, $r_C = \lambda_{dB}/2\pi$, and r_C is the fine structure constant times the Bohr electron radius makes one wonder about some sort of divine power-series expansion in 'α' :

$$r_e = \alpha r_C = \alpha^2 a_0. \tag{5}$$

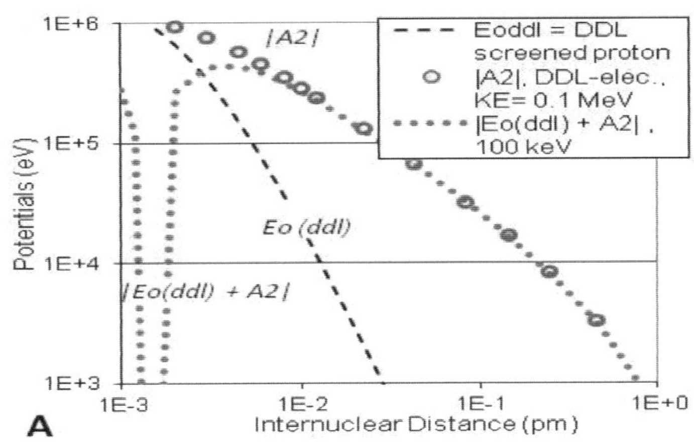

Figure 3. Log-log information for the screened Coulomb repulsion and the exchange potentials of femto-H_2^+ and their sums.

4. Summed Potentials of H_2^+ Femto-molecular-ions

Figure 3 is a log-log plot of the femto-molecular ranges (thus complementing the atomic-molecular range of Fig. 2). It has three curves that give an indication of the parameters to consider. The nuclear potential is ignored, since it would dominate if the spacing between protons reached that low. The unscreened Coulomb barrier (not shown) between the protons is the starting point. The presence of the electron provides a self-screening of that potential and this screened potential, E_0(ddl), is the one that is to be worked with. It is the femto-atom equivalent of E_0 in Feynman's model (and in Figs.1 and 2).

Comparison of the E_0 curves in Figs. 2 and 3 shows the effect of strong screening by a DDL electron vs that of an atomic electron. $|A_2|$ and the sum of the two potentials, $|E_0$ (ddl) + $A_2|$ in Fig 3, are displayed against the E_0 (ddl) curve to indicate explicitly their relative magnitude.

The absolute value of the exchange potential, $|A_2|$, is now much larger than the screened Coulomb potential E_0 (ddl) until the protons are within the multi-femto-meter range. The maximum attractive potential for the sum is at about 5 fm for the 100 keV electron kinetic energy in Fig. 3. In Fig 4, the electron KE is raised to $\sim$ 1.5 MeV (an expected value for a DDL electron) and the results are displayed on a semi-log plot for ease of interpretation. For this higher KE (A_2'), with a smaller average orbital radius that is in agreement with projected DDL electron orbits, the attractive summed potential is 'smeared' into the nucleus ($R < 1$ fm) and actual fusion of the protons to a deuteron could be unavoidable. However, there are things missing in this analysis, such as spin–spin coupling, which is very strong at this range and will alter the results; so what is shown is only indicative of expectation. Detailed calculations will have to await a more complete understanding of the proximity effects of the bound electron and proton.

A potentially important point in Fig. 4 is the solid curve (Yukawa exchange potential). While the nuclear potentials have been postulated to result from virtual pion-exchange (shorter range with much higher particle mass), an intermediate-range (100 fm) force has long-been proposed and sought (by one of the present authors, AM, and by Jacque Dufour [6]) to explain cold fusion. This mid-range force can be associated with the Yukawa potential for a

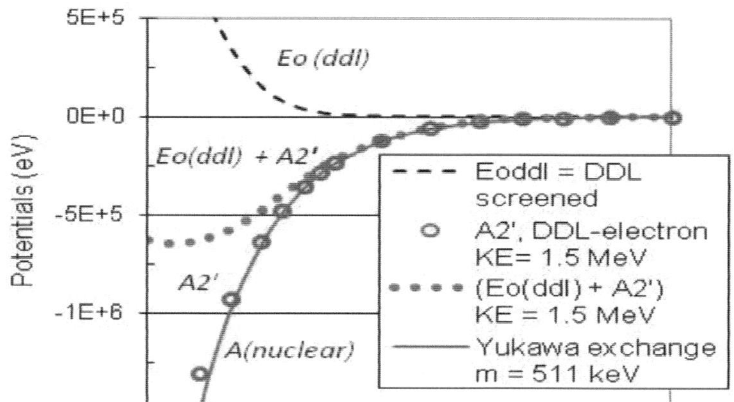

Figure 4. Semi-log information for the screened Coulomb repulsion and the exchange potentials of femto-H_2^+ and their sums. The Yukawa potential for a virtual-electron exchange is included.

real-electron-exchange mass of 511 keV/c^2. The A'_2 curve is coincident with the Yukawa exchange potential until the low femto-meter range. This should not be a surprise. The contributions to both the A'_2 and the Yukawa-exchange potentials are nearly identical except for the values of R. The $1/R$ values in Eq. (3) are based on the same distance between nuclei centers. However, the R value used in the exponent of the present model is a weighted average of the distances 'tunneled' between potential wells and this is slightly less than the distance between proton centers used in the Yukawa potential.

Such an intermediate-range attractive force could appear to be nuclear in nature, but much more extended; it would also provide a basis for the metastable halo nuclei [7]. The 'odd' nucleons (protons or neutrons) that are bound in the 5–8 fm range of a halo nucleus could be femto-hydrides.

5. Femto-molecular Hydrides and Experimental Data

The final issue, addressed here only with a quick comment, is that of the neutral femto-molecule. With two protons and two electrons, the attractive potential is much greater. As a first-order approximation to the problem, just double the DDL-electron attraction between the protons. The Coulomb repulsion between the electrons may be eliminated by the spin–spin attraction of the electron pair with anti-parallel spin vectors. Therefore, we may not have to be concerned about that otherwise major issue. If the electron-exchange potential is doubled by addition of the second deep-orbit electron, then this effect may dominate the other effects and a deeper or broader potential well results. Whether the well is metastable, or immediately leads to fusion, is still to be determined even from the simple case modeled.

What happens when a femto-H atom or femto-H_2 molecule forms within a lattice? This theoretical issue has now been partially addressed for transmutation [8,9]. Nevertheless, there have been some experimental transmutation results that, if valid, would appear to be self-contradictory. The first was from Defkalion where four of the natural isotopes of nickel (^{58}Ni, ^{60}Ni,^{61}Ni, and^{62}Ni), were used to test for excess heat [10]. All but ^{61}Ni, produced excess heat. Why did ^{61}Ni not do so (even when the experiment was repeated)? The second was from Rossi's system tests at Lugano [11], where the residual nickel isotope, after a month of producing excess heat, was ^{62}Ni. While both results would indicate nuclear fusion results, the apparent contradiction of the results could cast doubt on the whole process.

An attempt to understand the two results has provided suggestions that may or may not be valid. In the referenced Defkalion work, the ^{61}Ni is the only isotope tested that does not have a zero isotopic spin ground state. (It has $J = 3/2$, with negative parity.) What does this have to do with cold fusion? It is known that only low-angular-momentum states can result for/from low-incident-energy collisions. If these $H^{\#}$ states cannot interact with the high total angular momentum of the ^{61}Ni nucleus, then a femto-hydride is the preferred state (a chemical rather than a nuclear bond). The result is that the energy of formation of the femto-H atom or molecule (about 0.5 MeV per H) would be the only excess energy that could result from this formation of the femto-hydride. While there is still excess energy, it is much less than would result from the multi-MeV fusion or transmutation energies. Since cold fusion reactions in the nickel system increase with lattice temperature, it is possible that the ^{61}Ni test, with a much lower excess power ability, never achieved a critical level under the same test conditions that worked for the other isotopes. The test results are proprietary (and Defkalion is no longer in business), so that it is not possible to get more information, or actual samples, from those tests.

The Lugano test results showed a strong depletion of the natural ^{58}Ni, ^{60}Ni, ^{61}Ni, and ^{64}Ni isotopes. The ^{62}Ni was the main residual nuclide measured. There is no ready explanation for this observation that is consistent with the ^{61}Ni 'anomaly' and the reported results of the earlier experiment. However, there is a possible, and testable, explanation. The experimental results showing ^{62}Ni could in fact be a 'misreading' of the isotopic test results. While the equipment used was capable of very fine mass separation (e.g., resolution of adjacent isotopes), there is a real possibility that a slight shift in a strong peak could have been overlooked even if the high resolution mode were employed in test. The ^{62}Ni peak and the ^{61}Ni + $H^{\#}$ peak are probably within 10 MeV/c^2 of each other. The shift of the peak from ^{62}Ni to

^{61}Ni + H$^{\#}$ would not normally be expected and therefore would not be noticed. Had there been a residue of ^{62}Ni, the doublet would have been obvious if measurements were taken at the full resolution of the test equipment. Again, this is proprietary data and material samples, so neither may become available for further testing and analysis until 'profit' warrants it.

6. Conclusion

Using a text-book derivation for the diatomic-molecular ion, we have derived the bound-state solution for a pair of protons bound together by a deep-level electron. The present approach combines that of the molecular-binding problem using real charged particles, the electron, with that of the nuclear-binding problem using the Yukawa potential and virtual particles, the pion. The high binding energy of the deep-level electrons ($\sim$ 500 keV), is closer to that of the nuclear forces (multi-MeV) than that of the atomic forces (multi-eV); therefore, it is no surprise that the results for the femto-molecular ion appear to be that of a mid-range (i.e., $\sim$ 100 fm) Yukawa nuclear-binding potential. In fact, the results are almost indistinguishable from those for nuclear binding based on virtual-electron exchange.

This similarity of forces leading to the femto-molecular ion with the nuclear forces and the conclusions from such similarity can lead to two quite different responses. First, the statement that "nuclear forces can be mediated only by gauge bosons" is too limiting. Second, "the proposed electron deep levels (EDLs) cannot exist;" therefore, "Feynman's molecular-ion derivation cannot be applied in this case." We cannot accept either part of the second statement. Just as the anomalous solutions of the relativistic Schrodinger and Dirac equations predict the existence of the EDLs [5,12,13], and thus femto-atoms [8], the analysis in this paper predicts the possibility of femto-molecules, femto-molecular ions, and medium-range nuclear binding by real electrons. Evidence of the existence of all these effects is strongly suggested by experimental evidence provided by results of the cold fusion work over the last 25+ years.

Acknowledgement

This work is supported in part by HiPi Consulting, New Market, MD, USA; by the Science for Humanity Trust, Bangalore, India; and by the Science for Humanity Trust, Inc, Tucker, GA, USA.

References

[1] J.A. Maly and J. Vavra, Electron transitions on deep Dirac levels I, *Fusion Technol.* **24** (1993) 307–381, http://people.web.psi.ch/pruchova/Vavra_d_dirac_l.pdf.
[2] R.P. Feynman, R.B, Leighton and M. Sands, *The Feynman Lectures on Physics*, Vol. 3, Chapter 10, Sections 1–3, Addison–Wesley, Reading, MA, USA, 1965.
[3] A. Deltuva, Coulomb force effects in low-energy α-deuteron scattering, *Phy. Rev.* C **74** (2006) 064001.
[4] R.P. Feynman, R.B, Leighton and M. Sands, *The Feynman Lectures on Physics*, Vol. 3, Chapter 3.1, Eq. (3-7), Addison–Wesley, Reading, MA, USA, 1965.
[5] Jean-Luc Paillet and Andrew Meulenberg, Arguments for the anomalous solutions of the Dirac equations, http://viXra.org/abs/1506.0177, *J. Condensed Matter Nucl. Sci.* **18** (2016) 50–75.
[6] J. Dufour, An introduction to the pico-chemistry working hypothesis, *J. Condensed Matter Nucl. Sci.* **10** (2013) 40–45.
[7] https://en.wikipedia.org/wiki/Halo_nucleus.
[8] A. Meulenberg, Femto-atoms and transmutation, *ICCF-17*, Daejeon, South Korea, 12–17 August, 2012, *J. Condensed Matter Nucl. Sci.* **13** (2014) 346—357.
[9] A. Meulenberg, Femto-helium and PdD transmutation, *ICCF-18, 18th Int. Conf. on Cond. Matter Nucl. Sci.*, Columbia, Missouri, 25/07/2013, http://hdl.handle.net/10355/36500.
[10] J. Hadjichristos, M. Koulouris and A. Chatzichristos, Technical characteristics and performance of the Defkalion's hyperion pre-industrial product, *17th Int. Conf. on Condensed Matter Nucl. Sci.*, Daejeon, South Korea, 12–17 August, 2012.

[11] G. Levi et al., Observation of abundant heat production from a reactor device and of isotopic changes in the fuel, http://amsacta.unibo.it/4084/, ecat.com/wp-content/uploads/2014/10/ECAT-test-report-2014.pdf, 2014.
[12] J.L. Paillet and A. Meulenberg, The basis for electron deep orbits of the hydrogen atom, *ICCF-19, 19th Int. Conf. on Cond. Matter Nuclear Science*, Padua, Italy, 15/05/2015, to appear in *J. Condensed Matter Nucl. Sci.* **19** (2016).
[13] A. Meulenberg and J.L. Paillet, Nature of the deep-Dirac levels, *ICCF-19, 19th Int. Conf. on Cond. Matter Nucl. Sci.*, Padua, Italy, 15/05/2015, to appear in *J. Condensed Matter Nucl. Sci.* **19** (2016).

J. Condensed Matter Nucl. Sci. 19 (2016) 210–218

Research Article

Excerpts from Martin Fleischmann Letters

Melvin H. Miles*

University of LaVerne, LaVerne, CA 91750, USA

Abstract

There were many letters exchanged between Martin Fleischmann and this author that reveal Fleischmann's scientific thoughts about various cold fusion topics. These topics included possible critical factors for excess enthalpy effects in the palladium/deuterium system, errors in the CalTech, MIT and Harwell calorimetry, helium-4 production, critics such as Morrison, Taubes and Jones, and the advantages of using integrated forms of the calorimetric equations. Furthermore, Fleischmann mathematically shows that there is never a true steady state in the cell temperature for an open isoperibolic calorimetric system because both the cell heat transfer coefficient and the heat capacity of the cell change with time. Therefore, an important power term involving the change of the cell temperature with time has often been neglected by various groups using isoperibolic calorimetry. This power term is especially important when there are large changes in the cell temperature with time such as when the cell is first turned on.

Keywords: Calorimetry, Cell temperature, Critics, Excess enthalpy, Helium

1. Introduction

Martin Fleischmann was an outstanding scientist with a detailed grasp of chemistry, physics and mathematics that is seldom found in one individual. This is shown in a recent book about areas of science inspired by Martin Fleischmann [1]. Cold fusion is just one of the 19 chapters in this book on various topics, where Fleischmann made major contribution [1]. Eventually, Fleischmann's genius in cold fusion will be recognized by mainstream science despite the brutal criticism that he endured along with Stanley Pons. Fleischmann and Pons will then be in good company with Galileo, Joule, Semmelweis and other great scientists whose accomplishments were very slow in becoming accepted.

Fleischmann knew that James Prescott Joule had similar problems in gaining acceptance for his work some 150 years earlier. Joule studied the nature of heat and its relationship to work which led to the development of the first law of thermodynamics. At one point, Joule had to present his ideas in a church because they were rejected by his fellow scientists. In 1999, when there was much opposition to a cold fusion session that I had organized for an American Chemical Society (ACS) meeting in Ontario, California, Fleischmann proposed that we follow Joule's example and hold our session in a church. Fortunately, cooler heads eventually prevailed and our cold fusion session was permitted at this ACS meeting. However, we were moved to a room at a remote corner of the building.

*E-mail: melmiles1@juno.com

Only a small portion of these Fleischmann letters can be presented here, but a book is in preparation by the author that will present fully many of these letters. It is hoped that these letters will show Fleischmann's character and thoughts on various topics during the early history of cold fusion.

2. Possible Critical Factors for Cold Fusion

Prior to their 1989 announcement, the Fleischmann and Pons cold fusion work was funded by the US Office of Naval Research (ONR), managed first by Jerry Smith and later by Bob Nowak. I was also funded at China Lake by ONR for several years that included research in 1991 to test palladium materials made by the Naval Research Laboratory (NRL). None of these electrodes produced any evidence for the anomalous excess heat effect that I had previously observed using palladium rods purchased from Johnson-Matthey. In frustration, I sent Bob Nowak a list of questions to give to Fleischmann about possible critical factors for producing excess heat in the palladium/deuterium system. Fleischmann's answers to my questions were included with other topics in a confidential letter dated 21 January 1992. The following shows my 10 questions followed by Fleischmann's answers:

(1) *Heat treatment of palladium prior to the experiment?* "We do not heat treat the palladium prior to the experiment".

(2) *Anodic treatment of palladium prior to deuterium loading?* "We do not polarize the electrodes anodically prior to making calorimetric measurements. However, we have now carried out extensive investigations on the factors which control the deuterium loading and this includes anodic treatment. I wished I could say that we understand the complexities of this topic but in truth we do not".

(3) *Current density to use for deuterium loading?* "We always load at low to intermediate current densities, which is less than 100 mA cm^{-2} before we raise the current density. We believe this to be a critically important point and to the best of our knowledge, there are only three groups of people in the world who understand the reason for doing so which are Stan and I, EPRI/SRI and the National Fusion Centre in Nagoya.

The reason why we have talked about this matter the least is that our protocol is based on a surmise and we do not want to polarize people's thoughts. Equally, it affects the whole of our research strategy".

(4) *Effect, if any, of atmospheric CO_2?* "I do not know what the effects of atmospheric CO_2 might be. At the very least, it will complicate the calorimetry if CO_2 gets reduced at the cathode".

(5) *Effects of D_2O impurities such as Cu, H_2O?* "The impurities in the D_2O are probably of key importance and this goes for metals, as well as the borates and silicates. I wished I could tell you that we had this sewn up but we have not. However, we do believe that the blocking of the surface either by Under Potential Deposition (UPD) layers or by insoluble precipitates is very important. As you will see, I am sending a copy of this letter to Bob so I would just expand my comments here by telling you that it is Stan's and my view that as far as the bulk of the palladium is concerned, it is advantageous to produce a set of microelectrodes by suitable blocking of the surface".

(6) *Effects of impurities in lithium metal such as Hg, Sn, Zn, and Pb?* "The third impurity on your list gets into the D_2O anyway and we believe that this is one of the species which is implicated!"

(7) *Effects of sudden voltage or current changes?* "Sudden changes in the electrochemistry are very important. You will want to know that our strategy in Utah was oriented at developing a reproducible set of electrodes and cells on which we were then going to investigate systematically the effects of perturbations in temperature and perturbations in the electrochemistry. Unfortunately, our work there was terminated because the people at National Cold Fusion Institute (NCFI) did not buy any more D_2O. This is one topic which we have to start anew".

(8) *Effect of bubble patterns at the cathode?* "There is clearly a wide variation in the formation of bubbles at the cathode. If bubble evolution is irreproducible then this leads to noise in the electrode potential: the noise

levels for different cells are widely variable. You will be interested to know that this affects our data evaluation because the errors in fitting the black box model to the data can become totally dominated by the fluctuations in cell potential, rather than the time dependent change of temperature. We believe that Kalman filtering is therefore a better technique to use than the non-linear regression technique which we have so far used extensively".

(9) *Effect of impurities/additives in Pd such as Ce, Li, or Ag?* "I am sure you will know that our favored strategy now is to use alloys (which you have listed! – no prizes for guessing)".

(10) *Effect of Li-6 versus Li-7?* "We do not know this as yet."

After responding to my 10 questions, Fleischmann made some additional comments: "I would like to make some more general comments with regard to electrode materials. You will know that in our original measurements with palladium alone, we got good reproducibility with 1, 2 and 4 mm diameter rods but never had any excess heat with our original 8 mm rods. We believe that this showed Johnson Matthey that cold working was very important and in fact, we also believe that large diameter electrodes "do not work" because they crack. We have subsequently managed to get excess heat with specially prepared 8 mm rods (I presume that Johnson Matthey changed their procedures so that the rods were prepared in an analogous way to those having smaller diameters but our measurements with these rods proved to be irreproducible).

It is also true that a subsequent batch of 1, 2 and 4 mm diameter rods gave us irreproducibility and low levels of excess heat but we managed to recover more or less our original position with yet further batches of electrode materials.

Our promising results with alloys have been based on single batches of material and in view of our experience with the palladium electrodes, we now feel that we have to carry out a major investigation with repeated batches of alloy electrodes. Bob (Nowak) will tell you that Stan and I are in favour of making these materials widely available if and when we find that they give us reproducible results (but the decision is not up to us).

There is another important point which I must raise with you and that is most of our results are based on the first charging of the electrode. You may recall that at the Como conference, I said that people should give this information in future research. As of now, we do not know whether the first charge or any subsequent charge of the electrode will be optional. This information is obviously critically important but it will take several years to decide this point. I only hope that someone somewhere will provide the resources so that these long-term experiments can be initiated and maintained".

Note: Regarding Fleischmann's last comment above, several electrodes that gave excess heat in my experiments were cleaned and polished and then gave similar large excess heat effects in new experiments conducted both at China Lake and at the New Hydrogen Energy laboratory (NHE) in Sapporo, Japan.

3. Helium-4 Production in the Pd/D System

Fleischmann and Pons were actually the first to observe that helium-4 was produced in the Pd/D system. This was mentioned to me by Fleischmann in several conversations as well as in a letter dated 21 September 1993. Although this letter was mainly about other issues, including my problems with Steve Jones, Fleischmann states: "Next your Steve Jones saga. Of course he is terrified that people will come to accept heat and ^{4}He (incidentally, we had our first indication of ^{4}He in December 1988!)." In conversations with me, Fleischmann stated that because of the intense criticism following their March 23, 1989 announcement, they did not want to open up a second front about helium-4 production.

At a SPAWAR (Navy) meeting in San Diego in 2000, Fleischmann mentioned an experiment in France that produced excess heat for several months. According to Fleischmann, the excess heat production eventually ended because the palladium lattice became choked with helium-4. Fleischmann always considered that the excess heat in the Pd–D system was "a bulk effect mediated by surface reactions." My communications with both Fleischmann and Pons suggests that considerable work on helium-4 production was performed at their laboratory in France (IMRA Europe). It is hoped that the results of this research will someday be released.

My China Lake correlation of heat and helium was performed in the fall of 1990 with results first reported in 1991 [2,3]. I was not aware at this time of Fleischmann and Pons results for helium-4. In fact, I was expecting possible helium-3 based on Schwinger's suggestion of the D+H→ He-3 reaction that produces no neutrons or gammas. However, only helium-4 was detected. These experiments were conducted entirely at China Lake except for the helium-4 measurements on the gas samples sent to the University of Texas. I allowed the initial paper [3] to be released through the University of Texas because I was concerned about enemies of cold fusion research at China Lake who may have blocked this publication. Further research at China Lake confirmed our earlier heat/helium results [4–6].

Research by others followed that also showed helium-4 production in the Pd/D system. This includes results of McKubre, Gozzi, DeNinno, Arata and others [7]. In my opinion, any correct theory for cold fusion in the Pd/D system would have to explain the production of helium-4, and many theories do this. However, new theories are sometimes proposed that apparently consider the production of helium-4 to be some mistake [8]. After all these years, I do not find any errors in my heat and helium measurements. It is very unlikely statistically that my heat-helium results could have been produced by random errors [5,6].

4. Comments About Cold Fusion Critics

The most vocal critics following the 1989 announcement included Douglas Morrison, Gary Taubes and Steven Jones. Especially harsh personal statements were also made very early after the announcement by Nate Lewis and Steve Koonin of CalTech [9]. However, Fleischmann never expressed anger in any of his comments about these critics. Only a few comments from my Fleischmann letters will be given here.

4.1. Douglas R.O. Morrison

Douglas Morrison of CERN attended most of the early ICCF conferences and wrote biased reports about cold fusion that were sent to many scientists. The Fleischmann's letters of April and May of 1992 discussed possible ways to respond to Morrison's "Cold Fusion update No. 6" dated 5 April 1992. Fleischmann wanted to respond to Morrison in an accepted scientific manner and stated in one letter: "And come on Douglas: Write some papers for peer-reviewed journals". Morrison eventually submitted a paper to Physics Letters A, but this was rejected by the editors. Fleischmann sent me a copy of this submitted Morrison paper titled "Comments on Claims of Excess Enthalpy by Fleischmann and Pons Using Simple Cells Made to Boil". Fleischmann also sent me their reply to this Morrison paper that was included with his 21 September 1993 letter. Fleischmann stated in this letter: "With hindsight the outcome was predictable: the editors refused to accept Douglas Morrison's critique which they regarded as being too polemical and asked him to submit a shortened non-polemical comment. Naturally, this has once again deprived us of a "right to reply". (Note: These two papers have now been published: see *Infinite Energy* **21**(124) (2015) 17–27.)

4.2. Gary Taubes

In an article about Fleischmann's cold fusion seminar in 1991 at CalTech, Gary Taubes made some false statements about my China Lake helium-4 results (Science, News and Comments, 13 December, p. 1582, 1991). The editors limited my reply to a letter of only two paragraphs (Science, Letters, 13 March, p. 1335, 1992). Fleischmann states in his letter of 23 March 1992: "I find it really remarkable that people like Gary Taubes can have access to such Journals with uncorroborated statements (others might say lies) while reasoned arguments do not prevail". I had previously sent Fleischmann my full reply that covered several important topics, but only a small portion was published.

4.3. Steven Jones

Fleischmann obviously did not trust Steven Jones of Brigham Young University (BYU), and thought that Jones acted as a reviewer on a 1988 Fleischmann–Pons proposal and then used the information in the proposal to advance his own cold fusion experiments. Fleischmann made the following comments in his 21 September 1993 letter: "He (Jones) has a total blockage about the excess enthalpy measurements because of his disbelief of our results in 1988/Spring 1989 that precipitated the unfortunate chain of events leading to the Press Conference". Fleischmann also stated the following about the referee's comment: "Did we ever tell you what happened when we had the referees' comments on our application? I said to Stan: 'Hey Stan, this is Steven Jones and if we answer question X, then we tell him why there may be fusion in the earth and more likely Jupiter; if we answer question Y, we tell him how to set it up in the laboratory: Prophetic words." I also had problems with Steven Jones becoming a major critic of my work after inviting me to give a seminar at BYU that lasted more than three hours due to his many probing questions.

5. CalTech, Harwell and MIT Calorimetry

The failure of CalTech, Harwell and MIT to report any evidence for excess heat in their 1989 experiments on the Pd/D system was a major setback for the acceptance of cold fusion. In a 10 April 1992 letter, Fleischmann stated that the cold fusion publications from these institutions "were very influential in creating a negative climate of opinion – not because of their excellence but because of the reputations of CalTech, Harwell and MIT." In a 1 March 1995 letter, Fleischmann states "that the trio of results at MIT, CalTech and Harwell still pose a tremendous block to further work." In several letters in 1992, Fleishmann was trying to have a committee of scientists set up to carefully examine the calorimetry of various groups including that of Fleischmann and Pons along with CalTech, Harwell, and MIT. If this examination were carried out correctly, Fleischmann knew that his calorimetry would prove to be superior. Unfortunately, such a committee was never established. However, major errors for the CalTech, Harwell, and MIT calorimetry have been reported [10–12]. Furthermore, these three groups never reported obtaining the high D/Pd ratio required for producing excess heat or any other cold fusion effects [1].

Fleischmann's detailed letter of 2 April 1992 states: "As you will see we are in favour of reviewing the published material and we do have a set of questions which we believe should be considered". This letter lists 12 questions for the CalTech calorimetric papers and 29 questions for the Harwell calorimetry. These questions refer to possible errors based on the published papers.

One example of a calorimetric error will be presented here. An important term missed by CalTech, Harwell and MIT [10,11] is the power due to the rate of change of the cell temperature with time: $P_{\text{calor}} = C_{\text{p}}M\,\mathrm{d}T/\mathrm{d}t$, where $C_{\text{p}}M$ gives the heat capacity of the calorimeter (JK^{-1}) and $\mathrm{d}T/\mathrm{d}t$ gives the change in the cell temperature with time (Ks^{-1}). This term is especially important when the electrolysis is first initiated, when there is an applied internal heater, when there is a change in the applied cell current, when D_2O is added and when the cell is turned off. These important periods can be accurately modelled for isoperibolic calorimetry using $P_{\text{calor}} = C_{\text{p}}M\,\mathrm{d}T/\mathrm{d}t$ and should never be neglected. Several of Fleischmann's questions of the William's paper [13] relate to the strange endothermic behavior shown in William's Figs. 2(b–e) on p. 379. For example, Fleischmann's Question 18 for Fig. 2(e) in the William's paper states: "Why was the excess enthalpy for the dissolution of D in Pd not seen and what could be the cause of the large negative enthalpy over the first two hours?" The answer is William's neglect of the P_{calor} term which would be large and positive when the cell is turned on and the contents heat up.

This same error in neglecting P_{calor} has also frequently been made by noted cold fusion scientists. For example, Storms erroneously states that the temperature of an isoperibolic cell needs to be constant before the values have any meaning [14]. Actually, there is never a steady state where $\mathrm{d}T/\mathrm{d}t = 0$ for an open isoperibolic calorimeter, thus the P_{calor} term is always required. The positive and negative false excess power results shown in several of Storm's figures are due to the neglect of this P_{calor} term [14].

6. Temperature Versus Time for Isoperibolic Calorimetry

Fleischmann was foremost a consummate mathematician in developing models of electrochemical and other systems. Therefore, we must briefly enter this world. A long Fax was received from Stan Pons on 8 July 1992 that contained 27 pages of Fleischmann's modeling equations. In regard to Fleischmann's notes, Stan stated: "It should be possible to predict the temperature–time data for whole sequences of measurements (over periods of months) and for various modes of operation (including the Harwell type of experiments). This type of modelling should also really facilitate data evaluation and improve the accuracy."

Fleischmann's model of his open isoperibolic calorimetry where the electrolysis gases (D_2, O_2) escape along with D_2O vapor is given by the differential equation

$$\mathrm{d}\Delta\theta/\mathrm{d}t + (\gamma - yt)\Delta\theta/(1-\beta t) = \alpha/(1-\beta t). \tag{1}$$

Note that $\Delta\theta = T - T_\mathrm{b}$, thus $\mathrm{d}\Delta\theta/\mathrm{d}t = \mathrm{d}T/\mathrm{d}t$ if the bath temperature is constant. All other terms are defined in the Appendix. If the heat transfer coefficient and $C_\mathrm{p}M$ are assumed to be constant, then both $y = 0$ and $\beta = 0$. Eq. (1) then becomes

$$\mathrm{d}\Delta\theta/\mathrm{d}t + \gamma\Delta\theta = \alpha. \tag{2}$$

Equation (2) can be readily integrated to give

$$\Delta\theta = (\alpha/\gamma)[1 - \mathrm{e}^{-\gamma t}]. \tag{3}$$

If only the heat transfer coefficient is assumed to be constant, then only $y = 0$ and Eq. (1) becomes

$$\mathrm{d}\Delta\theta/\mathrm{d}t + \gamma\Delta\theta/(1-\beta t) = \alpha/(1-\beta t). \tag{4}$$

This equation can also be integrated using known integrals to yield

$$\Delta\theta = (\alpha/\gamma)[1 - (1-\beta t)^{\gamma/\beta}]. \tag{5}$$

However, neither the cell heat transfer coefficient nor the cell heat capacity remains constant in the open isoperibolic calorimeter used by Fleishman and Pons, thus Eq. (1) must be integrated in order to obtain the correct cell temperature–time behavior. There are no known integrals for Eq. (1), and Fleischmann's note contains many pages of mathematics to obtain the integrated equation for the correct temperature–time behavior. Fleischmann's final integrated solution to Eq. (1) is given by

$$\Delta\theta = \alpha[(\gamma - yt)^{-1} - \gamma^{-1}(1-\beta t)^{\varphi}\mathrm{e}^{-yt/\beta}], \tag{6}$$

where $\varphi = \beta^2(y - \gamma\beta)^{-1}$.

Equations (3), (5) and (6) all show that the cell temperature increases nearly linearly with time, $\Delta\theta = \alpha t$, when t is less than about 1000 s. The correct equation, Eq. (6), shows that there is never a true steady state for the cell temperature even over long time periods.

Figure 1 shows the cell temperature versus time for the first three hours for Eqs. (3), (5) and (6). The initial increase in the cell temperature is nearly linear with time with the slope $= \alpha$. The differences between these three equations are small over the first three hours and undetectable in Fig. 1.

The results are quite different for the first eleven days as shown in Fig. 2. The black line shows the correct behaviour corresponding to Eq. (6). There is never a steady state where the cell temperature remains constant. In most experiments, the cell would be refilled every two or three days (3 days = 259,200 s).

Although the results for Eqs. (3) and (5) shown in Fig. 2 are very similar, Eq. (5) never yields a true steady-state temperature. The difference between the cell and bath temperature, $\Delta\theta$, can eventually become zero or even negative

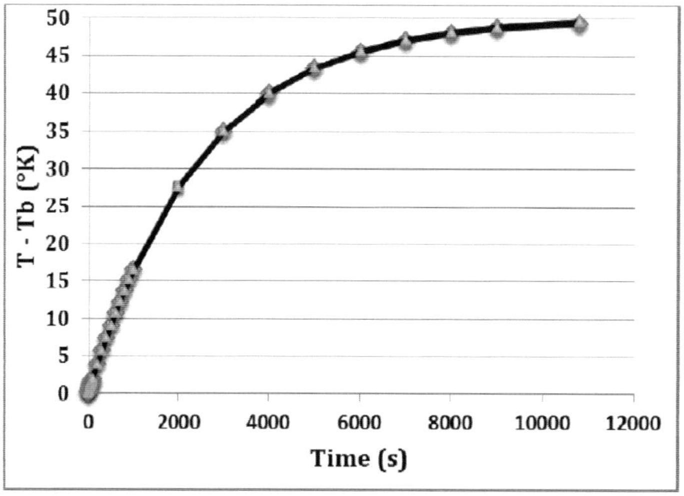

Figure 1. Cell temperature versus time for the first three hours of cell operation based on Eqs. (3), (5) and (6). The small differences in the three equations are not readily detectable for this time period.

based on Eq. (5). However, the electrolyte would become depleted during such large time periods, and Eq. (5) would no longer apply.

The very small effects of changes in the cell heat capacity, as shown in Fig. 2, likely led Fleischmann to start using

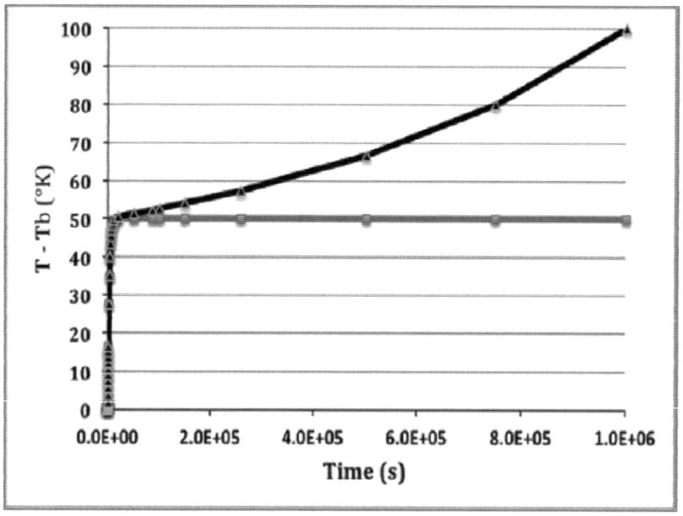

Figure 2. Cell temperature versus time for the first eleven days of cell operation based on Eqs. (3), (5) and (6). The black line gives the correct temperature-time behavior based on Eq. (6). The small differences between Eqs. (3) and (5) are not readily detectable for this time period.

the approximation of $C_p M\, d\Delta\theta/dt$ rather than the more exact term expressed by [15]

$$C_p d/dt(M\Delta\theta) = C_p M d\Delta\theta/dt + C_p \Delta\theta dM/dt, \quad (7)$$

where

$$M = M^o - (1+\beta')It/2F \quad \text{and} \quad dM/dt = -(1+\beta')I/2F. \quad (8)$$

Note that β' is a dimensionless term that accounts for the loss of water (H_2O or D_2O) due to evaporation or any causes other than electrolysis alone, and M is the equivalent moles of H_2O or D_2O that would give the total heat capacity of the calorimeter [15].

7. Summary

These excerpts from his letters show Martin Fleischmann's noble character and his scientific greatness. Fleischmann's mathematic skills are obvious from his modelling of the isoperibolic calorimetric system that should someday be recognized and applied to various electrochemical systems.

Acknowledgements

This work was supported from an anonymous fund at the Denver Foundation. An adjunct faculty position at the University of LaVerne is also acknowledged. The Author also thanks David L. Miles for the preparations of Figs. 1 and 2.

Appendix

The terms used by Fleischmann In Eq. (1) are defined as follows:

$$\alpha = Q_1/C_p M^\circ (\mathrm{Ks}^{-1}) \quad \text{where} \quad Q_1 = (E - E_H)I + Q_f, \quad (A.1)$$

$$\beta = I/2FM^\circ \ \mathrm{s}^{-1}, \quad (A.2)$$

$$\gamma = (4k'_R\theta_b^3 + \psi I/\theta_b)/C_p M^o \ (\mathrm{s}^{-1}), \quad \text{where} \quad \psi = E - E_{H}, \quad (A.3)$$

$$y = 2k'_R\theta_b^3(1+\lambda)I/FC_p(M^\circ)^2 \ (\mathrm{s}^{-2}), \quad \text{where} \quad \lambda = 0.0909, \quad (A.4)$$

$$\Delta\theta = T - T_b(\text{or } \theta - \theta_b)(\mathrm{K}). \quad (A.5)$$

Typical values used for the Fleischmann–Pons Dewar calorimetry in 1992 are $\alpha = 1$ to $2 \times 10^{-2}\mathrm{Ks}^{-1}, \beta = 1.0 \times 10^{-6}\mathrm{s}^{-1}$ for $I = 0.80\mathrm{A}, \gamma = 4 \times 10^{-4}\mathrm{s}^{-1}$, and $y = 2 \times 10^{-10}\mathrm{s}^{-2}$. Most of these terms have been defined elsewhere [10,11,15].

References

[1] M.H. Miles and M.C.K. McKubre, Cold fusion after a quarter-century: The Pd/D system, Chapter 13, in *Developments In Electrochemistry: Science Inspired by Martin Fleischmann*, D. Pletcher, Z.-Q. Tian and D.E. Williams (Eds.), Wiley, West Sussex, UK, 2014.

[2] M.H. Miles, B.F. Bush, G.S. Ostrom and J.J. Lagowski, Heat and helium production in cold fusion experiments, in *The Science of Cold Fusion, ICCF-2 Proceedings*, T. Bressani, E. Del Giudice and G. Preparata (Eds.), Italian Physical Society, Vol. 33, Bologna, Italy, 1991, pp. 363–372.

[3] B.F. Bush, J.J. Lagowski, M.H. Miles and G.S. Ostrom, Helium production during the electrolysis of D_2O in cold fusion experiments, *J. Electroanal. Chem.* **304** (1991) 271–278.

[4] M.H. Miles, R.A. Hollins, B.F. Bush, J.J. Lagowski and R.E. Miles, Correlation of excess enthalpy and helium production during D_2O and H_2O electrolysis using palladium cathodes, *J. Electroanal. Chem.* **346** (1993) 99–117.

[5] M.H. Miles, B.F. Bush and K.B. Johnson, Anomalous effects in deuterated systems, NAWCWPNS TP 8302, September 1996.

[6] M.H. Miles, Correlation of excess enthalpy and helium-4 production: a review, in *Condensed Matter Nuclear Science, ICCF-10 Proceedings*, P.L. Hagelstein and S.R. Chubb (Eds.), World Scientific, New Jersey, 2006, pp. 123–131.

[7] E. Storms, *The Explanation of Low Energy Nuclear Reaction*, Infinite Energy Press, Concord, NH, USA, 2014, pp. 30–40.

[8] G. Hubler, J. Aviles, V. Violante and A. Torre, On a possible cosmological explanation for the anomalous heat effect, *ICCF-19 Presentation, Abstract A011*, 13 April 2015.

[9] C.G. Beaudette, *Excess Heat: Why Cold Fusion Research Prevailed*, 2nd Edn., Oak Grove Press, South Bristol, Maine, 2002, pp. 73–76.

[10] M.H. Miles, B.F. Bush and D.E. Stillwell, Calorimetric principles and problems in measurements of excess power during Pd–D_2O electrolysis, *J. Phys. Chem.* **98** (1994) 1948–1952.

[11] M.H. Miles, Examples of isoperibolic calorimetry in the cold fusion controversy, *J. Condensed Matter Nucl. Sci.* **13** (2014) 392–400.

[12] M.H. Miles and P.L. Hagelstein, New analysis of MIT calorimetric errors, *J. Condensed Matter Nucl. Sci.* **8** (2012) 132–138.

[13] D.E. Williams et al., Upper bounds on cold fusion in electrolysis cells, *Nature* **342** (1989) 375–384.

[14] E. Storms, *The Science of Low Energy Nuclear Reaction*, World Scientific, New Jersey, 2007, pp. 160–164.

[15] M. Fleischmann, S. Pons, M.W. Anderson, L. Li and M. Hawkins, Calorimetry of the palladium–deuterium heavy water system, *J. Electroanal. Chem.* **287** (1990) 293–348.

J. Condensed Matter Nucl. Sci. 19 (2016) 219–229

Research Article

High Energy Density and Power Density Events in Lattice-enabled Nuclear Reaction Experiments and Generators

David J. Nagel*

NUCAT Energy LLC and The George Washington University, Washington DC 20052, USA

Alex E. Moser

Nova Scientific Inc., Sturbridge, MA 01566, USA

Abstract

The rapid release of energy from Lattice Enabled (or Low Energy) Nuclear Reactions is of interest for three reasons. First, it constrains and challenges theories about the mechanism(s) active in producing LENR. Next, it might heavily influence the design of heat and electrical generators based on LENR, since they have to be safe for use by a wide variety of people. Finally, there has long been interest in whether or not LENR could be used to augment existing weapons or produce to entirely new weapons. This paper first reviews reports in the literature of meltdowns or explosions that might have been caused by LENR. Then, each of the three areas cited above is examined. It is clear that reported high energy or high power events will heavily impact theories about LENR and the development of safe products. It is unclear now if LENR will be weaponized in any form. Control of the initiation of explosive LENR events is obviously necessary for that possibility.

Keywords: Craters, Explosions, High-power, LENR, Meltdowns

1. Introduction

Meltdowns and explosions have already been observed in LENR experiments and tests of prototypes of products. Analyzing and understanding those events are relevant to scientific understanding of LENR, the operational safety of generators based on LENR and, possibly, the weaponization of LENR. The major scientific question is whether or not chain reactions can occur in LENR systems. The development of safe products based on LENR is mandatory, if they are going to be used widely. Using LENR in or for weapons has already been studied and, undoubtedly, will continue to get attention.

*E-mail: nagel@gwu.edu

© 2016 ISCMNS. All rights reserved. ISSN 2227-3123

Sections 2–5 review instances of the fast release of energy in LENR experiments. Then three sections deal with the influence of such events on the science, commercialization and weaponization of LENR. A few comments on research needed to address the fast release of energy by LENR are in Section 9.

2. Fleischmann–Pons Meltdown

Fleischmann and Pons started their experiments on loading deuterons into palladium long before the controversial press conference in March of 1989. They were already performing such experiments more than four years earlier. In or about February of 1985 they were running an electrochemical experiment in which the cathode was a cube of Pd 1 cm on a side. One morning, they came to laboratory to find that the experiment was destroyed. They described the situation in a journal article [1]. "We have to report here that under the conditions of the last experiment, even using D_2O alone, a substantial portion of the cathode fused (melting point 1554°C), part of it vaporized, and the cell and contents and a part of the fume cupboard housing the experiment were destroyed." Kevin Ashely was a graduate student with a laboratory nearby that of Fleischmann and Pons. He is quoted in the book by Charles Beaudette [2]. "The bench was one of those black top benches that was made of a very, very hard material. I was astonished that there was a hole through the thing. The hole was about a foot in diameter." Chase Petersen was President of the University of Utah at that time. He wrote in his autobiography [3]. "There was a hole in the lab counter top it had occupied but no evidence of an explosion. Beneath the countertop was a hole in the cement floor some eight inches wide and three or four inches deep." Many of us found that report to be incredible. Could one cubic centimeter of even superheated Pd liquid cause such damage to concrete after burning through the experimental cell and the benchtop? Later, Larry Forsely wrote [4]. "I last discussed this with Martin Fleischmann at his home in 2007. He stated that the damage to the concrete floor in the bottom of the cabinet was the size of the tip of his thumb and about 1/2 inch deep. He acknowledged that the structure was destroyed, with a fist size hole in the benchtop... Both he and Stan Pons were scared that they would be shut down and worse." This level of damage seemed more reasonable. However, it was still remarkable.

Jed Rothwell offered his view of the event in his book "Cold Fusion and the Future" [5]. "Unfortunately we know nothing about the nature of the February 1985 explosion, because Fleischmann and Pons did not keep any physical evidence from this event, such as a piece of the burned table or scraps of the exploded cathode. In my opinion, this was irresponsible and unprofessional. Fleischmann ruefully agrees." Despite variations in reports of the effects of the meltdown, and the lack of either physical evidence or photographs, it happened and is significant.

As can be seen from these reports, there was confusion concerning the nature of the event in the Fleischmann–Pons laboratory. Both melting (fusing) and exploding were mentioned. However, it is clear now that the event was a meltdown. An explosion would not have produced damage to the bench and floor only directly under the experiment. It is understandable why Fleischmann and Pons did not want photographs of the damage to their laboratory circulated. However, it is very regrettable that such photographs are not available for scientific analysis. Further, it is odd that no one seems to have tried to replicate the experiments using cubic cathodes of Pd. Doing experiments with cube-shaped cathodes of varied size, say 2, 4, 8 and 16 mm, could be very instructive. Of course, they would have to be done in a safe place, probably on the floor of a bare concrete room. Full-time video monitoring would be easy, with a high speed camera available to be triggered by any sudden temperature rise.

It seems surprising that there are no published analyses of the Fleischmann–Pons meltdown, given its importance. It is not difficult to make some simple estimates. The cathode in the experiment was a cube of Pd 1 cm on a side (11.9 g). It contained 6.7×10^{22} Pd atoms, so if it was fully loaded with deuterons, there would be that number of deuterons in the cube. Assume that deloading the deuterium atoms from the Pd cube takes no energy. The oxidation of that number of deuterons would produce 12.8 kJ. It takes 0.53 kJ/g to melt Pd, taking into account the specific heat and heat of fusion. Hence, 12.8 kJ would melt 24.2 g of Pd. However, the dynamics of the situation would prevent that from happening. That is, the release of the deuterium might have been too slow, at least initially, to get the Pd to its melting point. And, if the deuterium were burned on the surface of the cube, only a fraction of the heat of combustion would have been transferred into the Pd cube.

The 1985 meltdown should be simulated using the temperature dependent properties of palladium, with various assumptions about the location, duration and magnitude of the LENR energy release. An estimate of the total amount of energy required to produce the observed effects, and possibly an upper limit on the duration of the release, might result.

3. Explosions Attributed to LENR

LENR experiments often involve high pressures. This is true of both electrochemical experiments and gas loading approaches to LENR. Hence, it is possible for explosions to occur because of the sudden release of pressure, an entirely mechanical event. In electrochemical experiments accumulation of gaseous hydrogen and oxygen above an operating cell can produce a dangerous mixture. Ignition leads to a rapid, explosive chemical reaction. Neither of these types of events is due to rapid release of energy by LENR. Two such events have been reported, one at SRI International [6] and the other at the University of Hokkaido [7].

There are three published reports of explosions during experiments that were attributed to LENR. The earliest was published in 1992 by Xinwei Zhang and six Collaborators [8]. Their setup is shown in Fig. 1. They were using a hollow Pd cathode 8 cm long with a 1.67 mm OD and 500 μm wall thickness. Three explosions in April of 1991 occurred with no one present. In two cases, the rubber stopper and electrodes were blown 1.5–2 m away, and the bottom of the cell was blown out. The water bath temperature increased 5°C. The authors wrote "... the real explosion cannot be caused by chemical reaction" and "The heat burst in the explosion reached MW per cm^3 Pd." That is, their estimate of the power density from the explosions was very high. In the abstract of a recent paper on thermal analyses of these explosions, five of the scientists state "The average power was greater than 6.7 W (65W cm^{-3} Pd or 430% of input power). Thermal analysis indicates that the power of the explosion was 5.1–5.5 kW (or 50–53 kW cm^{-3} Pd), and the event developed in 2–17 s" [9].

A second explosion that could only be attributed to LENR was reported by Mizuno and Toriyabe in 2006 [10]. They were using a tungsten cathode 1.5 mm in diameter and 29 cm long. A length of 3 cm was in 0.2 M K_2CO_3 H_2O electrolyte. The experiment is shown schematically in Fig. 2. The cell with 700 ml of electrolyte heated from 30°C to 80°C in 10 s, and then exploded. The authors took into account the heat capacity of the electrolyte and the measured hydrogen production. They stated that "The estimated heat output was 800 times higher than the input..." This is the highest reported energy gain in an LENR experiment. Gains in the range from 80 to 415 were reported by Focardi and Rossi [11]. Other, lesser reports of energy gains are available and were reviewed recently [12]. However, even the smaller reported energy gains are adequate for the commercial success of LENR heating units.

Another explosion attributed to LENR was reported by Biberian in 2009 [13]. He was using a hollow Pd cathode 10 cm long, 2 mm OD and 200 μm wall thickness. The experiment had been running for 30 days with 0.1 M LiOD in D_2O prior to explosion. The setup and the aftermath of the explosion are shown in Fig. 3. Biberian attempted twice to replicate the observed damage by inducing chemical explosions of hydrogen and oxygen in a similar Dewar, but did not shatter it. He concluded that "It is very likely that under some not yet understood conditions, chain reactions occur in highly loaded palladium samples giving rise to an explosion." Biberian's experience was consistent with, but not proof of chain reactions.

It is curious that two of the three explosions, which were attributed to LENR by the experimenters, occurred when hollow Pd cathodes were being used. This might be nothing more than an odd coincidence. However, it does suggest that more experiments be performed with tubular Pd cathodes. Such geometries have the added advantage of being able to insert sensors inside of an active cathode during electrolysis.

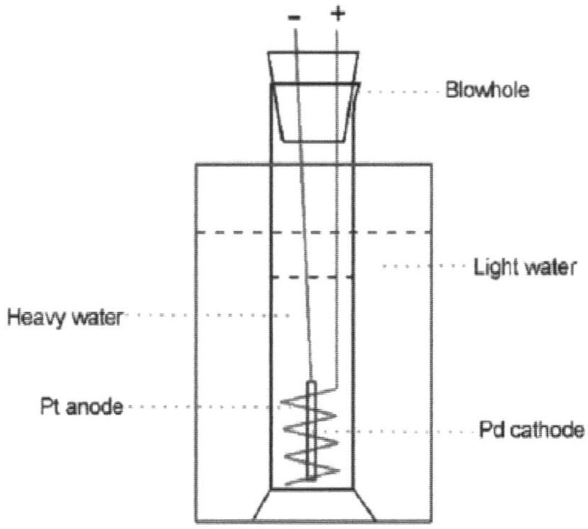

Figure 1. The experimental setup of Zhang and his colleagues .

The three explosions just discussed are not the only ones reported during LENR experiments. There have been other reported explosions, although not with detailed documentation. Hence, it is unclear whether the events were due to LENR or other causes. Krivit wrote [14] "...a source who wishes to remain anonymous states that the Lawrence Livermore National Laboratory had a Fleischman–Pons type explosion in 1989, as well." This leaves open the question of whether the event at that laboratory was a meltdown or, as reported, an actual explosion. Recently, there was an explosive event in the laboratory of the MFMP [15]. "An attempt by the Martin Fleischmann Memorial Project to

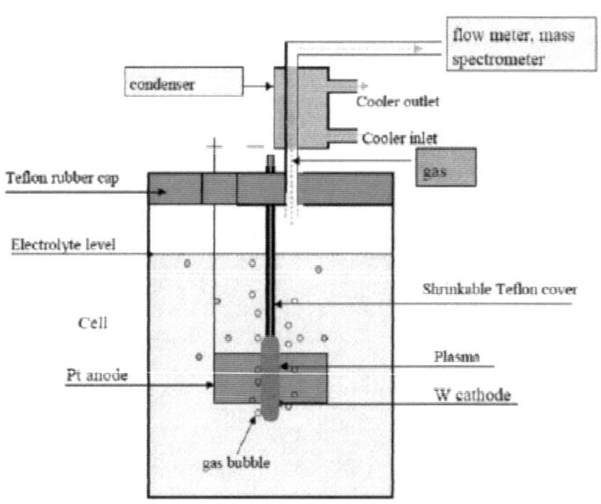

Figure 2. Drawing of the experiment of Mizuno and Toriyabe.

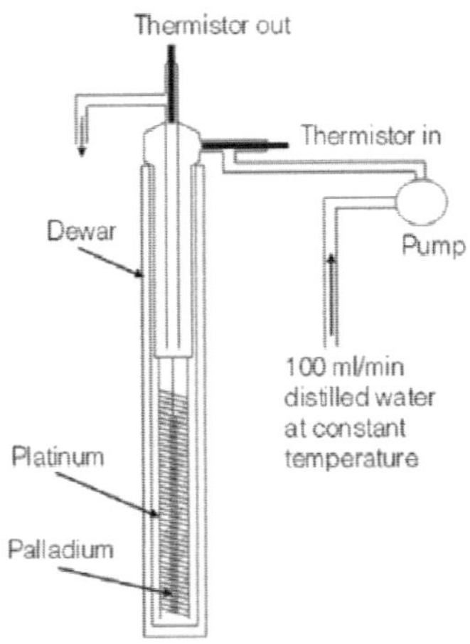

Figure 3. Schematic of Biberian's experiment and a photograph after it exploded.

replicate E-Cat, Andrea Rossi's alleged cold fusion reactor ended explosively yesterday after the reactor heated to over 1000°C." It is unclear whether this was due to LENR or high pressure or simply thermal stresses in the material. There have been unpublished reports of runaway events in LENR and related experiments by Francesco Piantelli, Stanislaw Szpak, Mark Snoswell, Vince Cockeram and Brian Ahern. It is very possible that other unreported meltdowns and explosions have occurred in LENR experiments. However, the fact that competent scientists have had explosive events occur during LENR experiments, and that they attributed them to LENR, is significant for scientific, commercial and, possibly, military reasons.

4. Explosions During Tests by Rossi

It appears that Rossi ran numerous tests of his E-Cat devices roughly during the time period from 2007 to 2009. There have been reports of explosions during such tests. In an interview [16], Christos Stremmenos stated that there were explosions during E-CAT tests in Bologna, fortunately all at night. Hank Mills wrote [17] "Andrea Rossi has stated many times in the past a self-sustaining system is dangerous, and there is a chance of explosions. He actually indicated that during stress testing of systems he has witnessed dozens of explosions". In response to a question by Hank Mills, Rossi wrote in his blog "We have seen explode hundreds of reactors..." [18]. It is clear that Rossi has paid a great deal of attention of control of his various E-Cat systems. In the first Levi report [19], control was achieved by the use of a square electrical waveform. When power was applied, the LENR reaction rate took off. Then, when the power was turned off, the reaction rate and heat production diminished. Chapter 12 of the book by Ventola and Nikolova is entitled "Control of Hot-Cat Reactions". It is the best available discussion of approaches to the willful variation of the output of Rossi's systems [20]. The absence of recent reports on E-Cat explosions indicates either they have been kept secret or have not occurred. The latter possibility seems the more likely.

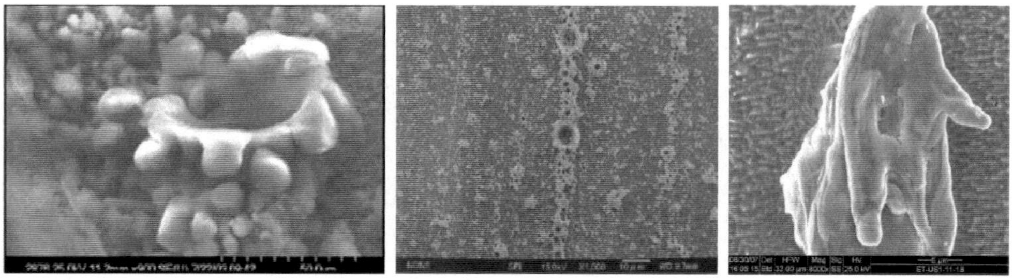

Figure 4. Examples of craters in the cathodes after electrochemical LENR experiments, which show evidence of melting. Sources of these images are cited in the text.

5. Crater Observations, Analyses and Simulations

Small craters in the surfaces of cathodes from LENR electrochemical experiments have been observed by several investigators. They generally range from about 1 to 100 μm in diameter. Many craters show evidence of melting and, hence, high temperatures. Three examples are exhibited in Fig. 4. They are, left to right, co-deposition of Pd with D [21], a gold surface with craters aligned along scratches [22], and Pd subjected to simultaneous super-wave and ultrasonic excitations [23].

Recent papers have provided analyses of the characteristics of craters. The first, which reviewed the literature on craters, determined that the energy required to form craters varies from 1 nJ (for 1 μm diameter) to 1 mJ (for 100 μm diameter) [24]. The next advance was by Ruer, who provided an analytical model for the dynamics of crater formation [25]. He found that crater formation times in the range of nanoseconds to microseconds.

It is instructive to consider a crater with an intermediate diameter of 10 μm. The energy required for formation of such a crater is about 1 μm J, and the eruption time is near 100 ns. These values give an averaged power density of 8×10^{10} W/cm^3. This is about one order of magnitude greater than the power density due to the explosion of TNT, namely 7×10^9 W/cm^3 [26]. It must be noted that the size of the power release volume could be *much* smaller. That is, using the entire volume of the final crater gives a very low power estimate. Taking the eruption time rather than the (unknown) energy release time gives a very conservative (low) estimate of the power density. More realistic estimates of the geometries, energetics and dynamics of crater formation can only be made using modern materials simulation tools.

The first full dynamical simulation of crater formation was conducted using the CTH Shock Physics Code [27] from Sandia National Laboratories. In the simulation, 680 μJ were released in 10 ns within 0.5 μm radius sphere 10 μm beneath a Ag surface. The simulation was conducted on a Dell Z800 high performance computer using 22 cores. A simulation spanning 520 ns in time took 115 h of computer time to complete. The results of the simulation for three times are shown in Fig. 5. The images show that it took almost 40 ns for the effect of the energy release to propagate to the surface of the material. The event was far from complete by 360 ns. Pressures as a function of location and time can also be obtained from CTH calculations. It is also possible to determine the ultimate disposition of the initially released energy using that code.

Simulations using the CTH code should be done as a function of several parameters. They include the amount of energy and the release time, that is, the initial power. The size (volume) of the energy release region is an important parameter since it determines the initial power density. Probably, the precise shape of the initial energy release volume is not important, as long as it is relatively equiaxed. That is, long and thin, or else flattened, energy release volume shapes could lead to different simulation results, but might not be physically realistic. The location of the release

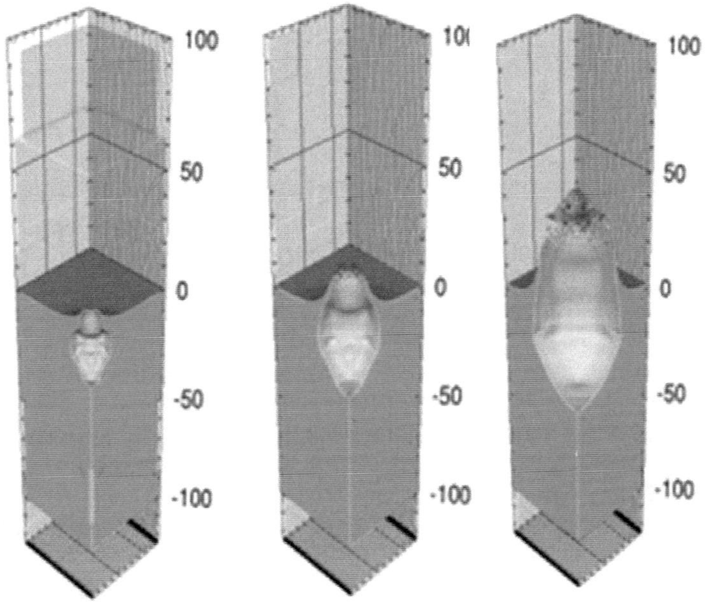

Figure 5. Three snapshots at 40, 120 and 360 ns during the evolving dynamics after release of energy in a small volume under the surface of a sample of silver. Only one-fourth of the entire simulation volume is shown for clarity. See text for details and discussion.

volume, either on the surface of the material or at some depth into the material, is an important variable. A basic issue in LENR remains whether energy is released only on the surface of materials, or only in the bulk, or in either type of location. It is hoped that comparison of the results of simulations with the energy release depth as a parameter, might contribute to the solution of that question. Clearly, several parameters should be exercised with CTH simulations, most of them over quite wide ranges of values. Variables include the material in which the craters are formed.

This completes the review of high energy and power density events in LENR experiments and prototype generators, and some work on their computational simulation. We now turn to the implications of these events for the LENR science, commercialization and potential weaponization.

6. Scientific Considerations

The scientific implications of fast LENR energy releases can be considered on the basis of some simple calculations. We start with crater formation. For a 10 μm crater, 1 μJ $= 6 \times 10^{12}$ eV or 3×10^5 20 MeV LENR events. Consider 10^6 LENR events on a cubic grid, 10^2 events on a side. The density of energy and power depend on the spacing between events, which can be measured in Pd lattice constants. Results for a wide variation in spacing are given in Table 1.

It is seen that, even for widely spaced events (3.9 μm) the energy density is very high. This can be appreciated by noting that the energy density of TNT $= 0.74 \times 10^4$ J/cm^3 [28]. The LENR energy density reported in the second report on E-Cat testing by the Levi et al. team about 2×10^9 J/cm^3. The LENR energy densities within craters might be dramatically high, if the energy release volumes are small, as indicated by the estimates in Table 1.

Table 1. Computed energy densities as a function of the spacing between LENR events.

Pd lattice constants between LENR	Side cube (nm)	Volume of cube(cm^3)	Energy density (1 μ J) (J/cm^3)
1	0.39	6×10^{-23}	1.6×10^{16}
10	3.9	6×10^{-20}	1.6×10^{13}
100	39	6×10^{-17}	1.6×10^{10}
1000	390	6×10^{-14}	1.6×10^{7}
10 000	3900	6×10^{-11}	1.6×10^{4}

One of the most fundamental questions about LENR is whether or not they can occur in "chain reactions", similar to fission reactions, where one reaction induces others. If not, then the fast, high-density release of LENR energy implied by crater formation and other events must be due to many reactions being nearly simultaneous in space and time. Curiosity about LENR chain reactions is old. In 1989, P.K. Iyengar wrote "Occasionally, nuclear events do appear to take place wherein over 100 neutrons are generated in a single sharp burst…this leads to the intriguing conclusion that a chain reaction involving as much as 10^{10} fusion reactions occurs within a time span of 100 μs"[29]. In 1996, Arata and Zhang provided a schematic sequence for nuclear reaction processes in solids [30]. It included "Generation and Propagation Process ('chain reaction')". At ICCF-15, Srinivasan reviewed the early work on "cold fusion" at the Bhabha Atomic Research Center, including "chain events" [31].

The core question about LENR chain reactions is simple: does the occurrence of one LENR at a specific point in space and time make it more or less likely that other LENR will occur nearby in both space and time, or are all LENR independent of each other? A cartoon relevant to this issue is given in Fig. 6. It indicates how the energy released from a LENR event in an array of unit cells in a Pd lattice would propagate outward. The distance (3.9 nm) at the bottom is determined by the Pd lattice constant. The time (0.13 ns) is gotten using the speed of sound in Pd. At issue is what happens in the cell labeled with ? when the results of the initial LENR event reach that cell. Does the passage of the energetic wave increase or decrease the likelihood of a LENR event at that location, assuming conditions favorable to such an event are otherwise satisfied.

7. Development of Safe Products

It is clear that, if LENR generators for heat and electricity are to be commercially successful, they must be safe to use. So, there has been a lot of attention to the safe operation of such potential units. Can commercial LENR generators be made safe, maybe even fail safe? If early LENR products are not highly safe, there will be regulatory intervention and the widespread adoption of LENR technologies could be long delayed. An extended discussion of the safety of LENR products is available [32]. The possibilities are widely varied.

Two types of unsafe operation are possible: (a) failure of critical LENR generators, which is a matter of their reliability, and (b) failure due to loss of control of the rate at which energy is produced. The high energy gains that are possible with LENR generators make loss of control a significant concern. Meltdowns and explosions are possible. LENR generators might be made fail-safe, so they automatically stop if over-heated. Rossi has stated that his units will be fail-safe. If they start of overheat, the high temperatures will destroy the efficacy of the fuel and shut them down automatically [33]. LENR units might also be engineered to prevent catastrophic runaway operation by some method, including venting of hydrogen gas, poisoning or by quenching by an auxiliary cooling system. However, it is more likely that LENR generators will be designed with active control systems, including feedback loops, to insure that they operate within bounds [20]. Such control is common in engineering systems. It might require the inclusion in the overall system of batteries to insure maintenance of positive control in the event of interruptions of power from the grid.

It is noted that nickel nano-materials pose health hazards [34], so if they are part of LENR generators, it will be necessary to insure that they are properly contained.

8. Weaponization of LENR

Historically, mankind has generally used new sources of energy for military purposes. This includes operations such as heating and transportation as well as production of weapons. It seems certain that LENR generators of heat and electricity will be used by the military for operational purposes. But, will the energy from LENR be used also in weapons? The central issue here is whether or not LENR explosions can be induced on demand.

There has been significant, albeit unpublished work on the possibility of using LENR in weapons. Experiments were conducted in the mid-1990s at a testing range in the desert west of the US to see if hydride materials imploded by conventional explosives would undergo nuclear reactions. The work was not conclusive. This author and his colleagues discussed impacting heavily hydrided materials, such as TiH_2 and TiD_2, onto targets at hyper-velocities in the early 1990s, but those experiments were never done. Possibly, the use of hydrided bullets has not been explored yet. Oddly, there was a publication on a "Metal Hydride Explosive System" in 1988 [35].

9. Conclusion

The meltdowns and explosions that have happened in LENR experiments and tests are important for three reasons: (a) they impact theoretical explanations of LENR mechanisms, (b) they must be controlled, even if not understood, for development of safe products, and (c) they raise questions about the potential weaponization of LENR.

There are two main questions. The first is: can the power density of LENR systems be turned up controllably? More pointedly, can explosions caused by LENR be willfully initiated? And, the second is: are there ways to limit and turn down LENR power output densities by use of either intrinsic effects like internal melting or by use of automatic external control systems?

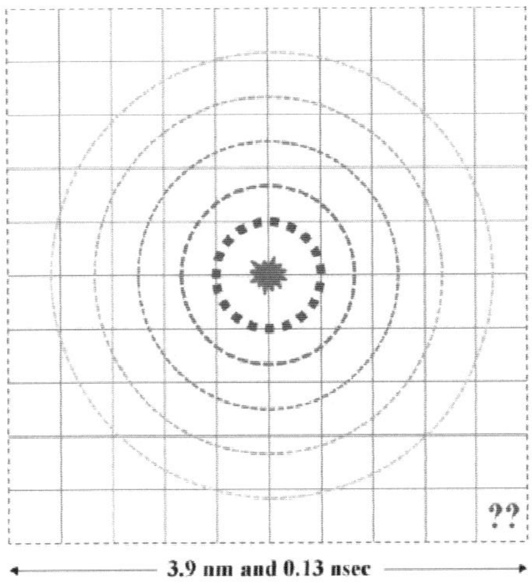

Figure 6. Schematic of a 10 × 10 of Pd lattice constants, where a LENR event occurs near the center. This array could be either on the surface of Pd or in its bulk.

Experimental research programs to address these questions would be expensive. For example, replication of the Fleischmann–Pons meltdown with at least four full electrochemical setups with significant amounts of Pd in an adequately safe laboratory might cost considerably more than $100K. Research on the controllable production of craters, and on their scaling to larger sizes is desirable. However, it will probably be difficult due to reproducibility issues, as is much of the experimental research on LENR even to this day.

Theoretical research on fast releases of LENR energy would not be so costly, but would be very challenging. Each of the concepts for mechanisms that cause LENR can and should be examined for whether or not it is compatible with the possibility of LENR chain reactions.

Acknowledgement

David Knies performed the initial CTH simulations of crater formation, which lead to the work reported in this paper. Comments by the reviewer are appreciated.

References

[1] M. Fleischmann et al., electrochemically induced nuclear fusion of deuterium, *J. Electroanaly. Chem.* **261** (1989) 301–307 and Erratum **263**.

[2] C.G. Beaudette, *Excess Heat: Why Cold Fusion Research Prevailed?* Oak Grove Press LLC, South Bristol, Maine, USA, 2002, pp. 35–26.

[3] C.N. Petersen, *The Guardian Poplar*, University of Utah Press, South Bristol, Maine, USA, 2012, p. 218.

[4] L. Forsley to CMNS Google Group in an email on 5 December 2014.

[5] J. Rothwell, *Cold Fusion and the Future*, 2004, p. 106, http://lenr-canr.org/.

[6] http://newenergytimes.com/v2/archives/fic/F/F199201.PDF.

[7] http://newenergytimes.com/v2/news/2005MTExplosion/2005MizunoT-AccidentReport.pdf.

[8] X. Zhang et al., On the explosion in a deuterium/palladium electrolyte system frontiers of cold fusion, *Proc. of ICCF-3,* 1992, pp. 381–384 and lenr.org.

[9] W.-S. Zhang et al., Thermal analysis of explosions in an open palladium/deuterium electrolytic system, *J. Condensed Matter Nucl. Sci.* **17** (2015) 116–123.

[10] T. Mizuno and Y. Toriyabe, Anomalous energy generation during conventional electrolysis, *Proc. of ICCF-12 on Condensed Matter Nuclear Science,* 2006, pp. 64–74 and lenr.org.

[11] S. Focardi and A. Rossi, A new energy source from nuclear fusion, *J. Nucl. Phys.* 2010; http://www.journal-of-nuclear-physics.com/?p=6.

[12] D.J. Nagel, Energy gains from lattice-enabled nuclear reactions, *Current Science* **108** (2015) 641–645.

[13] J.-P. Biberian, Unexplained explosion during an electrolysis experiment in an open cell mass flow calorimeter, *J. Condensed Matter Nucl. Sci.* **2** (2009) 1–6.

[14] S. B. Krivit, Explosion in Hokkaido University Cold Fusion Laboratory, http://newenergytimes.com/v2/news/2005MTExplosion/explosion-net.shtml (7 Nov 2009 Update).

[15] D. Hambling, Cold Fusion E-Cat Experiment Ends Explosively http://www.wired.co.uk/news/archive/2015-02/06/cold-fusion-reactor.

[16] http://www.e-catworld.com/2012/12/22/english-transcript-of-svt-cold-fusion-program/.

[17] H. Mills, Rossi's Self Sustaining One Megawatt Reactor. Pure Energy Systems News, http://pesn.com/2011/07/21/9501874_Rossis_Self_Sustaining_One_Megawatt_Reactor/ (21 July 2011)/.

[18] Quoted in Hot-Cat 2.0 by Eng. Roberto Ventola and Vessela Nikolova, ISBN: 8894003299 (2015), pp. 79–80.

[19] G. Levi et al., Indications of Anomalous Heat Energy Production in a Reactor Device Containing Hydrogen Loaded Nickel Powder, arXiv, 2013, 1305.3913.

[20] Roberto Ventola and Vessela Nikolova, Hot-Cat 2.0, ISBN 8894003299 (2015).

[21] S. Szpak et al., Evidence of nuclear reactions in the Pd lattice, *Naturwissenschaften* **92**(8) (2005) 394–397.

[22] Y. Toriyabe et al., Elemental analysis of palladium electrodes after Pd/Pd light water critical electrolysis, in A. Takahashi et al. (Eds.), *Proc. of ICCF-12 Condensed Matter Nuclear Science* ,World Scientific, Singapore, 2006, pp. 253–263.

[23] I. Dardik et al., Ultrasonically-excited electrolysis experiments at energetics technologies, in David J. Nagel and Michael E. Melich (Eds.), *Proc. of the 14th Int. Conf. on Condensed Matter Nuclear Science*, 2010, pp. 106–122.

[24] D.J. Nagel, Characteristics and energetics of craters in LENR experimental materials, *J. Condensed Matter Nucl. Sci.* **10** (2013) 1–14.

[25] J. Ruer, Simulation of crater formation on LENR cathode surfaces, *J. Condensed Matter Nucl. Sci.* **12** (2013) 54–68.

[26] http://webbook.nist.gov/cgi/cbook.cgi?ID=C118967&Units=SI&Mask=2.

[27] D.A. Crawford et al., CTH User's Manual and Input Instructions, version 10.2, Sandia National Laboratory, 1012.

[28] G. Levi et al., Observation of abundant heat production from a reactor device and of isotopic changes in the fuel, ecat.com/wp-content/uploads/2014/10/ECAT-test-report-2014.pdf, October 2014.

[29] P.K. Iyengar, Cold fusion results in BARC experiments, *Proc. 5th Intl. Conf. on Emerging Nuclear Energy Systems*, Karlsruhe, Germany, 1989.

[30] Y. Arata and Y.-C. Zhang , Deuterium nuclear reaction process within solid, *Proc. Japan Academy Ser. B* **72** (1996) 179–184.

[31] M. Srinivasan, hot spots, chain events and micro-nuclear explosions, *Proc. of ICCF-15*, V. Violante and F. Sarto (Eds.), Italian National Agency for New Technologies, Energy and the Environment, ISBN 978-88-8286-256-5, 2010, pp. 240–245, http://iccf15.frascati.enea.it/docs/ICCF15-PROCEEDINGS.pdf.

[32] D.J. Nagel, Questions about lattice enabled nuclear reactions: engineering, commercialization and applications, *Infinite Energy* No. 120 (March/April 2015) pp. 18–38.

[33] A. Rossi, unpublished.

[34] http://safety.fsu.edu/training/hazwaste/Occupational

[35] W.W. Lee et al., Metal hydride explosive system, U.S. Statutory Invention Registration H464 3 (May 1988).

J. Condensed Matter Nucl. Sci. 19 (2016) 230–243

Research Article

Basis for Electron Deep Orbits of the Hydrogen Atom

Jean-Luc Paillet*
Aix-Marseille University, France

Andrew Meulenberg†
Science for Humanity Trust Inc., USA

Abstract

In this paper, we look into the difficult question of electron deep levels (EDLs) in the hydrogen atom. Acceptance of these levels and, in particular, experimental evidence of their existence would have major implications for the basis for cold fusion and would open up new fields of femto-physics and -chemistry. An introduction shows some general considerations on these orbits as "anomalous" (and usually rejected) solutions of relativistic quantum equations. The first part of our study is devoted to a discussion of the arguments against the deep orbits and for them, as exemplified in published solutions. We examine each of the principal negative arguments found in the literature and show how it is possible to resolve the questions raised. In fact, most of the problems are related to the singularity of the Coulomb potential when considering the nucleus as a point charge, and so they can be easily resolved when considering a more realistic potential with finite value inside the nucleus. In a second part, we consider specific works on deep orbits, named Dirac Deep Levels (DDLs), as solutions of the relativistic Schrödinger and of the Dirac equations. The latter presents the most complete solution and development for spin 1/2 particles, and includes an infinite family of DDL solutions. We examine particularities of these DDL solutions and more generally of the anomalous solutions. Next, we analyze the methods for, and the properties of, the solutions that include a corrected potential inside the nucleus, and we examine the questions raised by this new element. Finally, we indicate, in the conclusion, open questions such as the physical meaning of the relation between quantum numbers determining the deep levels and the fact that the angular momentum seems two orders-of-magnitude lower than the values associated with the Planck constant.

Keywords: Deep electron levels, LENR, Relativistic Schrödinger and Dirac equations, Relativistic quantum physics, Singular solutions

*Corresponding author. jean-luc.paillet@club-internet.fr
†E-mail: mules333@gmail.com

© 2016 ISCMNS. All rights reserved. ISSN 2227-3123

1. Introduction

For many decades, the question of the existence of electron deep levels (EDLs) for the hydrogen atom led to numerous works and debates. Why once more a study on this subject? For several reasons:

- the arguments in favor of the deep orbits have become progressively more mature by the use of relativistic quantum tools for a full three-dimensional description of the system;
- by accepting the reality of a non-singular central potential within a nuclear region, many mathematical arguments against anomalous solutions of the relativistic equations no longer pertain;
- numerical evaluation of the relativistic equations are now detailed and available for interpretation of the models, their implications, and their predictions;
- hydrogen atoms, including deuterium, with electron deep orbits (femto-atoms) can contribute to the understanding of: processes of cold fusion inside condensed matter, the avoidance of nuclear fragmentation in D–D $\rightarrow^4$He fusion reactions, and a means of increasing the rate of energy transfer between an excited nucleus and the surrounding lattice;
- and, above all, recognition of these levels opens up a whole new realm of atomic, nuclear, and subatomic-particle physics as well as nuclear chemistry.

There are various theoretical ways to define a state of the hydrogen atom with electron deep level (EDL) or deep Dirac level (DDL) orbits. In the following, we denote $H^\#$ as any state of hydrogen atom with EDL orbits. Some authors use the term *hydrino* for denoting the $H^\#$ states owing to the work of [1] on the hypothetical existence of H atoms with orbit levels under the Bohr ground level and where the values of orbit radii are fractional values of the Bohr radius. Here, we do not use this term, a physical concept specifically attached to the cited work, because it is not deduced from (standard) quantum equations, while we essentially consider the states $H^\#$ obtained by the methods of relativistic quantum physics. With the quantum equations habitually used in the literature for computing the bound states of the H atom, we can note that there is in general a crossroad with a choice of value or a choice of sign for a square root in a parameter. According to which path is chosen, the resolution process leads either to the usual solution or to an unusual one called an "anomalous" solution; one that is rejected in the Quantum Mechanics textbooks.

In our present study, we note that a $H^\#$ solution is always an anomalous solution but *every anomalous solution is not a $H^\#$ solution.* For example, the anomalous solution also contains the regular energy levels of *anti-hydrogen.* We will see below that it is easy to recognize $H^\#$ solutions, if we have an expression of the anomalous solutions obtained by an analytic method.

A solution provides the eigenvalues of the Hamiltonian, representing the total energy of the electron, in the form of a family of quantized energy levels depending on quantum numbers. We consider only relativistic equations because, in the deep level orbits considered here, the electrons are relativistic. Indeed, we can make a quick computation:

> In [2] the authors plot the curve of the normalized electron density of the deep orbit wavefunction corresponding to the ground-state DDL orbit (–2s), and this curve has a peak for a radius equal to ~ 1.3 fm. By using the formula of the Coulomb energy potential CP $= -\alpha c\hbar/r$, we can calculate $|\text{CP}| \sim 1.09$ MeV for this deep orbit. On the other hand, an electron on this orbit has a potential energy equal, in absolute value, to $\gamma mv^2 = |\text{CP}| \sim 1.09$ MeV, where v is the electron velocity. From this, we can deduce $v = 2.75 \times 10^8$ m/s, $\beta \sim 0.91$ and $\gamma \sim 2.5$ confirming that the electron is actually relativistic.

The total energy E corresponding to a regular solution in a non-relativistic form for a bound state electron is expressed in negative value and $|E| \ll mc^2$. In relativistic form, the rest mass of the electron is included and, for atomic electrons, $E \sim mc^2(1-\varepsilon)$. Here $\varepsilon \ll 1$ and depends the coupling constant α and on quantum numbers. We can recognize a $H^\#$ solution, if the relativistic total energy is of the form $E \sim mc^2\varepsilon$.

As the movement of the electron is in a central field, its eigenstate equation is usually written in spherical co-ordinates and so it can be decomposed into a part depending on angular parameters θ,φ, and another part, a radial equation. The equation on angular parameters has for solutions the spherical harmonics $Y(\theta,\varphi)$ and the wavefunctions $\psi(r,\theta,\varphi)$ verify $\psi(r,\theta,\varphi) = R(r)Y(\theta,\varphi)$, where $R(r)$ is solution of the radial equation (here we omitted the usual quantum numbers indexes). The expression of the wavefunction can take a more complex form, as e.g. for the Dirac equation, but in any case only the radial part can raise questions. Thus, we consider the reasoning on the radial equations and on the radial wavefunctions.

2. Discussion on the Arguments Against the H$^{\#}$ States

Here, we classify our reflections according to the arguments found in the literature against the existence of these special states of the hydrogen atom and we discuss these arguments.

2.1. The wavefunction can have a singular point at the origin

This argument is rising in all known cases of H$^{\#}$ states with a $1/r$ Coulomb potential. The spatial part of the solutions of the radial equation, in the most general form, has several factors:

- one factor is a decreasing exponential $\exp(r)$ such that $\exp(r) \to 0$ when $r \to +\infty$,
- another one is $\propto 1/r^s$ with s a real number, due to the form of the Coulomb potential,
- and there can be a further one in polynomial form.

In the case of the "anomalous" solutions, the exponent s of the factor in $1/r^s$ is $s > 0$, then the radial function $R(r) \to \infty$ when $r \to 0$ and the wavefunction $\psi(r,\theta,\varphi)$ does not obey a boundary condition. This problem comes from the expression of the Coulomb potential in $1/r$. Some authors of H$^{\#}$ solutions remove this trouble by saying that the classical expression of the central potential is a good approximation for the bound state of a single electron atom, but considering the nucleus as a mathematical point is an unphysical abstraction. In fact, the Coulomb approximation in $1/r$ of the central potential generated by the nucleus is suitable if the electron is not too near the nucleus.

At this point, many authors do not consider a non-singular potential and stop without further development of the anomalous solution; but others work on this subject in expressing corrected potentials in the close vicinity of the nucleus, as e.g. in [2].

That is an actual enhancement of the theory, but it unfortunately entails extra difficulties. Indeed, there are no serious difficulties for defining the weakening of the electrical potential very near the charge radius of the proton and inside it. But problems arise when wanting to find solutions of the equations by taking into account the chosen potential in the neighborhood of the nucleus. Let *Pn* be this potential, then there are two possible procedures to use *Pn* :

(1) To solve the equation with the Coulomb potential and to find a first solution S1, then to solve the equation with *Pn* and find a second solution S2 near and inside the nucleus, and finally carefully connect S1 and S2 in a suitable manner: continuity conditions and even conditions on derivatives at the interface of both potentials.
(2) In the case where the solution S1 implies the bound electron is at the contact of the proton or deeper, we can think S1 is rather erroneous. Then, one should solve the equation directly with the whole rectified potential from 0 to $+\infty$.

2.2. The wavefunction 'cannot be square integrable'

In this case, the wavefunction cannot be normalized in the entire space and it does not obey a boundary condition for the bound states. This case results essentially from the behavior of the wavefunction ψ at the origin and not for

$r \to \infty$. To be normalized, the wavefunction has to satisfy $||\psi(r,\theta,\varphi)|| < +\infty$. As the Jacobian of the transformation from Cartesian to spherical coordinates is $J = r^2 \sin\theta$, we have

$$||\psi(r,\theta,\varphi)|| = \int |\psi|^2 \sin\theta \, r^2 \, \mathrm{d}\theta \ \mathrm{d}\varphi \, \mathrm{d}r = \int |Y(\theta,\varphi)|^2 \, \mathrm{d}\Omega \int |R(r)|^2 r^2 \, \mathrm{d}r, \quad \text{where } \Omega \text{ is the solid angle.}$$

Since the spherical harmonics are normalized, we have only to verify $\int |R(r)|^2 r^2 \, \mathrm{d}r < +\infty$. In fact, the behavior of $|R(r)|^2 r^2$ at infinity does not make any difficulty, because the leading factor, which induces ψ to vanish, is a decreasing exponential factor. Thus, only the behavior of $|R(r)|^2 r^2$ at the origin can be a problem.

Here, we can cite the work of Naudts [3], where a $H^{\#}$ state is found by using the Klein–Gordon (K–G) equation and the corresponding solution is square integrable. The author derives in one step the K–G equation for the bound electron of the hydrogen atom from the time-dependent Schrödinger equation by introducing the relativistic formulation of the energy. We recall this process in a more explicit way in order to more clearly see what are the implications and the limitations of the K–G equation.

- The relativistic total energy of a free particle of mass m is given by the equation $E^2 = \mathbf{p}^2 c^2 + m^2 c^4$.
- If we consider that an electron of charge e is submitted to an exterior electromagnetic field defined by an scalar electric potential φ and a vector potential $\mathbf{A}$ in covariant form, then the momentum vector $\mathbf{p}$ becomes $\mathbf{p} - e\,\mathbf{A}$ and the energy scalar E becomes $E - e\varphi$. By substituting in the previous energy equation we obtain $(E - e\varphi)^2 = (\mathbf{p} - e\mathbf{A})^2 c^2 + m^2 c^4$. However, as the electron is in a static Coulomb potential generated by the central proton, we have $\mathbf{A} = 0$ and $e\varphi$ is equal to $V = - e^2/r = -\alpha c\hbar/r$, where α is the coupling constant, also called the fine-structure constant, equal to $\sim$1/137.
- Finally, we apply the canonical quantization principle, by expressing E and $\mathbf{p}$ by differential operators:

$$E \to i\hbar\partial_t, \qquad \mathbf{p} \to i\hbar\nabla.$$

Then the equation energy becomes the K–G equation for the electron in the hydrogen atom, as written in the cited paper:

$$(i\hbar\partial_t - V)^2\psi(r,t) + \hbar^2 c^2 \Delta\psi(r,t) = m^2 c^4 \psi(r,t),$$

where $\psi(r,t)$ is a time-dependent radial wavefunction and m the rest mass of the electron. Historically, this equation was called the *relativistic Schrödinger equation*. In fact, the gauge invariance has been applied in the hypothesis of the Fock minimal coupling [4], so this equation describes the evolution of a relativistic electron in an exterior electromagnetic field [5].

Naudts then tries an ansatz with a function

$$\psi(r,t) = \mathrm{e}^{i/\hbar E t} r^{-l} \mathrm{e}^{-r/r_0}$$

solution of the equation with the hypothesis $l < 3/2$ and $r_0 > 0$ (here the parameter l does not represent the angular momentum). The condition on the l guarantees the behavior of $|\psi(r,t)|^2 r^2$ at the origin and $r_0 > 0$ its behavior at the infinity, in order for the wavefunction solution to be *square integrable*. The resolution leads to an equation to be satisfied by the parameter l : $l^2 - l - \alpha^2 = 0$ with solutions

$$l = \frac{1}{2}\left(1 \pm \sqrt{1 - 4\alpha^2}\right).$$

We can see a possible choice of sign $\pm$ before the square root. In both cases, the constraint on l is satisfied, i.e. $l < 3/2$. The choice of the negative sign leads to a regular solution that corresponds to the ground energy level associated with the usual quantum number $n = 1$. Indeed the value of the total energy is

$$E \sim mc^2\left(1 - \frac{\alpha^2}{2}\right)$$

and thus the binding energy is equal to

$$\mathrm{BE} \sim -mc^2\frac{\alpha^2}{2} = -13.6\ \mathrm{eV}$$

and the orbit radius r_0 is equal to 53 pm. If the positive sign is chosen, the obtained solution is a $H^{\#}$ state, with total energy very low, $E \sim mc^2\alpha \sim 3.73$ keV. That means a high value for the binding energy, since BE$\sim mc^2(\alpha - 1) = -507.3$ keV.

Naudts calculates excited states by using an analogous ansatz, but all his excited states are found to correspond to regular states. The only $H^{\#}$ state obtained is the above one, because he considers only spherically symmetric states. Since the exponent $l > 0$, the origin is a singular point for the wavefunction. Naudts argues against this problem by saying that the nucleus is not a point, but its charge is "smeared" over a distance of about 1 fm. Solving the equation with a smeared out Coulomb potential would produce a solution not diverging at the origin, but with certain minor changes on the $H^{\#}$ state.

Some criticism can be raised about the application of the K–G equation for a question concerning a bound electron.

- The most classical criticism concerns the fact that the electron is a fermion, spin $\pm 1/2$, whereas this equation does not take into account the spin and there is no way to introduce the Pauli spin matrices without destroying the Lorentz invariance. On one hand, this trouble is slight in comparison with the benefit of finding a square integrable $H^{\#}$ solution. On the other hand, it's the same problem for the classical Schrödinger equation and its use is well accepted for finding the energy levels for the light atoms. Moreover, it is not yet time to worry about fine structure for the $H^{\#}$ states.
- A more subtle criticism concerns the conservation equation

$$\partial_t\rho + \nabla\cdot\mathbf{J} = 0, \quad \text{where } \mathbf{J} = \frac{\hbar}{2im}\left(\psi^*\nabla\psi - \psi\nabla\psi^*\right).$$

For the non-relativistic equation, where $\rho = |\psi|^2$ and represents a probability density, ρ satisfies this equation, so its space-integral is time-independent and $\mathbf{J}$ represents a probability current density. However, from the solutions of the relativistic equation, the only possibility is that ρ be proportional to

$$\frac{\hbar}{2imc^2}\left(\psi^*\partial_t\psi - \psi\partial_t\psi^*\right)$$

and $\mathbf{J}$ be proportional to the same expression as above. The occurrence of time derivatives in ρ is due to the presence of a second-order time-derivative in the K–G equation. It can be shown that this expression of ρ can be reduced to $\rho = |\psi|^2$ in the non-relativistic limit. Nevertheless, the expression for ρ, when including time derivatives, is not necessarily positive and thus it cannot be considered as a probability density. However, it is possible to interpret it as a charge density in inserting $e\varphi$ (and $e\mathbf{A}$ for $\mathbf{J}$) as indicated in [6,7], because a charge can be positive or negative. Note that this problem does not exist when using the Dirac equation because it contains only first-order time-derivatives. Regardless, this question does not remove the interest in the $H^{\#}$ solution found by [3]. Finally we note that, if the singular point at the origin is suppressed, the wavefunction is automatically square integrable.

2.3. The orthogonality criterion cannot be satisfied

The Hamiltonian, which represents the total energy, has to be a Hermitian operator in order for its eigenvalues to have real values. This property leads to the following needed condition: eigenfunctions corresponding to distinct values have to be orthogonal. In [8], de Castro examines the asymptotical behavior of the solutions of the non-relativistic Schrödinger, of the K–G and of the Dirac equations by using the Frobenius series method and by considering the variations of this behavior as a function of formal variations of the coupling constant α. Then, from the orthogonality condition, he indicates deduced conditions under two different forms for the radial solutions, according to which equation is used:

For the K–G equation:

$$(u_k * (\mathrm{d}u_{k1}/\mathrm{d}r) - u_{k1}\,(\mathrm{d}u_k * /\mathrm{d}r)) \to 0 \quad \text{when } r \to 0, \text{ where } u_k(r) = rR(r).$$

For the Dirac equation, we have a condition on the components

$$f, g : (f_k^* g_{k1} - f_{k1} g_k^*) \to 0 \quad \text{when } r \to 0.$$

So, for the K–G case, only the solutions with u "less singular" than $\sqrt{r}$ can satisfy the orthogonality criterion. For the Dirac solution, he finds that only the regular solutions for the components f, g can satisfy the orthogonality. In fact, we can see the orthogonality condition is closely related to the behavior of the radial function at the origin [9], itself related to the behavior of the Coulomb potential at the origin. Therefore, this problem can be resolved by a corrected potential, without singular point at $r = 0$, that corresponds in fact to the physical reality. This point in developed in Section 3.3. We can also note several works [10,11] on self-adjoint extension of operators for potentials with singularity. In particular, the first cited work explicitly shows that, for the K–G equation in the case of the Coulomb potential, the "singular" solutions satisfy the orthogonality condition and satisfy also the boundary condition, i.e.: when $r \to 0, \lim u(r) = u(0) = 0$.

2.4. The strength of the binding seems to increase when the coupling strength decreases

It may seem absurd to make changes of a physical constant whose value is in principle given by 'Dame Nature'. But, it can be very instructive to make this "thought experiment": to imagine variations of the coupling constant α and to examine the consequence of such variations on the energy parameters of the hydrogen atom. In [12], Dombey points out a very strange phenomenon concerning the $H^{\#}$ solutions of the relativistic equations: when α decreases and tends towards 0, the binding energy of the electron increases and tends towards its maximum. The author solves the K–G equation in an analytic way leading to a Whittaker's (second-order differential) equation [13,23] on the radial function and the solution is achieved by transformation into a Kummer's equation [14,24]. So the radial function has the general form with three factors as we noted in Section 2.1, where polynomials are obtained by fixing some parameters of the confluent hypergeometric series [25] solution of the Kummer's equation. There is a choice of sign for a parameter occurring in the expression of the energy levels. One choice leads to the regular energy levels, while the other sign leads to anomalous ones, E_N, where N is a positive quantum number:

$$E_N = mc^2 \left[1 + \frac{\alpha^2}{\left[N + {}^1\!/_2 - \sqrt{({}^1\!/_4 - \alpha^2)} \right]^2} \right]^{-1/2}.$$

We note the presence of a minus sign before the square root in the denominator of the fraction $\ldots 1/2 - \surd \ldots$ For the anomalous ground level E_0, the author obtains the same expression as the $H^{\#}$ state energy found by Naudts [3], i.e. $E_0 \sim mc^2\alpha = 3.73$ keV. We can see the total energy E_0 of the electron decreases when α decreases, which means its binding energy |BE| increases in absolute value. And finally $|\mathrm{BE}| \to mc^2$, the *whole rest mass energy*, when $\alpha \to 0$.

This strange phenomenon seems to concern only the ground state E_0 because, for $N \neq 0$ we have $E_N \sim mc^2(1 - \alpha^2/2N^2)$ for $\alpha \ll 1$, thus E_N increases when α decreases, which is the "normal" behavior. Of course, we have the same result as the ones obtained in [3] by means of an ansatz, but as we have an algebraic expression resulting from an analytic process, we can see at least a mathematical reason for discriminating the $H^{\#}$ solution among the set of the anomalous solutions. The explanation lies in the expression Δ in the denominator,

$$\Delta = N + \frac{1}{2} - \left(\frac{1}{4} - \alpha^2\right)^{1/2}.$$

Indeed if $N = 0$, then Δ can be reduced to α^2, for $\alpha \ll 1$, leading to the expression for E_0 above. Nevertheless, in some cases [15] there is an infinite family of $H^{\#}$ states and all these states have the strange behavior with respect to the coupling constant. Note in [12] the author also examines the solutions given by the Dirac equation in two-space-dimensions and finds the same strange behavior of the ground state, which is a $H^{\#}$ state, with respect to the coupling constant. In fact, we think this result is obtained in a context of an ill-defined system, uniquely on a pure mathematical basis. From a physical point of view, we can see the coupling constant α is actually entangled with several fundamental constants, in particular the Planck constant, the velocity of the light, and the elementary electric charge. So, modifying α without caution can certainly lead to paradoxical physical results (e.g., letting α go to zero means that the charge does also; thus, there are no bound states and no BE).

Another example, extracted from [16], of this kind of problem concerning a physical constant, in a case where the constraints are simpler: the non-relativistic limit of a relativistic theory can be obtained if one lets c tend to infinity, and thus the relativistic coefficient γ becomes 1 for any speed v. However, if doing this on the Dirac operator in an electro-magnetic field, one has to proceed carefully because of terms as such as mc^2, which would tend to infinity, and as the term $(e/c)\mathbf{A}$ that would turn off the vector potential $\mathbf{A}$ if c tends to infinity. Then the author is led to develop specific techniques and to define some concepts needed on account of the nature of the so-called *c-dependence* of the Dirac operators.

3. Deep Dirac Levels (DDL). The Deep Orbits Obtained as Solutions of the Relativistic Quantum Equations

In [15], Maly and Va'vra publish their first article on this subject where they define the concept of Deep Dirac Levels (DDL) for the electronic orbits of the Hydrogen-like atoms. Here we consider only hydrogen atoms. In fact, they use two methods for obtaining these deep orbit levels: the former by means of the relativistic Schrödinger equation, and the latter by the Dirac equation. For both equations, they follow the solution method indicated in [6]. We emphasize the use of the Dirac equation.

3.1. Solutions obtained with the relativistic Schrödinger equation

After separation of the radial equation, one introduces an ansatz $R(\rho) = \rho^s e^{-\rho/2} L(\rho)$ into the radial equation, $L(\rho)$ is a series of powers of ρ, s is a real parameter and ρ is a real numerical parameter, without physical dimension but proportional to the radius r. New parameters are defined to obtain a pure numerical second-order differential equation in $L(\rho)$. One shows that the eigenvalue energy E of the Hamiltonian is defined by the following expression: $E = mc^2[1 - (\gamma^2/\lambda^2)]^{-1/2}$, where λ is a numerical parameter of the equation and $\gamma = Z\alpha$, α being the coupling constant. It is in fact the Sommerfeld relation.

Two conditions must be satisfied: $s(s+1)+\gamma^2 - l(l+1) = 0$, where l is the angular momentum quantum number, and $\lambda = n' + s + 1$, for obtaining the convergence of the series $L(\rho)$, where n' is an integer number ≥ 0. The first condition, a quadratic equation, has two roots:

$$s = -\frac{1}{2} \pm \left[\left(l + \frac{1}{2}\right)^2 - \gamma^2\right]^{1/2}.$$

It is the "crossroad" condition indicated at the beginning of this paper. When taking the positive sign in the expression of s, we obtain the usual solution for the electronic energy levels. This choice is taken because with the negative sign we have $s < 0$ for any $l > 0$ and thus the radial wavefunction R tends to infinity when ρ tends to 0 because of the exponential term ρ^s. Note that for $l = 0$, we also have $s < 0$, even when taking the positive sign. But in this case, for small Z and as $\gamma \approx Z/137$, $\gamma^2 \ll 1$ and s is close to zero. Moreover, considering that the nucleus has a size $a \neq 0$, the potential has no singular point near 0 and is finite everywhere. Then, one can show [6] that the solution R is finite at $r = \rho = 0$ and approaches that solution with a singular-point-Coulomb potential when a tends to 0. Nevertheless, as noted in [15] for heavy atoms, the value of γ^2 becomes great enough that s (a negative value) has a non-negligible absolute value for $l = 0$. For example, in the Cs atom, $Z = 55$ and then $s \approx -0.2$. One can observe that for an even-higher-Z hydrogen-like atom, i.e. Fr with $Z = 87$, we have s with an imaginary part; but this fact goes beyond the subject of our paper. As we consider only the H atom, it is sufficient to let $Z = 1$. Anyway, at this point we can consider there is no serious reason for systematically eliminating the so-called "anomalous" solutions obtained with a negative sign in the root s. The argument concerning the physical reality of the finite ($\neq 0$) size of the nucleus can be applied in this case too.

The energy levels corresponding to the "anomalous" solutions are provided by the following expression [15]:

$$E{=}mc^2\left[1+\frac{\alpha^2}{(n'+1/2-[(l+1/2)^2-\alpha^2]^{1/2})^2}\right]^{-1/2},$$

where n' is the radial quantum number and l is the angular momentum quantum number. For hydrogen-like atoms, α^2 is simply replaced by $Z^2\alpha^2$. As for the usual solutions, the authors define the total quantum number $n = n'+l+1$. Then they compute the new energy levels E for all the possible combinations of the quantum numbers $n = 1, 2, \ldots, n' = 0, \ldots, n-1$; and $l = n - n' - 1$. The values represented by E are the total energy of the electronic orbitals. The corresponding binding energies, i.e. the values $|\mathrm{BE}| = E - mc^2$, are quoted in several Tables T1 [15] together with the energies of the regular solutions (positive sign in the expression of s) and with the non-relativistic Schrödinger levels for comparison. The results for $Z = 1$ are reproduced in our companion paper [17]. Each table is built-up for a different hydrogen-like atom of the class of the alkali metals. In fact, *every anomalous solution is not a deep* orbit: such *a deep orbit appears only for* $\boldsymbol{n' = l}$, as emphasized by the authors and as we can see in the tables (where $n = N$, $n' = M$, and $l = L$).

As we have an analytic expression of E, it is possible to find the "secret" of this difference between the three solutions. For doing this, we start from the formula

$$E = mc^2[1 + (\gamma^2/\lambda^2)]^{-1/2}, \quad \text{where } \lambda = n' + s + 1 \text{ and } s = -\frac{1}{2} - \left[(l+1/2)^2 - \alpha^2\right]^{1/2}.$$

For small Z (here $Z = 1$), we can first show that $\lambda \sim n' - l + \alpha^2/(2l+1)$. Then we can see that the condition $n' = l$ drastically reduces the expression of λ : $\lambda \sim \alpha^2/(2l+1)$, and so $\lambda \ll 1$. Carrying this into the expression of E, one can deduce $E \sim mc^2\alpha/(2l+1)$. As the fraction $\alpha/(2l+1) \ll 1$, the binding energy $|\mathrm{BE}|$ is very high and that means the orbit is very deep. So for every l, there is an infinite series of these very deep energy levels with $|\mathrm{BE}| > 507$ keV and an orbit radius of order fm. Concerning the energy levels corresponding to other combinations of the quantum numbers, the tables gives two kinds of results: values similar to the usual energy levels, and values annotated by the authors as "negative energy states, not observable", appearing for $n' > l$. Finally, we can note that for $n' = l = 0$, for the relativistic Schrodinger case, the binding energy $\mathrm{BE} = -507$ keV, in agreement with the result found in [3] for the energy of the anomalous solution. Moreover the necessary condition, $n' = l$ for the EDL orbits, explains the negative result of the same author concerning the "excited" states ($n' > 0$), because Naudts considered only the case of null angular momentum for the deep-levels.

3.2. Solutions obtained by means of the Dirac equation

The Dirac equation is certainly a cornerstone of modern physics: reconciling quantum physics and special relativity with success, accounting for spin of particles, and having the historical source of the concept of anti-particle even before their actual discovery. It is copiously handled in the literature, e.g. in [4, 6,7,18]. Essential features of the Dirac equation and its solutions can be found in [9].

3.2.1. *Determining the DDL solutions*

Maly and Va'vra use the method developed in [6], where the "crossroad" condition is the sign of a parameter

$$s = \pm\sqrt{(k^2 - \gamma^2)}$$

appearing during the solution process. If choosing the sign "–", one obtains the expression E of the energy levels for the anomalous solutions

$$E = mc^2 \left[1 + \frac{\alpha^2}{\left(n' - \sqrt{(k^2 - \alpha^2)}\right)^2} \right]^{-1/2}.$$

The radial quantum number n' can take any positive value 0, 1, 2, ...The quantum number k, eigenvalue of the specific Dirac operator K, is related to the total angular momentum (now including the electron spin). It can take values $\pm 1, \pm 2, \ldots$ but not the value 0. Indeed, as K satisfies $\hbar^2 K^2 = \mathbf{J}^2 + \frac{1}{4}\hbar^2$ we can see that k cannot be null, but we can give also a more "physical" argument to understand this fact. In the radial equation, there is a term in k/r playing the role of a repulsive angular momentum barrier that prevents the "fall to the center." It is like the "effective potential" $l(l+1)/r^2$ appearing in the non-relativistic radial Schrödinger equation. One also defines the main quantum number to be $n = n' + |k|$.

The authors built tables of the binding energies for combinations of the Dirac quantum numbers n' and k appearing in the expression of E, plus the main (or total) quantum number n and the orbital quantum number l. This latter is connected to k by the relation $l = k - 1$, if $k > 0$, else $l = -k$. These Tables T2 also include the hydrogen-like atoms of the alkali class. Here we consider only the hydrogen tables and $k > 0$. ($n = N$, $n' = M$, $k = K$, and $l = L2 = -k$ (used with $k < 0$) for the Dirac levels.) Here again, *every energy level is not a* DDL, but *only those computed for* $\mathbf{n}'(= M) = \mathbf{k}$. The mathematical explanation is similar to that for the Schrödinger equation: we can show that the energy values E_D of the DDL orbit satisfy $E_\mathrm{D} \sim mc^2\alpha/2k$. So, we can see that there is an infinite series of DDL solutions.

Now, we consider the expression $E_\mathrm{S} \sim mc^2\alpha/(2l+1)$ obtained in the Schrödinger case. Here, the indices D and S refer to the Dirac equation and Schrödinger equations, respectively. While comparing the expressions of E_D and E_S, we can verify the slight shift between the values of E_D and E_S, i.e. $E_\mathrm{D} > E_\mathrm{S}$ and thus $|\mathrm{BE}_\mathrm{D}| < |\mathrm{BE}_\mathrm{S}|$, for the equivalent levels determined by n' and thus by $k = l$. For example, for $n' = 1$, the tables displays $\mathrm{BE}_\mathrm{S} - 509.8$ keV while $\mathrm{BE}_\mathrm{D} - 509.1$ keV. Note that the first deep orbit energy $\mathrm{BE}_\mathrm{S} \sim -507$ keV, for $n' = 0$, has no equivalent in the "Dirac table" of the same atom, because the Dirac number k cannot be null. The differences between E_D and E_S is certainly due to the corrective term of *spin–orbit* energy included in the Dirac Hamiltonian and associated with the spin precession. For the regular solutions, there is a smaller total spread in energy of fine structure levels [6] than for the Schrödinger solutions. In fact this corresponds to a slightly bigger TE and thus implies a slightly smaller $|\mathrm{BE}|$. Moreover, we can think this energy shift is much more appreciable at the deep level.

3.2.2. *Some particularities of the DDL solutions*

In considering $E_D \sim mc^2\alpha/2|k|$ for the total DDL energy, we can see that E_D decreases when k increases. Since the total quantum number $n = n' + |k|$, and $k > 0$, we can write $E_D \sim mc^2\alpha/n$. Thus when n increases, the binding energy $mc^2[1 - \alpha/n]$ increases. So, the variation of the binding energy as a function of the principal quantum number n is the *inverse* of the case for the regular solutions. This fact raises a question: what is the variation of the mean orbit radius as a function of the quantum number n (or k)? It seems this question has never been mentioned, much less addressed. We think the most logical answer, based on the results of [15], should be the following: when n increases, the binding energy increases. That is possible only if the electron moves nearer to the nucleus, so the mean orbit radius decreases. A coarse computation seems to lead to the same hypothesis and a remark in ([2], p.61), the next paper of the authors cited here, corroborates this hypothesis. Under these conditions, we assume that the mean radius corresponds to a charge-accumulation area in orbitals close-about the charge volume of the nucleus in more circular (or spherical) orbits. Of course, only a computation based on the variations of the quantum electron density could determine the correct result. This can be the object of a further paper. Another question leading to further study is: how to physically interpret the fact that DDL orbits appear only when the quantum number n' and k are equal?

3.3. The deep orbits obtained by considering a corrected potential near the nucleus

3.3.1. *DDL orbits with a finite potential inside the nucleus*

After their earlier work [15], Maly and Va'vra produced a second paper [2] on DDLs where, in particular, they estimate the size of the DDL atoms. They used another method ([19], p.195) for the Dirac equation solution. This method, using similar ansatz, transforms the system of coupled first order differential equations on the radial functions into a second-order differential equation, a Kummer's equation. The general solutions of this equation take the form of confluent hyper-geometrical series, requiring suitable convergence conditions. Of course, there is always the same "crossroad" parameter s that determines the DDL solutions when $s < 0$. In order to make accurate calculations of the size of the DDL atoms, the authors consider that the nucleus has finite dimension, taking into account a finite specific potential inside the nucleus, and they look for the wavefunctions inside the nucleus. Then they show that it is possible to normalize and to "connect" both solutions (outside and inside the nucleus) at a conventional value R_0 of the nucleus, not indicated in the paper. Because of the wavefunctions inside the nucleus, we can show that the global wavefunctions satisfy the orthogonality condition (Section 2.3).

From the couple of radial functions f and g found outside the nucleus, Maly and Va'vra [2] compute the electron density (Eld) outside the nucleus by the formula

$$\mathrm{Eld} = 4\pi r^2(|f|^2 + |g|^2)$$

and they deduce the mean orbit radius

$$\langle r \rangle = A_0 \int_0^\infty r\,\mathrm{Eld}\,\mathrm{d}r,$$

where A_0 is a normalization constant. They plot curves of Eld for various atoms, for regular and for DDL orbits. In particular, in their Fig. 2 [2], they give the curves of Eld for DDL orbits corresponding to the Dirac quantum number $k = +1$, for H and Li (as hydrogen-like) atoms. For H, the energy level is ~ -509.1 keV, while for Li (with one electron on DDL) it is ~ -505.4 keV. The authors say that the peak of Eld corresponds to the radius of the nucleus. The peak occurs for $r \sim 1.3$ fm. After this, they propose mechanisms of atomic transitions to the DDLs and suggest chemical behaviors of the DDL atoms that, in fact, could behave almost as neutral particles. This would explain the difficulty in detecting them. Finally, they report experimental results such as calorimetry and radiation detection. This is beyond of the scope of our present paper.

3.3.2. *Techniques used when considering a finite potential inside the nucleus and criticism*

When considering a finite potential inside the nucleus, there are three stages for finding the solution.

First, the solution is computed outside the nucleus with the Coulomb potential, but with considering the radius $r > R_0$, where R_0 is near the "charge radius" of the nucleus. For example, if we consider the hydrogen atom, the nuclear-charge radius is ~ 0.87 fm. In [20], R_0 is computed for a nucleus of mass number A by means of the formula $R_0 = r_0 A^{1/3}$, where $r_0 = 1.2$ fm.

Next, the solution is computed inside the nucleus, with a chosen potential which is an approximation physically suitable for the problem. For example, one can use, as in [2,18], the Smith–Johnson potential that corresponds to a uniformly distributed spherical charge, i.e.

$$V(r) = -\left[\frac{3}{2} - \frac{1}{2}(r^2/(R_0)^2)\right] Ze^2/R_0,$$

or simply a constant potential, or more complex ones.

Finally, let $g_\mathrm{i}(r)$ be the inside solution and $g_\mathrm{o}(r)$ be the outside one, both have to be correctly "connected" at $r = R_0$. More precisely, if the initial equation(s) is (are) of differential order 2 as the Schrödinger equation, we have to satisfy the continuity condition for the functions $g_\mathrm{i}(R_0) = g_\mathrm{o}(R_0)$, and also for their first order derivative, i.e. $g_\mathrm{i}'(R_0) = g_\mathrm{o}'(R_0)$. In fact both conditions can be combined into a "matching" equation of the form $g_\mathrm{i}'(R_0)/g_\mathrm{i}(R_0) = g_\mathrm{o}'(R_0)/g_\mathrm{o}(R_0)$. Of course, the normalization of the whole wavefunctions defined for $r \in [0, +\infty)$ has to be carried out after matching.

An interesting criticism is found in [20] where the Dirac "anomalous" solution is not rejected, but is in a way combined with the regular one in a linear combination with coefficients to be computed to satisfy the continuity conditions at the matching radius. From the result, we can see the anomalous solution is involved with a very small ratio, as a little perturbation. We observe that the authors use a solution method based on the use of a Whittaker's [13] (second order) differential equation. The solutions include confluent hypergeometric series as factors of exponential functions, in particular ρ^s, which does not diverge at $\rho = 0$ thanks to the finite chosen potential. Nevertheless, if considering both the regular and the anomalous solutions at the same time, a difficulty appears for the convergence of the series when $\rho \to +\infty$: the convergence condition depends on the sign of the crossroad parameter s. To resolve this difficulty, the authors have to combine both kinds of series in order for the divergences to be exactly balanced, when using asymptotic forms of the series. Then the authors have to unify a parameter E representing the energy, when they verify the continuity conditions at $r = R_0$. Here we can note that, if using the solution method proposed in [6] or in [21], then the convergence of the series is independent of the crossroad parameter s, which greatly simplifies the problem.

The criticism of Deck, Amar, and Fralick [20] about the method used in [2] concerns the lack of dependence on the potential inside the nucleus and on the boundary conditions at the nuclear radius. In fact, matching in a simple way the wavefunction outside the nucleus with a solution inside the nucleus is an approximation that only allows removal of the singularity at the origin. We think that the form of the nuclear potential can have a significant effect on the energy levels, particularly for deep orbits. So, in order to improve the precision of the DDL levels, we suggest the following method: to start from a corrected global potential built by connecting in a smooth way the Coulomb potential outside the nucleus with a chosen nuclear potential [22], then to numerically solve the radial equations with this global potential and to compute the corresponding DDL energies. This could be the object of further work.

Note finally a method indicated in [16,18], which allows one to "regularize" the Coulomb potential without the arbitrariness of the cut-off near the nucleus (in particular with the choice of a radius R_0). This method consists of taking into account the "anomalous" magnetic moment of the electron in the unmodified Coulomb potential, and it can be generalized to other potentials with singularity at the origin. Thaller shows that, in doing this, one modifies the radial Dirac operator written in matrix form to have an additional term $\mu_a\alpha/r^2$ not diagonal, where in natural units $\mu_a = 0.00058$ determines the anomalous magnetic moment of the electron. We cite the author: " this term acts as a repulsive interaction that forces the wavefunction away from the singularity ". In fact, this factor (one of several physical mechanisms that would prevent an electron from reaching $r = 0$ in a $1/r$ potential) is extremely small; nevertheless, it becomes dominant against the attractive diagonal Coulomb term α/r for an electron near the origin. Here, the term "regularize" means that the modified Dirac operator has all the "good" properties for providing counter arguments to all the criticisms analyzed above.

4. Summary, Conclusions, Open Questions, and Future Work

In the first part of this paper, we discussed the principal arguments against deep orbits (DDL) for the H atom and we showed how it is possible to resolve the questions raised. Next we looked at the computational results of Maly and Va'vra [2,15] that produced an infinite set of the anomalous solutions, usually rejected, of the relativistic Schrödinger equation and the Dirac equations.

- We observed that only a subset of these solutions, but an infinite one, corresponds to deep orbits: the ones satisfying the equality between the quantum numbers that determine the values of energy levels, i.e. the radial number n' and the angular number (l for the Schrödinger equation and k for the Dirac equation).
- We saw that the electron binding energy on these DDL orbits, of order 509–511 keV, increases when the angular quantum number increases. This result seems to indicate that the mean radius of the DDL orbits decreases when n' increases. Though it is not explicitly stated by the authors, some remarks in their second paper clearly corroborates this hypothesis, and it seems there is an accumulation zone of the orbits in the neighborhood of the nucleus, near a radius of order 1 fm. Of course only a precise quantum computation of the mean radius as function of n' could confirm this hypothesis.
- Another feature that begs for explanation is the predicted 'angular momentum' quantum numbers that are for quantities, which are two orders-of-magnitude lower than the values associated with the Planck constant.
- We also noted that another infinite subset of solutions give energy values very near the regular atomic-electron levels, while the corresponding wavefunctions are not the ones of the regular solutions. This result and the previous one about the quantum numbers, raise questions about their possible physical interpretation.

In seeking the source of the deep orbits and their unique structure, we thought of the situation of the DDL orbits in an extreme **B**- and **E**-field from the binding nucleus. This implies a major strengthening of the several "special" known effects that affect the regular atomic-electron orbits. These include the spin–orbit and spin–spin interactions, the Zitterbewegung, and the Lamb shift. For example, the spin–spin interaction, responsible for the hyperfine structure, and the corresponding quantum number associated could play an important role in the determination of energy shifts. However, the deep levels, discrete and asymptotically approaching a binding energy of 511 keV, are also predicted by the relativistic-Schrödinger (K–G) equation, which does not include spin effects. Therefore, any spin-related contributions must be limited to the Dirac equations.

Since the relativistic-Schrödinger equation does not include spin effects, the remaining feature in common (and one not shared with the non-relativistic-Schrödinger equation) is the relativistic formulation that includes the rest mass of the electron. Both equations predict deep orbits with spin effects as a minor perturbation. We might suggest that there is a 'hidden' variable within quantum mechanics that may be associated with relativity (or its formulation in quantum mechanics). Perhaps there are possible new quantum numbers associated with known physical effects, such as relativistic and field-induced precession and nutation of the electron-spin vector. Nevertheless, the latter are only the perturbation effects, not the main story.

From the analysis of the work [2] on the DDL orbits obtained with a corrected potential near the nucleus and the questions raised about these solutions, we think that a more accurate estimation of the DDL orbit energies and radius should result from a numerical "direct" computation of the radial wavefunction with an improved potential from $r = 0$ to infinity. An alternative to this computation could take into account the anomalous magnetic moment of the electron with unmodified Coulomb potential. Introduction of a Hamiltonian including the nuclear-spin interaction with the DDL electron spin would be necessary to include the spin–spin effect. This has an extremely small effect on the atomic-electron orbits, but should be a major effect on electrons in femto-meter orbits. These questions will be the object of further work.

The conclusions from this paper are as follows:

- the anomalous solution to the relativistic Schrodinger and Dirac equations cannot be rejected if a physically real potential is used;
- even though the electron deep orbits do not have sufficient angular momentum to allow standard quantized orbits, the mathematical solutions to these equations include an infinity of discrete orbits;
- by similarity with the atomic-electron orbital solutions, it would appear that there is another set of electron waves (perhaps only when bound) at the higher frequency associated with the deep orbits.

The final conclusion is a series of questions and comments. Is the major effect producing the electron deep orbits from relativistic-velocity effects or is it some form of symmetry-breaking for deep orbits from mass effects? Is the effect real? What are the equations and their solutions trying to teach us? The anomalous solutions have been rejected because there is no data to prove that they exist. Because the solutions have been rejected, data that might prove the deep levels are rejected, explained away, or used to build different models. Cold fusion provides hundreds of experiments that can be more readily explained by the existence of the electron deep levels. Cold fusion will probably provide the physical material from which the deep levels can be verified. Such material probably already exists (even in 2014) that can uniquely and incontrovertibly prove both cold fusion and the electron-deep-level models.

Acknowledgement

This work was supported in part by HiPi Consulting, New Market, MD, USA, by the Science for Humanity Trust, Bangalore, India; and by the Science for Humanity Trust Inc., Tucker, GA, USA. The authors would like to thank Prof. Jan Naudts for his useful comments on the manuscript.

References

[1] R.L. Mills, *The Grand Unified Theory of Classical Physics*, 2011, edn. Black Light Power. Vydrino.

[2] J. Maly and J. Va'vra, Electron transitions on deep dirac levels II, *Fusion Sci. Technol.* **27**(1) (1995) 59–70. http://www.ans.org/pubs/journals/fst/a_30350.

[3] Jan Naudts, On the hydrino state of the relativistic hydrogen atom, http://arXiv:physics/0507193v2.

[4] David Tong, *Lectures on Quantum Field Theory*. University of Cambridge, UK. http://www.damtp.cam.ac.uk/user/tong/qft.html.

[5] H. Kleinert, *Particles and Quantum Field*, Lectures at the Freie Universität Berlin. http://users.physik.fu-berlin.de/~kleinert/b6/psfiles/qft.pdf.

[6] L.I. Schiff, *Quantum Mechanics*, Third Edition, McGraw-Hill, Nrew York, 1968.

[7] S. Weinberg, *The Quantum Theory of Field*, Vol. 1, *Foundations*, Cambridge University Press, Cambridge, 1995.

[8] Antonio S.de Castro, Orthogonality criterion for banishing Hydrino states from standard quantum mechanics, *Physics Letters* A **369** (2007) 380–383.

[9] J.-L. Paillet and A. Meulenberg, Arguments for the anomalous solutions of the Dirac equations, http://viXra.org/abs/1506.0177. *J. Condensed Matter Nucl. Sci.* **18** (2016) 50–75.

[10] T. Nadareishvili and A. Khelashvili, *Some Problems of Self-Adjoint Extension in the Schrodinger equation*, http://arxiv.org/pdf/0903.0234.pdf.

[11] Guy Bonneau, Jacques Faraut and Galliano Valent, Self-adjoint extensions of operators and the teaching of quantum mechanics, http://arXiv:quant-ph/0103153v1.

[12] Norman Dombey, The hydrino and other unlikely states, http://arxiv.org/pdf/physics/0608095.pdf.

[13] E.T. Whittaker and G.N. Watson, *Modern Analysis*, Fourth edition, Chapter XVI, Cambridge University Press, 1940.

[14] https://en.wikipedia.org/wiki/Confluent_hypergeometric_function.

[15] J.A. Maly and J. Va'vra, Electron transitions on deep Dirac levels I, *Fusion Sci. Technol.* **24**(3) (1993) 307–3018. http://www.ans.org/pubs/journals/fst/a_30206

[16] Bern Thaller,The Dirac operator, in *Relativistic Electronic Structure Theory: Part* 1, Theory (Theoretical and Computational Chemistry # 11), Chapter 2, P. Schwerdtfeger (Ed.), Elsevier, New York, 2002.

[17] A. Meulenberg and J.-L. Paillet, Nature of the deep-Dirac levels, *J. Condensed Matter Nucl. Sci.* **19** (2016).

[18] Bern Thaller, *The Dirac Equation*, Springer, Berlin, 1992.

[19] S. Fluegge, *Practical Quantum Mechanics*, Vol. 2, Springer, Berlin, 1974.

[20] R.T. Deck, Jacques G. Amar and Gustave Fralick. *Nuclear size corrections to the energy levels of single-electron and -muon atoms, J. Phys. B: At. Mol. Opt. Phys.* **38** (2005) 2173–2186.

[21] J.J. Sakurai and Jim Napolitano. *Modern Quantum Mechanics*, Second Edition, Addison-Wesley, San Francisco, Pearson Education, 2011.

[22] D. Andrae, Finite nuclear charge density distributions in electronic structure calculations for atoms and molecules, *Physics Reports* **336** (2000) 413-525.

[23] G. Arfken, *Mathematical Methods for Physicists*, Third Edition, Academic Press, New York, 1965, p 756.

[24] M. Abramowitz and I. Stegun. *Handbook of Mathematical Functions with Formulas, Graphs, and Mathematical Tables, Ninenth Edition,* 1970.

[25] John W. Pearson. *Computation of Hypergeometric Functions,* School of Mathematics, The University of Edinburgh, 2009/9/4. Link= ox.ac.uk .

J. Condensed Matter Nucl. Sci. 19 (2016) 244–256

Research Article

Research into Heat Generators Similar to High-temperature Rossi Reactor

A.G. Parkhomov* and E.O. Belousova

Lomonosov Moscow State University, Moscow, Russia

Abstract

Devices similar to a high-temperature Rossi reactor were made. Excess heat at the temperature of about 1100°C and higher was demonstrated. No nuclear radiation above the background level was observed during the excess heat production.

Keywords: Element composition, High temperature, Hydrogen, Isotope composition, Nickel powder, Nuclear radiation, Power consumption, Power production, Reactor

1. Introduction

According to a report by experts observing a test of the high-temperature Rossi reactor in Lugano [1], we assume that reactor itself is just a ceramic tube sealed using a heat-resisting cement, also containing nickel powder with lithium aluminum hydride addition. To initiate the process, it is necessary to heat the tube from 1200 to 1400°C. Based on this assumption, we made two versions of a device which is similar to the high-temperature Rossi reactor. In the devices of the first type [2], the amount of heat produced was measured using the mass of the evaporated water. In devices of the second type [3], the amount of heat produced was measured by comparing the power consumed by the reactor with fuel and without fuel.

2. Design of First Type Device

Alumina ceramic tubes 120 mm length with outer diameter 10 mm and inner diameter 5 mm were used (Fig. 1). Nichrome wire as electrical heater was wound around the tube. One gram of Ni powder with 0.1 g $Li[AlH_4]$ are placed inside the tube. A thermocouple was placed in contact with the outer surface of the tube. The ends of the tube were sealed with the heat-resistant cement. The entire reactor surface is coated with the same cement.

*E-mail: alexparh@mail.ru

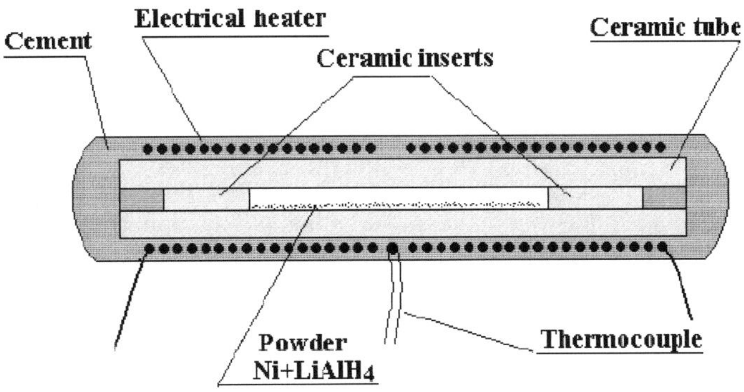

Figure 1. Schematic of the first type device.

3. Heat Output Measurements

The measurement method used during the Rossi's reactor test is very complicated. In our experiment, we used a method based on the amount of water boiled during the operation. This method was verified repeatedly during various experiments, including experiments with plasma electrolysis.

The reactor is placed in a closed metal vessel (Fig. 2), which is submerged in water. Some amount of water is steamed away when the reactor is in operation. Measuring the amount of water evaporated and using the heat of vaporization (2260 kJ/kg), we calculated the heat produced. We can also calculate the heat loss through the thermal isolation taking into account of the cooling rate after the reactor is switched off.

The reactor was placed either in the air on the alumina supports or immersed into alumina powder and put it into a metal box. The latter method requires 2–3 times less power to heat up the reactor to a given temperature, but in this mode, reactor operation is less stable.

4. Equipment for Heater Power and Diagnostics

The equipment used to measure and control power consumption and temperature measurements is shown in Fig. 3. We used a transformer with switching coils. The coils are switched manually and automatically by the regulator which is controlled by the thermocouple signal. When the temperature rises above a threshold, the regulator switches to the lower voltage. When temperature goes below another threshold, the voltage is increased. This allows long-duration operation at a preset temperature, which makes reactor's operation more stable. To measure the power consumption, we use a voltmeter and an ammeter, and a wattmeter that allows data recording to a computer.

To monitor the radiation level, we used a Geiger counter, a dosimeter DK-02 and neutron-activation technique with indium. Counter SI-8B has a thin mica input window that allows it to detect not only beta and gamma radiation, but also alpha particles and soft X-rays. The dosimeter DK-02 is a capacitor-based ionization chamber with measurement range 200 mR (beta and gamma rays). Indium plates, placed in the calorimeter water, are used to monitor neutrons. To measure indium activity, we used two Geiger counters. Impulses from the counters are recorded by the computer. The large surface of the indium plate (18 cm^2) allows the detection of slow neutrons that have a flux density more than 0.2 neutron/cm^2 s. In addition, the computer recorded impulses from the counter mounted on the reactor's top and impulses from the wattmeter. Another computer running PCLAB-2000 was coupled with a data logger. It displayed

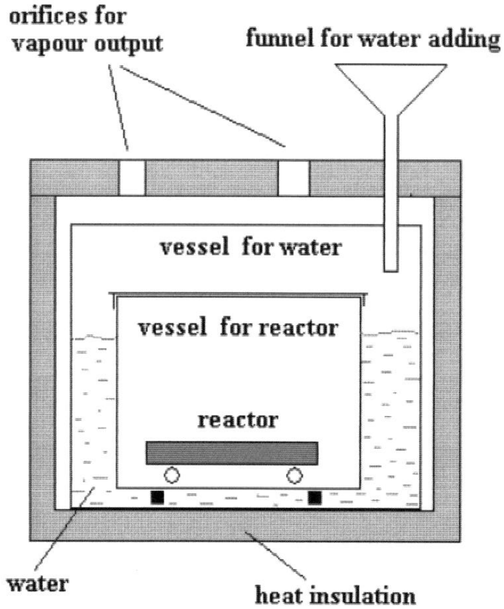

Figure 2. Schematic of the calorimeter with vaporized water.

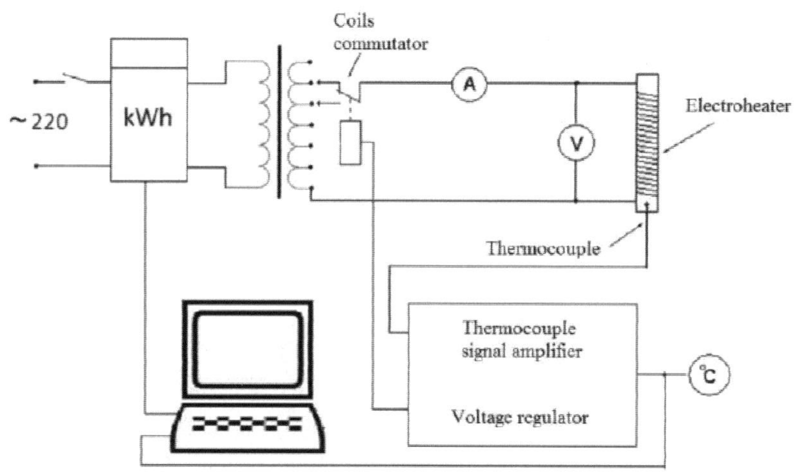

Figure 3. Schematic of power supplies and the resistance heater power regulation.

Figure 4. First type of reactor during operation.

and recorded the reactor temperature, and the signal from the DK-02 counter.

5. Temperature Changes during the Heating

To operate the device with heat production exceeding the input electric power, the following conditions must be met: Li[AlH_4] decomposition to hydrogen release, clarification of nickel granules surface, absorption of the emitted hydrogen by nickel, and finally the device must be operated at a high temperature. The production of energy is a result of processes in the fuel mix. It is important to note that the speed of heating has to be low in order that the emitted hydrogen can be absorbed by nickel to avoid an excessive increase of hydrogen pressure.

As an example, we show diagrams of the temperatures rising for experiment 20-12-2014. We slowly increased the power of the heater from 25 up to 500 W (Fig. 5). The temperature of 1000°C was reached after 5 h of heating. In the same diagram, we also show the count rate of the Geiger-counter SI-8B. As one can see during the entire heating process, the radiation level was similar to the background. The DK-02 dosimeter did not show the dose above the accuracy limit (5 mR) during the experiment. No noticeable activation of indium was detected.

Figure 6 shows the temperature changes with heating powers 300, 400, and 500 W. One can notice that with constant heating power the temperature gradually increases, especially in the last interval. It should be pointed out that excess heat occurs in addition to the electrical heating. At the end of the interval, at the maximum temperature, some temperature oscillations occurred. This interval ended when the electrical heater burned out, terminating the electrical heating. After that, the temperature became constant at 1200°C for 8 min, and afterwards begins to decrease. It should be noted that the reactor at this time was producing heat at kilowatt levels without electrical heating.

Thus, from the heating diagram itself we can see that reactor is capable of generating large amounts of heat in addition to the electrical heating.

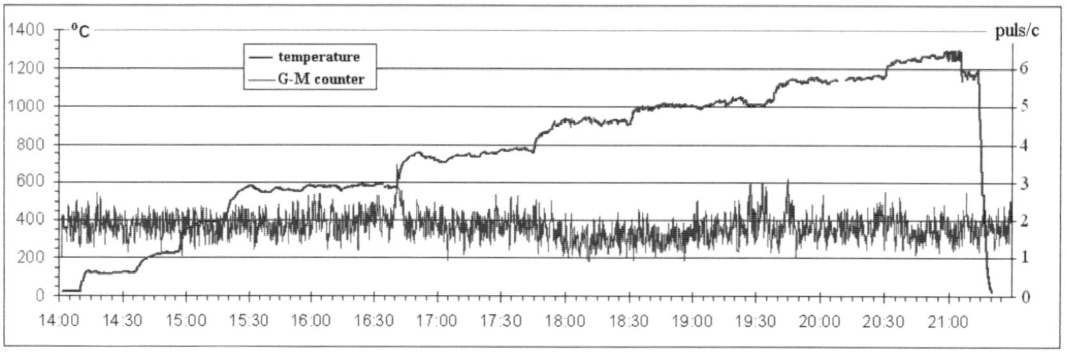

Figure 5. Dynamics of temperature and data from the Geiger counter during the experiment on 20-12-2014.

6. Produced Heat and COP Calculation

In Table 1 calculations are made for three reactor's operation modes: with temperature about 1000°C, about 1150°C, and 1200–1300°C. With temperatures 1150°C and 1200–1300°C the heat produced is meaningful exceeds the consumed energy. While operating in these modes, for about 90 min, about 3 MJ or 0.83 kWh was produced. The energy would be produced by burning about 70 g of gasoline.

Table 2 shows results obtained in all the experiments from December 2014 to January 2015. Besides the experiments with reactors loaded with Ni + Li[AlH_4] mixture, we also did experiments with a mock-up reactor without fuel. In cases with the mock-up reactor, the ratio of produced to consumed energy was close to 1, as it was in the experiments with fuel but with temperature below 1000°C. Significant excess heat was observed only when the reactor is loaded with the Ni + Li[AlH_4] fuel and held at a temperature about 1100°C or higher.

7. The Problem of Uncontrolled Local Overheating

The maximum duration reached with this type reactor operating in excess heat mode was 1.5 h. The causes of such short operating time are the failures due to local overheating. A view of the reactor during this kind of overheating

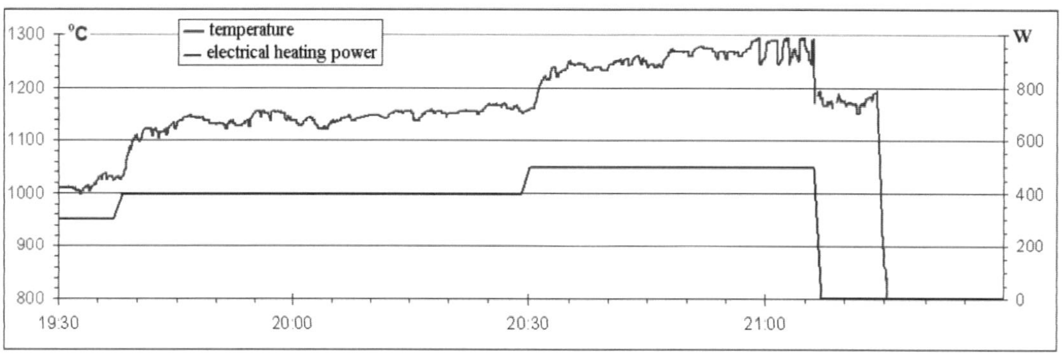

Figure 6. Dynamics of temperature and the power of electrical heating at high temperature for the experiment on 20-12-2014.

Table 1. COP calculation for the experiment on 20-12-2014.

Average temperature (°C)	970	1150	1290
Duration of regime (min)	38	50	40
Electrical input power (W)	300	394	498
Electrical input energy (kJ)	684	1182	1195
Mass of evaporated water (kg)	0.2	0.8	1.2
Energy to heat water to boiling (kJ)	63	251	377
Energy spent on evaporation (kJ)	452	1808	2712
Heat leakage through insulation (W)	70	70	70
Heat leakage through insulation (kJ)	159	210	180
Total emitted heat energy (kJ)	674	2269	3269
Ratio of the emitted heat energy to input energy	0.99	1.92	2.74

Table 2. COP in experiments in December 2014 and January 2015 (Reactor loaded with fuel).

Date	Temp. °C	Duration min	Cons. W	Prod. W	COP
20-12-2014	970	38	301	297	0.99
20-12-2014	1150	50	395	758	1.92
20-12-2014	1290	40	499	1365	2.74
04-01-2015	940	131	304	305	1.00
04-01-2015	1020	75	377	407	1.08
10-01-2015	1080	73	161	284	1.77
18-01-2015	800	90	308	293	0.95
18.01.2015	1080	38	78	135	1.73

Table 3. COP in experiments in December 2014 and January 2015 (Reactor loaded without fuel).

Date	Temp. (°C)	Duration (min)	Concentration (W)	Production (W)	COP
02-01-2015	210	56	211	227	1.07
02-01-2015	470	88	433	414	0.95
02-01-2015	1050	16	928	1035	1.12
21-01-2015	1000	69	297	296	1.00
21-01-2015	1080	43	306	297	0.97
28-01-2015	900	65	95.5	105	1.08
28-01-2015	1100	66	116	116	1.00
28-01-2015	1200	50	151	147	0.97

is shown in Fig. 7. The level the temperature reached was enough to melt the alumina (melting point 2040°C) (see Fig. 8).

8. Type-2 Reactor Works for Long Duration

The experiments with the above-described devices showed that the mixture of Ni + Li[AlH_4] heated in a hermetically closed ceramic tube at temperatures higher than 1100°C produce significantly more heat than the input energy. However, the operating time of these reactors before they fail is too short to produce measurable isotopic or atomic changes, and thus to short to show that the release of the excess heat is caused by cold nuclear transmutations.

To achieve longer continuous work durations, we had to make many changes to the design of the reactor. First of all, we had to abandon calorimetry based on the measurement of the quantity of evaporated water, because it is difficult

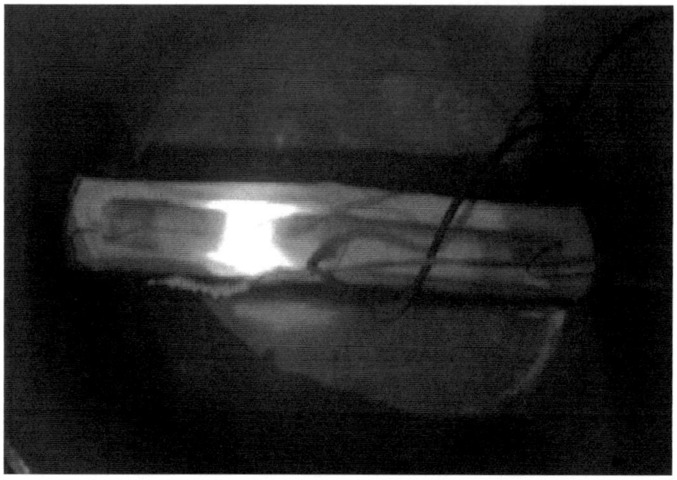

Figure 7. The local overheating that leads to reactor's failure.

to add make-up water all day long.

9. Construction of Type-2 Reactor

The reactor tube is designed for long duration operation is 29 cm long. Only its central part is heated. Due to the low thermal conductivity of the ceramic, the ends of the tube are not very warm (at 1200°C in the center, the ends are not

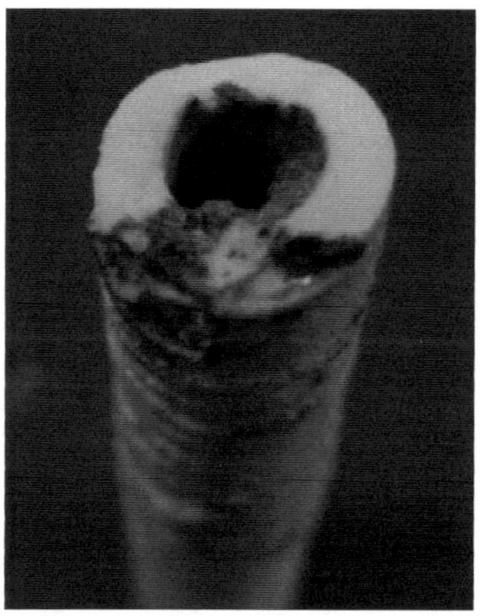

Figure 8. Fragment of the reactor destroyed due to local overheating.

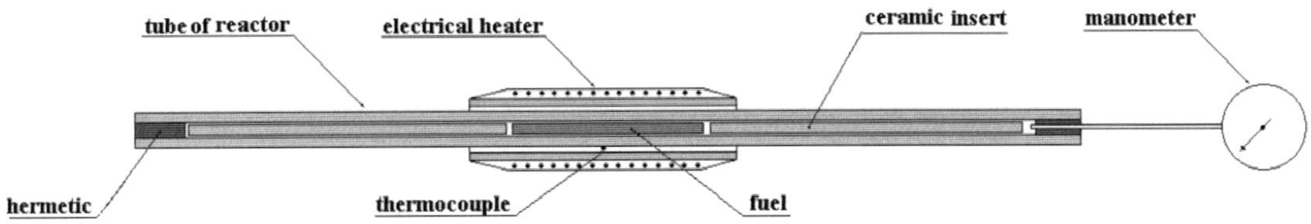

Figure 9. Construction of the reactor for long duration work.

warmer that 50°C); this allows the use of epoxy hermetic sealant to close the tube (Fig. 9).

The heater made from Kanthal A1 that works up to 1400°C. The fuel mixture (640 mg Ni + 60 mg $LiAlH_4$) is in a container of thin stainless steel. To displace the air from the tube, we used ceramic inserts. A manometer with a limit of measurement of 25 bar is connected to the reactor with a thin tube of stainless steel.

10. Equipment for Heater Power Supply, Measurement of the Consumed Energy and Control of Temperature

The electric heater is connected to the power supply through a thyristor regulator. An AC ammeter and a voltmeter are used to monitor and control the power consumption. Because the thyristor regulator distorts the sinusoidal form of current, such devices do not give correct measurements. To get correct measurement of the consumed electric power, the electronic watt-meter is used. This displays not only the amount of consumed electric power but it also sends consumption data to the computer.

To control of reactor temperature, a K-type thermocouple is used. The hot junction is placed on the surface of the reactor tube in the middle of the heating zone. The thermocouple data is displayed on a hand-held meter and also sent to the computer. The signal from the thermocouple is also used to thermostatically regulate the electric heater power so that a fixed temperature is maintained.

The computer used to record the temperature had to be switched-off periodically to recharge the battery. At this time, temperature control was maintained with the hand-held meter, and the temperature continued to be printed on paper. Figure 10 shows Type-2 reactor during testing.

11. Change of Temperature and Pressure during Reactor Operation

Figure 11 shows how temperature changed during heating, and it shows the power needed to reach the set temperature. The temperature of 1200°C at the surface of the reactor tube was reached after 12 h of stepwise increases of temperature, at heater power of 630 W. After 1 hour the power needed to maintain the temperature of 1200°C decreased to 330 W.

In Fig. 12, the change of pressure in the reactor chamber in the course of heating compared to the change in temperature is shown. The increase of pressure starts around 100°C. Maximum pressure of app. 5 bar was attained at 180°C. After this pressure starts to fall and at 900°C was lower than atmospheric pressure. The greatest decrease (–0.5 bar) was attained at 1150°C, then the pressure starts to increase slowly to atmospheric pressure.

In Fig. 13, the power of the electric heater for 4 days until burn-out of the heater wire as a result of its gradual oxidation is shown.

For almost 3 days the power necessary to maintain the temperature of the reactor tube at 1200°C was in the range of 300–400 W (Fig. 13). Before the power supply burned out, the power started to increase and at the time it burned out, power was 600 W. The burning out was caused by gradual oxidation of the resistor.

Figure 10. Type-2 reactor during testing.

12. Operation of the Reactor with the New Heater

One day after the heater burned out, the reactor was switched on repeatedly with the same reactor tube, but with a new heater. Figure 14 shows the power consumed by the reactor after replacement of the heater. Unlike the first heating, here after achievement of temperature 1200°C essential decrease in power consumption did not occur. The power which is required for maintenance of temperature 1200°C was in the range 600–700 W, i.e. was approximately the size it was at the time of burn-out of the first heater. Only at the end of repeated turning on the reactor power consumption decreased a little.

Approximately a day after repeated turning on of the reactor it was switched off by gradual reduction of heater power.

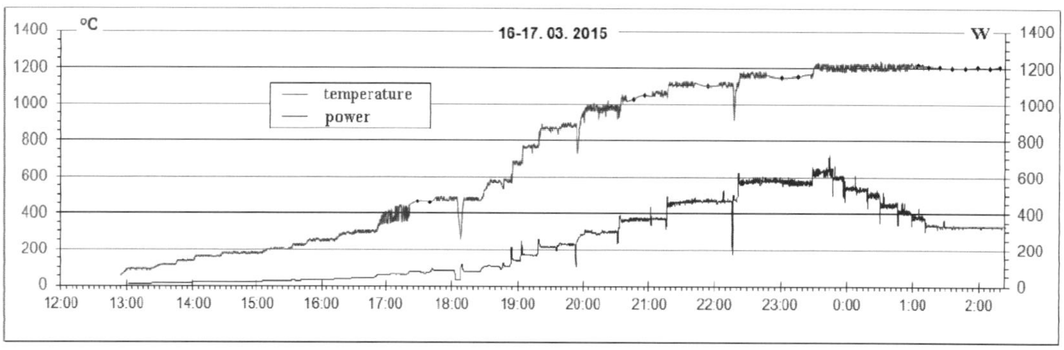

Figure 11. Heating of reactor to working temperature. Moscow time is shown on the horizontal scale.

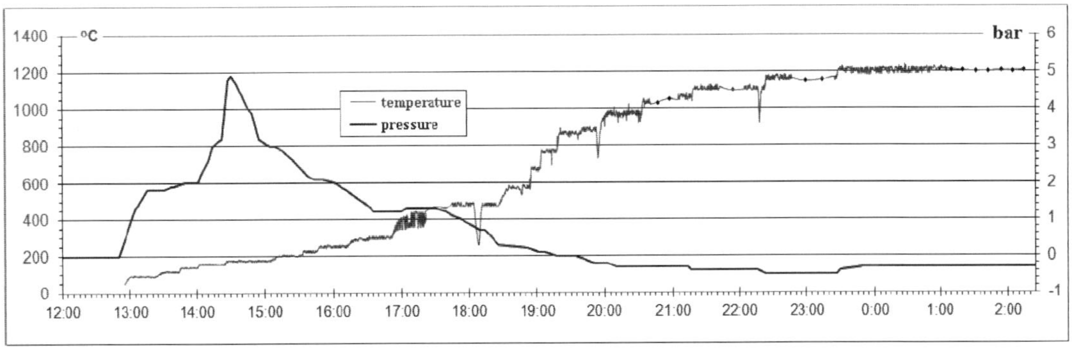

Figure 12. Heating of reactor to working temperature. Change of pressure during the course of heating.

13. Detection of Radiation

For radiation level detection, the same instruments were used as in the type-1 experiments: and SI-8B Geiger counter and the DK-02 dosimeter. For neutron measuring a neutron-activation technique with indium was employed. Measurements showed that the level of nuclear radiation in reactor operating time does not significantly exceed background.

14. Ratio of produced heat and the consumed electric power (COP)

As in the described experiment direct measurement of amount of the heat cannot be done, and reliable measurement of it is quite a complex challenge. This problem can be solved by comparing parameters of the reactor containing the fuel mix and the reactor without the fuel mix.

Figure 15 shows the power necessary to achieve the set reactor temperature without fuel and with fuel. It can be seen that at temperatures above 700°C the reactor with fuel consumes less electric power than the same reactor without fuel. This indicates the existence of a heat source besides the electrical heater.

About 1100 W of power is needed to maintain a temperature of 1200°C without fuel. In the presence of fuel (at the first switching) in the beginning 650 W is needed to maintain this temperature, and an hour after reaching it, only 300–330 W were required. Based on these data, it is possible to make an assessment of the COP. It is necessary to

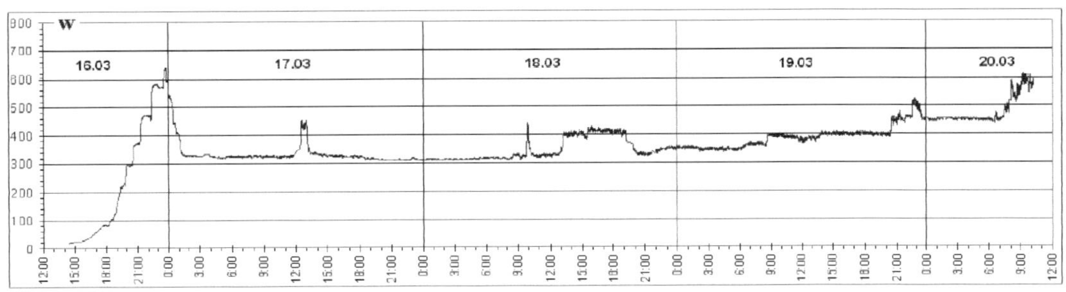

Figure 13. The power of heating during almost 4 days until the heater burned out. After reaching a temperature of 1200°C at 23:30 16-03-2015 this temperature was supported until 10:50 20-03-2015 by automatic adjustment of heater power.

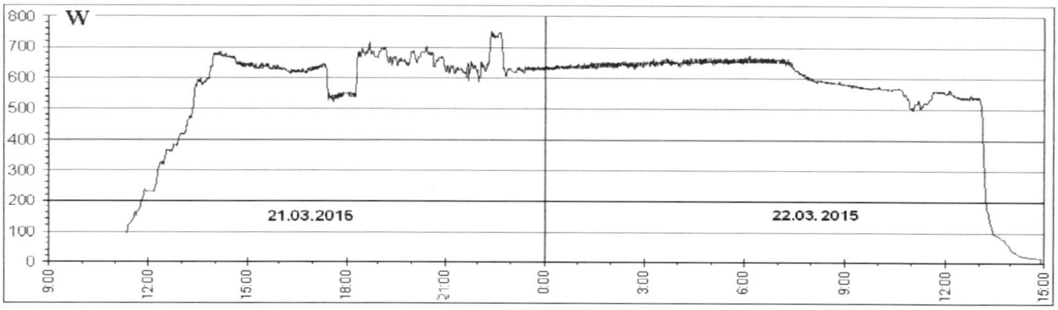

Figure 14. The power consumed by the reactor during repeated switching with the new heater. After reaching the temperature of 1200°C at 13:50 21-03-2015 this temperature was supported until 13:00 22-03-2015 by automatic adjustment of power of the heater.

consider the difference between processes in the reactor without fuel and with fuel when there is an additional thermal emission.

The intensity of heat exchange (produced power) depends on the temperature at the border of the environment, i.e. the external temperature of the heater. In the absence of an internal source of heat, the temperature outside and inside cannot strongly differ. In the presence of an additional thermal emission in the center of the reactor there is a thermal flux from inside to outside, which causes a temperature gradient. Therefore, the temperature measured by the thermocouple on the reactor tube surface (inside the heater tube) is higher than heater outer surface temperature (Fig. 16).

Measurements by the additional thermocouple show that with fuel at a temperature 1200°C on the reactor's inner tube heater outer surface temperature (at the first switching) was near 1070°C. As the produced power is defined by external temperature, the reactor overall makes a lot of heat, about 800 W, which is how much input was required

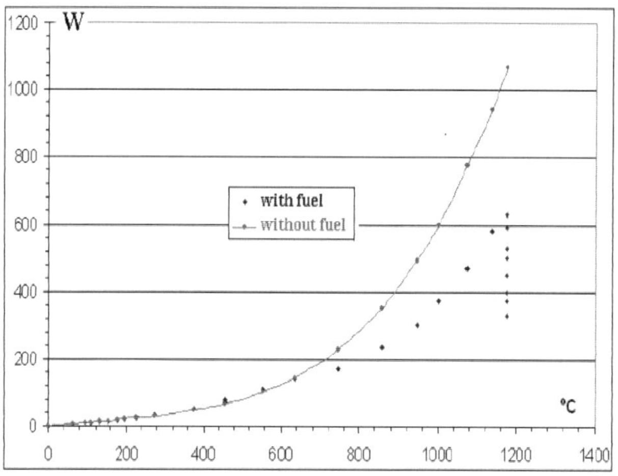

Figure 15. Power necessary for maintenance of the set temperature.

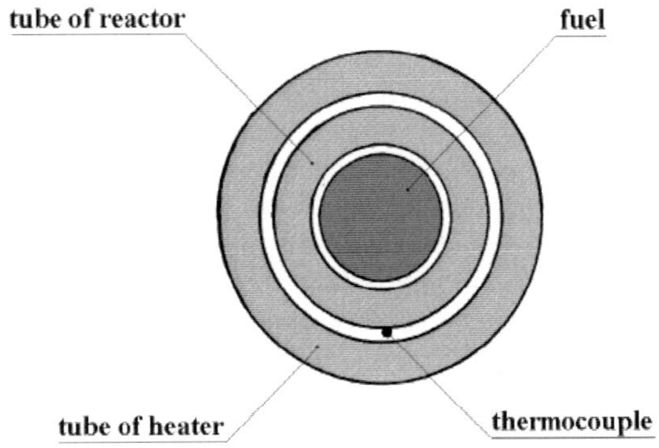

Figure 16. Cross section of the reactor central zone.

without fuel to maintain the temperature a 1070°C. Based on this, the COP is 800/330 = 2.4.

At repeated switching at a temperature 1200°C on a reactor tube surface, the heater surface temperature is 1130°C. The reactor makes apparent excess heat of about 950 W, which is how much input power was required without fuel to maintain the temperature 1130°C. Power consumption fluctuated from 550 to 700 W. Therefore, with repeated switching this COP is 1.3–1.7.

15. Research after the Test Run

After stopping and cooling the reactor, the tube of the reactor was opened, and the ceramic inserts and the container with fuel were removed from it. It was revealed that the container well remained though it and sites of ceramic inserts, close to it, became covered by a film of black color. The fuel extracted from the container represents caked substance of light gray color by the form very different from initial fuel mix (fine powder of black color). In optical microscope, it is visible that the used fuel has an appearance of baked small droplets of golden color with impurity of powder of gray color.

The images received on the electronic scanning microscope show that nickel in the initial fuel mix has an appearance of porous spherical clusters about 10 μm in size. Lithium aluminum hydride has an appearance of flakes with the size from 1 to 100 μm. In the fuel extracted from the reactor (after operation) we can identify two components: alloyed mass, consisting mainly of nickel, and flakes, consisting mainly of aluminum and oxygen.

16. Analysis of Compositions of the Initial and Used Fuel

Investigations of element and isotope composition of fuel mix are in progress. The analyses carried out so far did not reveal large changes in the isotopic composition of fuel. It is possible that it is connected with insufficient duration of this experiment. Lugano experiment which found strong isotope changes was 10 times longer at higher power.

Research *of element* structure on electronic microscope showed a very strong distinction for different places of test selection. Nevertheless, two fractions distinctly differ: in one aluminum and oxygen prevail, and in the other nickel prevails. The ratio of elements in fractions taken before and after operation of the reactor is approximately equal. After the reactor operates, in the fraction with nickel the contents of iron, chrome, silicon, sodium, potassium, titanium and some other elements considerably increased.

17. Conclusions

(1) Experiments with devices similar to the high-temperature Rossi heat generator loaded by mixture of Ni and Li[AlH_4] demonstrated that these devices produce more energy than they consume at the temperature of about 1100°C and above.
(2) Type-2 apparatus worked continuously for more than 3 days, producing more than twice as much heat as the applied electrical energy. More than 40 kWh or 150 MJ were produced in excess of the electrical energy consumed. This is the amount of energy obtained by burning several liters of petroleum.
(3) The reactor chamber pressure during slow heating was relatively low.
(4) No ionizing radiation above the background level was observed while operating the reactor.
(5) Preliminary conclusions from the analysis of fuel element and isotope composition indicate a minor change in isotopic ratios and the emergence of new elements in the used fuel.

Acknowledgments

The authors would like to thank K. Alabin, S. Andreev, E. Belousova, and E. Buryak for organizing the analysis of the fuel mixture, R. Grinyer for useful discussions, N. Samsonenko, L. Urutskoev, V. Zhigalov, G. Levi, P. Gluck and many other people in Russia and abroad for their support and valuable advice.

References

[1] G. Levi, E. Foschi, B. Höistad, R. Pettersson, L. Tegnér and H. Essén, Observation of abundant heat production from a reactor device and of isotopic changes in the fuel, http://www.sifferkoll.se/sifferkoll/wp-content/uploads/2014/10/LuganoReportSubmit.pdf.
[2] A.G. Parkhomov, Investigation of the heat generator similar to Rossi reactor, *Int. J. Unconventional Sci.* **7**(3) (2015) 68–72. http://www.unconv-science.org/pdf/7/parkhomov-en.pdf.
[3] A.G. Parkhomov, Investigation of new version of the device similar to high-temperature Rossi heat generator, *Int. J. Unconventional Sci.* **8**(3) (2015) 34–38, http://www.unconv-science.org/pdf/8/parkhomov-en.pdf.

J. Condensed Matter Nucl. Sci. 19 (2016) 257–263

Research Article

Search for Low-energy X-ray and Particle Emissions from an Electrochemical Cell

Dennis Pease*, Orchideh Azizi, Jinghao He, Arik El-Boher and Graham K. Hubler

The Sidney Kimmel Institute for Nuclear Renaissance (SKINR), Department of Physics and Astronomy, University of Missouri, Columbia, MO, USA

Sango Bok, Cherian Mathai and Shubhra Gangopadhyay

Department of Electrical Engineering, University of Missouri, Columbia, MO, USA

Stefano Lecci and Vittorio Violante

ENEA Research Center, via E. Fermi 45, 00044 Frascati (Rome), Italy

Abstract

Several theories to explain anomalous heat production predict the emission of low-energy X-rays and/or MeV alpha particles from PdD cathodes in electrochemical cells. Such radiation, however, is not detectable from outside of a standard electrochemical cell due to absorption in the electrolyte and cell walls. A custom cell was therefore assembled which permits X-rays of energy $\geq$ 1 keV to pass through a thin cathodic membrane and enter into an X-ray detector with minimal attenuation. This test cell geometry also potentially allows any emitted MeV alpha particles to be detected when they impact a Pd cathode and cause fluorescent emission of Pd-K_α (21.2, 23.8 keV) X-rays. The detection of X-ray emissions from a membrane electrolytic cell potentially permits the mechanism(s) for anomalous heat production to be investigated with great sensitivity. As an example, a typical X-ray detector allows 1 keV X-rays to be detected at emission rates of less than one per second and this level of sensitivity corresponds to a thermal resolution of < 0.2 fW. Time resolved X-ray spectral data ranging from 1 to 30 keV was collected for over a year using various types of membranes and different electrolytic solutions. None of these test cells, however, yielded any X-rays which were above ambient background levels.

Keywords: Anomalous heat effect, Electrolytic cell, Low energy, PdD cathode, X-rays

1. Introduction

Low energy emissions in electrochemical anomalous heat (AHE) experiments have never previously been thoroughly investigated by using time-resolved detectors with real time threshold analysis and continuous data collection. While

*E-mail: peased@missouri.edu

© 2016 ISCMNS. All rights reserved. ISSN 2227-3123

Table 1. A list of various theories which predict low energy X-ray or particle emissions associated with AHE.

Author	Theoretical basis for emissions	Predicted emissions
Hagelstein [5]	2D fusion	1–4 keV X-rays
Takahashi [1]	4D fusion	4 keV X-rays, MeV alphas
Kucherov [3]	Photon stimulation of nucleus	Alpha decay, fission
Mayer [6]	Tight molecular binding of light particles	4.7 keV X-rays
Meuhlenberg, Mills, others [2,4]	Collapsed electron states	~4 keV X-rays

many AHE theories predict the production of low energy X-rays or MeV alpha particles, there has always been a fundamental problem in trying to detect such emissions. This is because low energy radiation would normally be attenuated to undetectable levels by absorption processes acting within the electrolyte, electrodes, and/or calorimeter structures of a standard electrochemical cell. Table 1 summarizes prior theoretical predictions of nuclear emissions expected to emanate from active cathodes which are producing anomalous heat by various postulated processes.

2. Experiment Concept and Design

A series of experiments were performed using custom electrolytic cells with two types of thin membrane cathodes having 0.1 M LiOD electrolyte on one side and air on the other. The first type of cathodic membrane used a 1 μm silicon nitride film coated with 200 nm of platinum (Pt) and palladium (Pd). Three different electrolysis co-deposition methods were used during electrolysis using this type of cathode: D/Pd deposited on Pt cathode from a Pd anode; D/Pd from a Pd salt solution; and D/Pd from a Pd nanoparticle suspension. The second type of cathodic membrane used a 30 μm thick palladium foil. Both of the above cathode types were electrolyzed using constant current for periods of time ranging between 1 and 90 days.

An electrochemical cell was fabricated that permits emissions of low energy X-rays to be detected with very little attenuation by using a thin membrane cathode. (Note: Si_3N_4 membranes are obtained from Silson LTD.) X-ray absorption calculations indicate that for the thin Si_3N_4 membrane, absorption is 90% at 1 keV, and rapidly decreases to 10% at 2 keV. Since a Pd film on the membrane would delaminate when charged with deuterium or hydrogen, 100 or 200 nm Pt film was sputter deposited on the membrane for the anode.

Most calorimeters have sensitivity in the range of 1–10 mW. Suppose that AHE is active much of the time but at micro-, nano-, or picowatt levels. A standard calorimeter would be unable to detect such low levels of power output. If 2 keV X-rays are emitted at $\sim$1 Hz into a background $<$0.2 counts per second, the corresponding heat sensitivity is $<$100 Attowatts. (This sensitivity calculation includes transmission through 100 nm Pt, 100 nm Pd, 1 μm Si_3N_4, 2 mm Air, and measured background X-ray counts.)

Advantages of the above method are as follows.

- Potential sensitivity to detect nuclear emissions corresponding to $<$1 Femtowatt of anomalous heat.
- Ability to directly detect 1–30 keV X-rays.
- Detection of MeV alphas via X-ray induced fluorescence generated from particle impacts on Pd, Pt, and Si cathode structures.
- Enable direct testing for energetic emissions predicted by many different AHE theories including non-nuclear proposals.

Disadvantages of the above method are as follows.

- The novel thin film anode/cathode/electrolyte combinations have not previously been demonstrated to reliably produce excess heat. However, a Pd anode/Pt cathode combination did produce 10% excess at

"Energetics Inc." in 2005 (ref. unpublished internal data) and Pd nanoparticles in electrolyte produced excess heat at SKINR in 2014–2015 (ref. unpublished internal data).

- Pd foil loaded from just one side is not known to yield AHE but similar size and thickness foils certainly can yield anomalous heat when charged from both sides.

Figure 1 is a schematic representation of an electrochemical cell which shows a thin cathodic membrane and the relative location of the detector used to collect X-ray spectral data in recent experiments at SKINR and ENEA.

The X-ray detector used for most of the data collection at SKINR was an Amptek XR100-CR Si PIN diode connected to an Ortec multichannel analyzer running Maestro-32 software. This detector was sensitive to 1–30 keV X-rays and had a spectral resolution of <0.3 keV FWHM. Time resolved spectra were collected for 100-second time intervals, compared to nominal ambient background levels using a custom software program, and then archived into a file for future reference. The custom program compared the amplitude of the X-rays detected in each of 4096 energy bins relative to specific threshold values which were previously determined as being the nominal ambient background level expected at that energy. If ANY threshold value was found to be exceeded the program would flag the data set as a *possible* event, write a summary of the specific data which exceeded the threshold into a text file, and generate a real time audible alarm.

The X-ray detector used for data collection at ENEA was an Amptek X123 Si PIN diode containing an integrated multichannel analyzer and having a spectral sensitivity and resolution nearly the same as stated for SKINR's detector. Data collection using the X123 detector at ENEA was not time resolved but instead was time integrated over the entire data collection period.

3. Experimental Results at SKINR

Figure 2 shows a calibration spectrum of the X-ray detector which used a ^{241}Am radioactive source while Fig. 3 is a typical background spectrum collected by acquiring data for 200,000 s. The ambient background was found to be very repeatable and essentially free of any naturally occurring radioactive source emission lines. Figure 4 shows two 100-second interval runs, one using a 30 μm thick Pd cathode and the other using a PdCl co-deposition of Pd on a

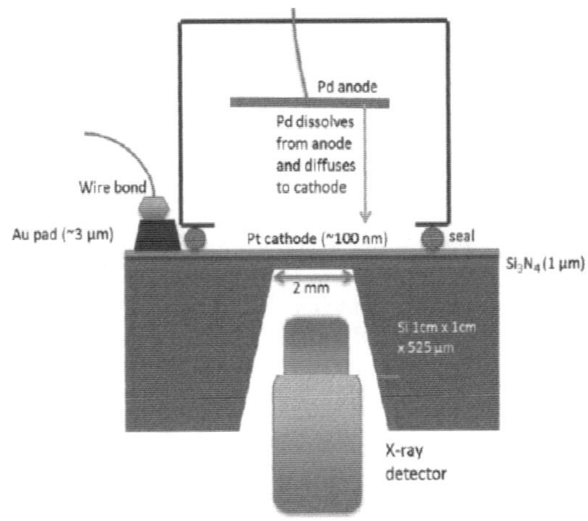

Figure 1. The cathodic membrane electrolytic cell used in SKINR and ENEA X-ray spectra tests.

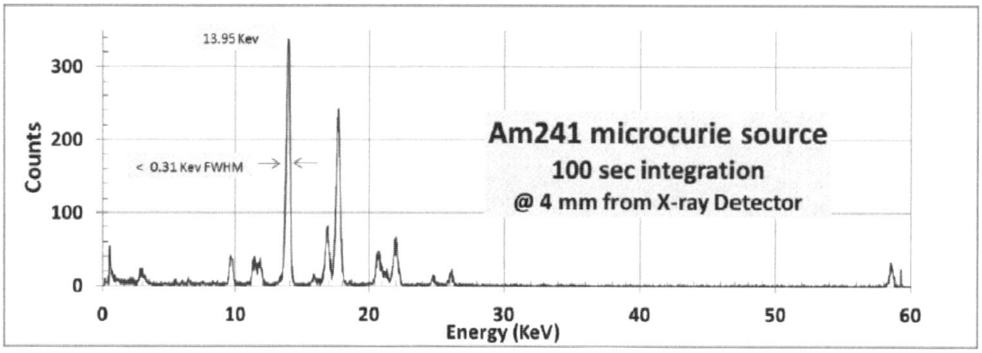

Figure 2. SKINR X-ray spectral calibration data acquired from an ^{241}Am microcurie test source.

SiN membrane. During each of the experimental test runs the custom search program was set to automatically trigger manual inspection if any channel exceeded five counts. At the end of each data run an even more sensitive check was performed by adding together all of the individual time-resolved test spectra and searching for the presence of any weak X-ray lines. Note that the above experimental technique would have also allowed any MeV alphas, if present, to have been detected by impact induced fluorescent X-rays from within the experimental setup materials including Pd $\sim$21.2, 23.8 keV $K_{\alpha,\beta}$; Pt $\sim$9.4, 11.1 keV $L_{\alpha,\beta}$; Pd $\sim$2.8, 3.0 keV $L_{\alpha,\beta}$; or Si$\sim$1.7,1.8 keV $K_{\alpha,\beta}$.

Table 2 contains a summary of the membrane characteristics and electrolytic cell parameters used in the various X-ray spectral data test runs at SKINR. For most of these data sets continuous X-ray data was collected and stored at 100 s intervals throughout the listed duration of the test so as to create a time-resolved view of each experiments X-ray spectral emissions. A re-occurring computer error (now fixed) caused some experiments to miss spectral data during portions of the total test run time. Even in those cases, however, there were many, many, 100 s long X-ray spectral data sets which were successfully acquired, stored, and reviewed.

Table 2. Membrane electrolytic cell parameters used in SKINR X-ray spectra tests.

Cathode		Anode	Electrolyte	Current density (mA/cm^2)	X-ray detector model/cell	Duration (days)
Membrane	Coated					
Si_3N_4	100 nm Pt	Pt	$PdCl_2$+LiCl	0.01–0.1	XR100/Glass cell	10
Si_3N_4	100 nm Pt	Pd	0.1 M LiOD	0.2–1	XR100/Glass cell	5
Si_3N_4	100 nm Au	Pd	0.1 M LiOD	0.5–2.5	XR100/Glass cell	11
Si_3N_4	100 nm Pt	Pd	0.1M LiOD	0.25–2	XR100/Glass cell	8
Si_3N_4	100 nm Pt	Pd	0.1 M LiOD	0.25–6	XR100/Glass cell	60
Si_3N_4	100 nm Pt	Pt	PdNp+0.1 M LiOD	0.1–6	XR100/Glass cell	7
Si_3N_4	100 nm Pt	Pt	PdNp+0.1 M LiOD	0.1–6	XR100/Glass cell	12
Si_3N_4	100 nmPd	Pt	PdNp+0.1 M LiOD	0.1–6	XR100/Glass cell	1
Si_3N_4	100 nm Pt	Pt	PdNp+0.1 M LiOD	0.1–6	X123/Glass cell	20
Si_3N_4	100 nm Pt	SS	PdNp+0.1 M LiOD	0.1–6	X123/Glass cell	90
Foil	Pd 30 μm	Pt	0.1 M LiOD	0.25–6	XR100/ Peek cell	30
Foil	200 nm Au on 25 μm Pd	Pt	0.1 M LiOD	0.5–10	XR100/ Peek cell	20

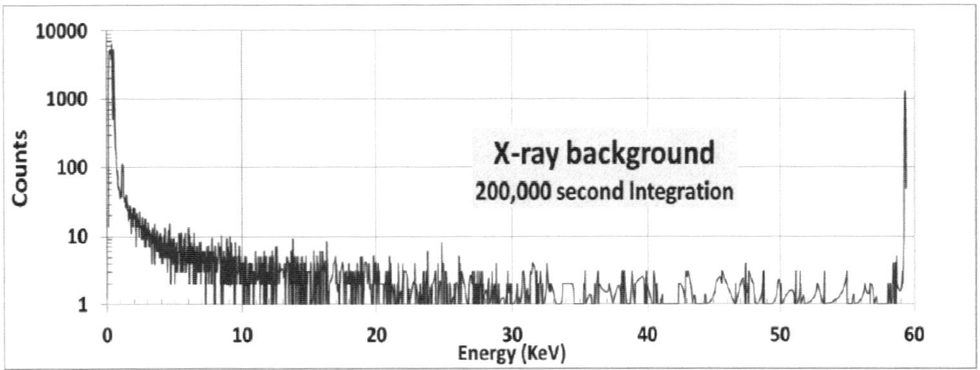

Figure 3. SKINR typical X-ray ambient background spectral data acquired and integrated for 200,000 s.

4. Experimental Results at ENEA

Additional tests using a similar cathodic membrane electrolytic cell and X-ray detector were independently conducted by ENEA, Italy. Figure 5 contains two representative spectrums from one of ENEA's experiments. One spectrum (blue trace) shows the nominal ambient background and has been scaled to match the same data acquisition period as was used in the collection of the active cell membrane data (red trace). This test data used a 25 μm thick Pd foil coated with 100 nm Au on the air/detector side and the plot shows X-ray spectra from 1 to 32 keV that was continuously accumulated for 5.8 days. The background and active cell spectra closely match one another at all wavelengths and are featureless except for an obvious line near 22 keV. The source of this line has been conclusively determined to be due to Pd $K_{\alpha,\beta}$ X-rays at 21.1 and 23.8 keV and it is created by fluorescence from 100 keV gamma rays emitted by a Bi radioactive isotope contaminant within the Pb shielding.

Table 3 contains a summary of the membrane characteristics and electrolytic cell parameters used in the various X-ray spectral data test runs at ENEA. For each of these runs a single time integrated X-ray spectrum was acquired

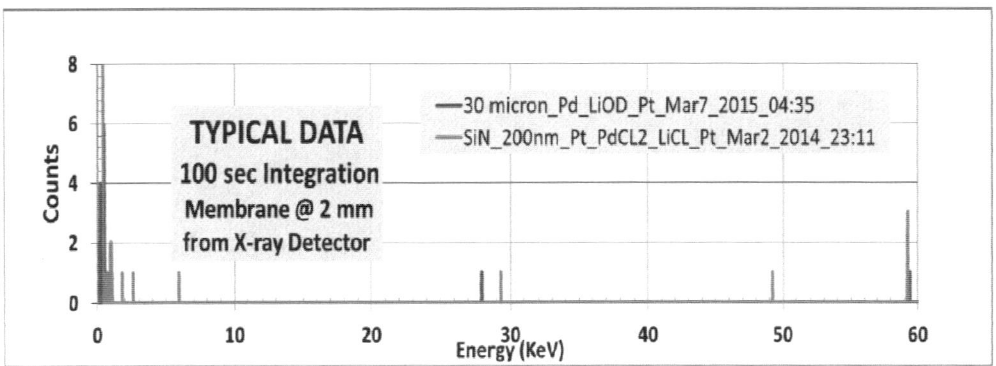

Figure 4. SKINR typical X-ray spectral data acquired from either a co-deposited Si_3N_4-Pd membrane or a thin Pd foil electrolytic cell.

Table 3. Membrane electrolytic cell test parameters used in ENEA X-ray spectra tests.

Cathode		Anode	Electrolyte	X-ray detector model	Duration (days)
Membrane	Coated				
Foil	100 nm Pt	Pd	0.1 M LiOD	Amptek X123	0.71
Foil	100 nm Au	Pd	0.1 M LiOD	Amptek X123	0.82
Foil	200 nm Au on 25 μm Pd	Pd	0.1 M LiOD	Amptek X123	5.8

which spanned the listed experimental duration.

5. Conclusions

It was hoped that these membrane X-ray experiments would reveal important information to help discriminate between the many different theoretical hypotheses proposed to explain the physical mechanisms which underlie and create anomalous heat effect (AHE) in electrolytic cells. No X-rays, however, were detected at SKINR above ambient background levels during several months of experimental runs. While ENEA data initially showed a couple X-ray peaks, these were later conclusively demonstrated to be due to Pd $k_{\alpha,\beta}$ fluorescence induced by 100 keV gamma rays emitted from Bi contaminants within external Pb shielding. While these detections showed that the experiment was definitely sensitive to fluorescence induced Pd X-rays the final result from ENEA was the same as from SKINR: *No X-rays were detected in any experimental runs.* The above null results at ENEA and SKINR imply the following derived conclusion.

If AHE involves the emission of low energy X-rays and/or MeV alpha particles then there are only two possibilities. One is that AHE was not active in any of the test runs at power levels between $\sim$1 Femtowatt and $\sim$1 mW and was therefore in an OFF state. The other possibility is that the cathode was actually in an ON state but was producing both imperceptible low levels of anomalous heat and undetectable levels of low energy radiation. In the latter case, our near individual particle sensitivity to radiation yields the conclusion that NONE of the theories in the table are likely to be viable.

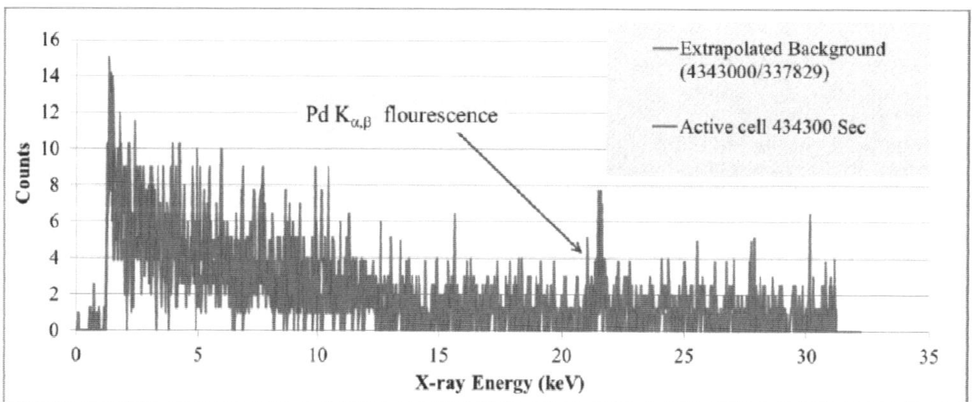

Figure 5. ENEA acquired X-ray spectra from a membrane electrolytic cell compared to an ambient background spectra scaled in intensity to match the same data acquisition time.

It was also concluded from the above experiments and their null X-ray results that the continuation and improvement of membrane electrolytic cell testing is warranted. While the fragile Si_3N_4-Pd co-deposited membranes only survived from 1 day to 3 months, the solid Pd cathode foils seemed to have an unlimited lifetime and could therefore be run for the longer times previously needed to produce substantial levels of anomalous heat. Several useful additions to the membrane experimental setup were also identified. These included placement of the membrane system in a calorimeter, searching for RF emission (RF was first detected by NRL [7] and later confirmed by ENEA [8], and/or stimulating the cathode with RF fields, acoustic waves, electrical pulses, or lasers.

Acknowledgement

This work was fully funded by Mr. Sidney Kimmel.

References

[1] A. Takahashi, Physics of cold fusion by TSC theory, *J. Phys. Sci. Appl.* **3**(3) (2013) 191–198.

[2] R. Mills, J. Lotoski, J. Kong, G. Chu, J. He and J. Trevey, High-power-density catalyst induced hydrino transition (CIHT) electrochemical cell, *Int. J. Hydrogen Energy* **39** (2014) 14512–14530.

[3] Y. Kucherov, Slow nuclear excitation model, *ICCF-6, Progress in New Hydrogen Energy,* M. Okamoto (Ed.), Hokkaido, Japan, Vol. 2, 1996, p. 502.

[4] A. Meulenberg and K.P. Sinha, Deep-orbit-electron radiation emission in decay from $^4H^{*\#}$ to ^4He, in *17th Int. Conf. on Condensed Matter Nucl. Sci.*, Daejeon, Korea, 12– 17 August 2012, *J. Condensed Matter Nucl. Sci.* **13** (2014) 357–368.

[5] P.L. Hagelstein and I.U. Chaudhary, Phonon models for anomalies, condensed matter nuclear science, Special Section: Low Energy Nuclear Reactions, *Current Science* **108**(4) (2015) 507–413.

[6] Frederick J. Mayer and John R. Reitz, Electromagnetic composites at the compton scale, *Int. J. Theoret. Phys.* **51** (2012) 322–330.

[7] D.A. Kidwell, D.D. Dominguez, K.S. Grabowski and L.F. DeChiaro Jr., Observation of radio frequency emissions from electrochemical loading experiments, *Current Science* **108** (2015) 578.

[8] V. Violante, E. Castagna, S. Lecci, G. Pagano, M. Sansovini and F. Sarto, RF detection and anomalous heat production during electrochemical loading of deuterium in palladium, EAI Energia, *Ambiente e Innovazione* **2-3** (2014) 63–77.

J. Condensed Matter Nucl. Sci. 19 (2016) 264–273

Research Article

Investigation of Enhancement and Stimulation of DD-reaction Yields in Crystalline Deuterated Heterostructures at Low Energies using the HELIS Ion Accelerator

A.S. Rusetskiy*, A.V. Bagulya, O.D. Dalkarov, M.A. Negodaev and A.P. Chubenko

P.N. Lebedev Physical Institute (LPI), Russian Academy of Sciences, Leninskii pr. 53, Moscow 119991, Russia

B.F. Lyakhov and E.I. Saunin

A.N. Frumkin Institute of Physical Chemistry and Electrochemistry, Russian Academy of Sciences, Moscow 119071, Russia

V.G. Ralchenko

Prokhorov General Physics Institute, Russian Academy of Sciences, Moscow 119991, Russia

Abstract

In this study, we present the results of studies of DD reactions in crystalline heterostructures at low energies using the ion accelerator HELIS. The results of measurements of the DD-reaction yields from the Pd/PdO:D_x and the Ti/TiO_2:D_x heterostructures in the energy range of 10–25 keV are presented. The neutron and proton fluxes are measured using a neutron detector based on ^{3}He-counters and a CR-39 plastic track detector. Comparisons with calculations show significant DD-reaction yield enhancement. It was first shown that the impact of the H^+ and Ne^+ ion beams in the energy range of 10–25 keV at currents of 0.01–0.1 mA on the deuterated heterostructure results in an appreciable DD-reaction yield stimulation. We also studied the neutron yield in DD reactions within a polycrystalline deuterium-saturated CVD diamond, during irradiation of its surface by a deuterium ion beam with energy of less than 30 keV. The measurements of the neutron flux in the beam direction are performed in dependence on the target angle, β, with respect to the beam axis. A significant anisotropy in neutron yield is observed, it was higher by a factor of 3 at $\beta = 0$ compared to that at $\beta = \pm 45°$.

Keywords: Enhancement of reaction yield, DD-reaction, Ion accelerator, Neutron detector, Plastic track detector

1. Introduction

The HELIS facility [1] at the P.N. Lebedev Physical Institute (LPI) operates with continuous ion beams with currents up to 40 mA and energies up to 50 keV. This multi-purpose accelerator addresses a wide spectrum of physics experiments,

*Corresponding author. E-mail: rusets@lebedev.ru

© 2016 ISCMNS. All rights reserved. ISSN 2227-3123

such as, e.g. light nuclei collisions at energies of several keV, investigation of elementary and collective processes in ion-beam plasma, and studies of the beam–target interactions using different materials with modification of the properties of the latter through ion-beam spattering of the thin-film coatings. In recent years, at HELIS, we studied the interactions of the deuterium beam with deuterium enriched crystalline heterostructures at energies of 10–25 keV.

Hitherto yields of DD-reaction in metal targets were studied mainly using only accelerators at energy $E_{\mathrm{lab}} \geq 2.5$ keV [2–7]. The desired further reduction of the accelerating voltage leads to serious problems related to the maintenance of current which makes it impossible to measure the products of the DD-reaction in a reasonable experiment duration, due to the extremely low yield. At the same time, a study of the behavior of yield (cross-section) DD-reaction at low energy deuterons (<1 keV) is of great interest in the study of processes and astrophysical screening effects in condensed media, leading to possible enhancement of the nuclear interaction of hydrogen isotopes in the metal.

Work [2] is very well known and frequently refereed, but the first results on dependence of DD-reaction yield on the target material was obtained in [3,4]. Experiments [8] demonstrated that the pulsed glow discharge capable of generating ions with energies of 0.8–2.5 keV and current densities of 300–600 mA/cm^2 at a deuterium pressure of 2–10 mm Hg. The current density used in the bombardment of the surface of the cathode (target) in a glow discharge is almost three orders of magnitude higher than that achieved with the use of accelerators. Analysis of the data showed that when $E_{\mathrm{d}} = 1.0$ keV Ti target bombarded by deuterons at a discharge current $I \sim 300$ mA, the gain of the reaction d(d,p)t was 10^9 in relation to the value of the yield at cross-section that defines the standard DD-reaction (approximation Bosch–Halle) [9]. Important factors that affect the enhancement of nuclear reactions in solids include a variety of external radiation (electrons, ions, X-rays, etc.). For example, in [10], it was shown that the effect of the electron beam with an energy of 30 keV and an X-ray beam with an energy up to 120 keV photons initiate in the systems Pd/PdO:D_x and Ti/TiO_2:D_x fusion of deuterium nuclei with the emission of 3 MeV protons.

Previous works [11–14] presented our results on dependence of DD-reaction yield on the target material and current beam density, as well as the results on enhancement of DD-reaction yield by ion beans.

2. Experimental Methods

2.1. HELIS facility

The HELIS facility (Fig. 1) is an ion accelerator of different gases to energy < 50 keV and includes: ion source with equipment, providing the power supply; beam focusing system; vacuum system; diagnostic apparatus for measuring current and ion beam energy.

2.2. Detector technology

Block diagram of the detector placement is shown in Fig. 2. A neutron detector based on ^{3}He-counters with organic glass and paraffin radiators is placed along and across the beam direction. CR-39 track detectors with various coatings surround the target irradiated by deuterium beam.

The CR-39 detector calibration by protons and alpha-particles was presented in [10]. The calibration of detectors 1 and 2 (placed along and above the sample) with the extended ^{238}Pu alpha source, simulating the sample position. Detection efficiency of CR-39 (in positions 1 and 2) to alpha-particles at 7 h etch was defined as $\varepsilon = 2.6\%$. The CR-39 detector was also calibrated by ^{252}Cf neutrons. Proton recoil spectrum after 7 h etch in 6 M NaOH at 70°C is at 4.5–8.0 μm track diameter (maximum at 5.2 μm) (Fig. 3). The neutron detection self-efficiency of CR-39 at 7 h etch was determined as $\varepsilon_{\mathrm{n}} \sim 10^{-4}$.

The ^{3}He detector was also calibrated by ^{252}Cf neutron source placed in the target position. The neutron detection efficiencies of ^{3}He detector groups at distance $R_1 = 85$ cm and $R_2 = 38$ cm were determined to be 0.1% and 0.4%, respectively.

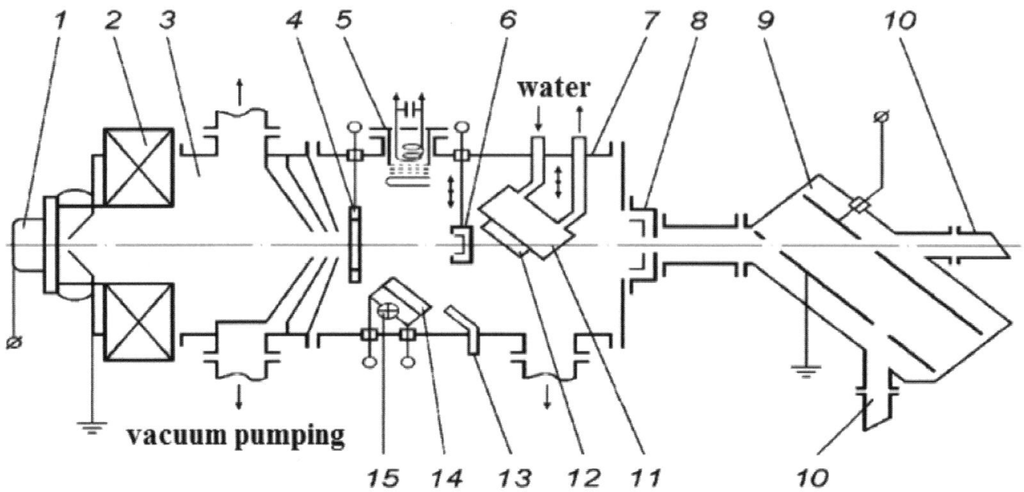

Figure 1. HELIS facility: 1 – ion source (duoplasmatron), 2 – electromagnetic lens, 3 – three-stage chamber of differential pumping, 4 – meter of a current of a transient-time type, 5 – auxiliary ion source, 6 – Faraday cap, 7 – chamber of targets, 8 – the device for calorimetric definition of a current of an ion beam, 9 – electrostatic analyzer, 10 – receivers of parsed fragments, 11 – water (or liquid gas)-cooled holder of the target, 12 – target, and 13 – feeder of gas in an vacuum chamber.

2.3. Pd and Ti sample preparation

The PdO/Pd/PdO samples were prepared by thermal oxidation from Pd foil (99.95% purity) of 50 μm thick with dimensions $S = 30 \times 10$ mm^2. Electrochemical loading in 0.3 M-LiOD solution in D_2O with Pt anode; $j = 10$ mA/cm^2. D/Pd ~ 0.73 (about 40 min required). The samples were rinsed in pure D_2O and were put in the Dewar glass to cool them down to 77 K. The cooled samples were then rapidly mounted (during $\sim$ 1 min) in sample holder in front of CR-39 detectors set, placed into HELIS chamber, vacuumed and irradiated by D^+ beam.

The Ti foils of 30 and 300 μm thick were loaded in a 0.2 M solution of D_2SO_4 in D_2O over 36 h at 10 mA/cm^2, in order to dissolve the TiO_2 oxide layer at the Ti-surface and to provide D-penetration. The average loading (D/Ti = 0.1 at depth of $\sim$ 1 μm) was determined by weight balance. Saturation of the sample can be carried out long before irradiation because the compound is absolutely stable at 300 K. The sample preparation procedure is described in more detail in [10].

2.4. The target of the polycrystalline diamond (CVD-diamond)

In our previous investigations of DD-reactions in the crystal targets (Pd, Ti), anisotropy was observed: the neutron flux along the beam direction was higher than that in the transverse direction [11,12]. This effect could be explained by the presence of narrow channels in the samples, where the bulk of deuterium, trapped during electrolysis, is concentrated. Particularly large anisotropy was observed using polycrystalline textured CVD diamond samples. The structure of the polycrystalline diamond film in cross section is shown in Fig. 4. The structure looks like columns growing from a few microns to about a hundred microns. More details about the CVD sample preparation procedure are described in [15]. The gaps between the columns are filled with deuterium atoms after electrolytic saturation.

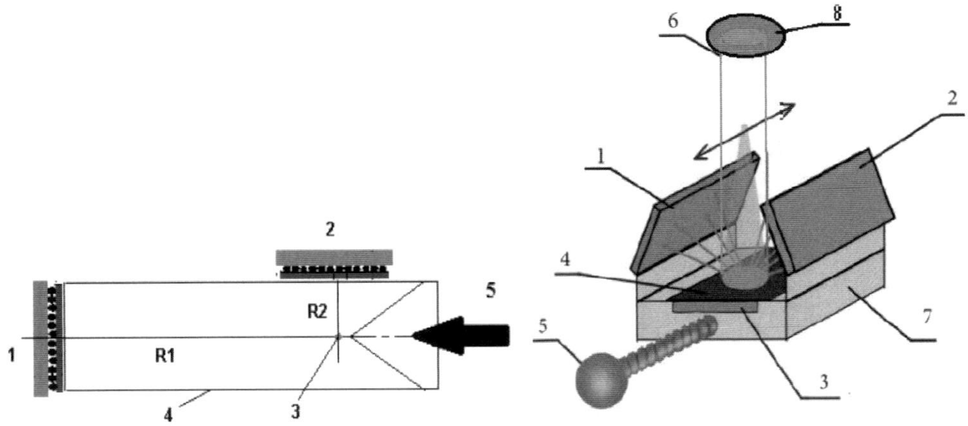

Figure 2. *Left panel*: the ^{3}He detector setup at HELIS, representing the first (1) and the second (2) ^{3}He-counter groups with radii $R_1 = 85$ cm and $R_2 = 38$ cm, respectively. The target is placed at (3) inside the HELIS beam pipe (4). The ion beam direction is indicated by (5). *Right panel*: Schematic representation of the target and track detector positions in the ion beam of the HELIS setup. (1)–(3) CR-39 track detectors with various coatings, (4) target, (5) manipulator, (6) ion beam, (7) water-cooled holder of the target, and (8) diaphragm.

3. Experimental Results

Thick target DD-reaction yield calculated using formula (1).

$$Y_{\mathrm{DD\,a.u.}} = Y_{\mathrm{DD}}/J_{\mathrm{d}} = N_{\mathrm{eff}}(T) \times \int_0^{E_{\mathrm{d}}} \sigma_{\mathrm{DD}}(E)(\mathrm{d}E/\mathrm{d}x)^{-1}\mathrm{d}E, \qquad (1)$$

where Y_{DD} is the DD-reaction intensity, J_{d} the deuteron current; $N_{\mathrm{eff}}(T)$ the effective concentration of bounded D in metal at temperature T, captured at depth x:

$$N_{\mathrm{eff}}(T) = N_0 \exp(-\varepsilon_{\mathrm{d}}\Delta T/k_{\mathrm{B}}T\,T_0),$$

where N_0 is the D concentration at $T_0 = 290$ K, ε_{d} the deuteron activation energy, k_{B} the Boltzmann constant, σ_{DD} the "bare" DD- cross-section, and $\mathrm{d}E/\mathrm{d}x$ is the stopping power in target calculated with Monte-Carlo code SRIM [16].

$$f(E) = Y_{\mathrm{exp}}E)/Y_{\mathrm{b}}(E) = \exp[\pi\eta(E)U_{\mathrm{e}}E] \text{ – enhancement factor},$$

where $Y_{\mathrm{exp}}(E)$ is the experimental yield of DD-protons, $Y_{\mathrm{b}}(E)$ is the yield at the same energy, determined according to the Bosch & Halle extrapolation, and $2\pi\eta = 31.29Z^2(\mu/E)^{1/2}$ is the Sommerfeld parameter (where Z is the deuteron charge, μ and E are the reduced deuteron mass and energy, respectively). U_{e} is the screening potential.

3.1. Ti/TiO$_2$:D$_x$ target and Pd/PdO:D$_x$ under the D$^+$ beam

The energy dependence of the fluxes of protons and neutrons emitted along and opposite to the beam for Ti/TiO$_2$:D$_x$ and Pd/PdO:D$_x$ targets is shown in Fig. 5. There is the anisotropy in the escape of DD-reaction products emitted along

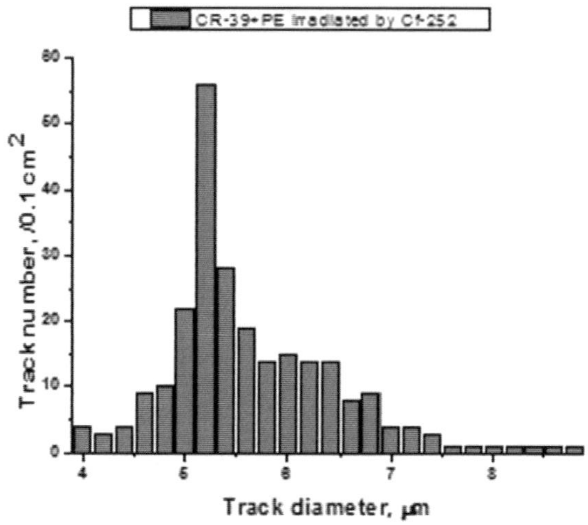

Figure 3. Proton recoil spectrum of CR-39 after 7 h etch in 6 M NaOH at 70°C.

and opposite to the beam. The dependence of the DD-reaction yield from the $Ti/TiO_2:D_x$ and $Pd/PdO:D_x$ target on the deuteron energy and the DD-reaction yields calculated by (1) for the given experimental conditions are shown in Fig. 6.

Accurate measurement of the temperature on the target is a difficult technical problem. The screening potentials were calculated for two "extreme" conditions:

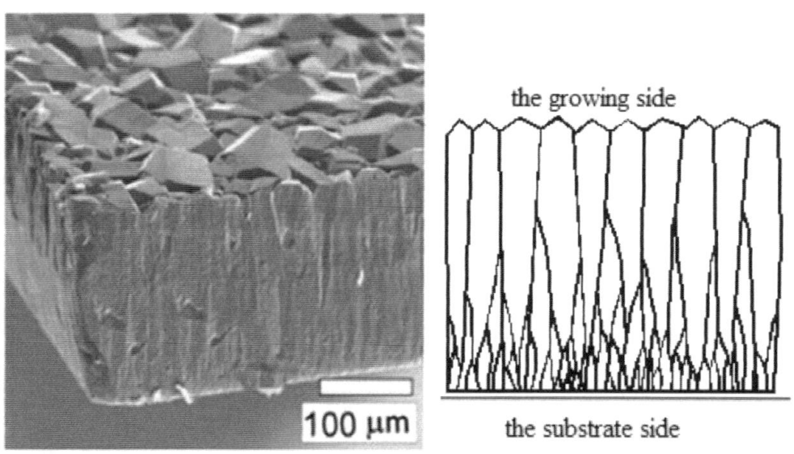

Figure 4. The structure of the polycrystalline textured diamond film in cross section.

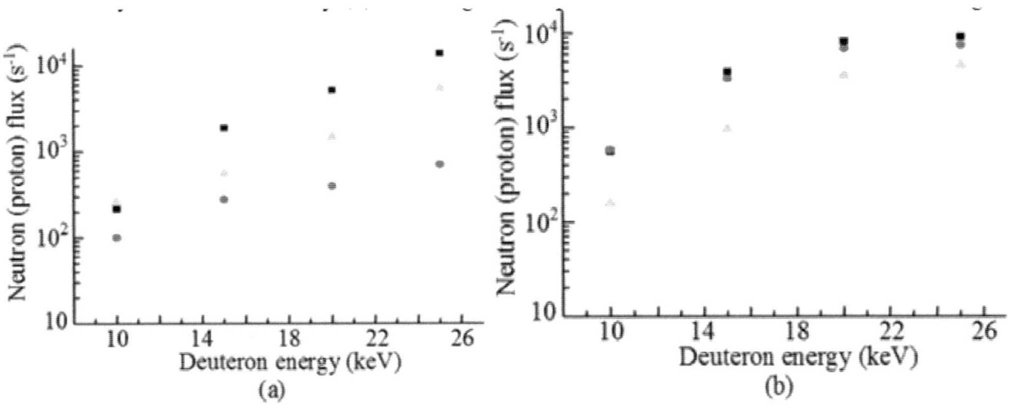

Figure 5. The energy dependence of the fluxes of protons and neutrons emitted along and opposite to the beam for Ti/TiO$_2$:D$_x$(left) and Pd/PdO:D$_x$ (right) targets. Neutron flux along the beam (■), neutron flux opposite to the beam (red filled circles), and proton flux opposite to the beam (green filled triangles). Measurements were performed by the CR-39 track detector.

(1) at $T = T_{\rm m}$, i.e., the metal melting temperature in the beam region, and
(2) at 350 K, i.e., the temperature measured by a thermocouple at the target edge.

Comparison with calculations shows significant effects of DD-reaction yield enhancement. The screening potential for Pd/PdO:D$_x$ heterostructure under experimental conditions was estimated as $U_{\rm e} = 630 - 980$ eV. The screening potential for Ti/TiO$_2$:D$_x$ heterostructure under experimental conditions was estimated to be in the range $U_{\rm e} = 160 - 750$ eV [11,12].

3.2. Ti/TiO$_2$:D$_x$ target and Pd/PdO:D$_x$ under the H$^+$ and Ne$^+$ beam

We studied the possibility of stimulating the DD-reaction yield from the Ti/TiO$_2$:D$_x$ and the Pd/PdO:D$_x$ heterostructures by ion beams using the HELIS setup. Instead of the deuteron beam, we used H$^+$ and Ne$^+$ ion beams in the

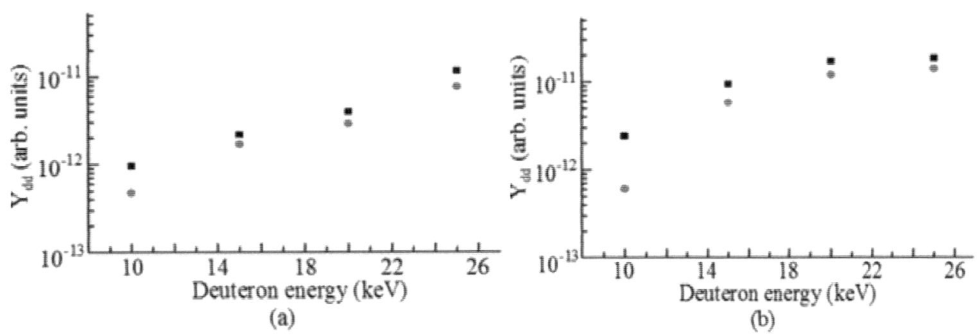

Figure 6. Dependence of the DD-reaction yield from the Ti/TiO$_2$:D$_x$ target (left) and from the Pd/PdO:D$_x$ target (right) on the D$^+$ beam energy: the measured DD-reaction yield along the beam (■) and the DD-reaction yield calculated for a given energy (red filled circles).

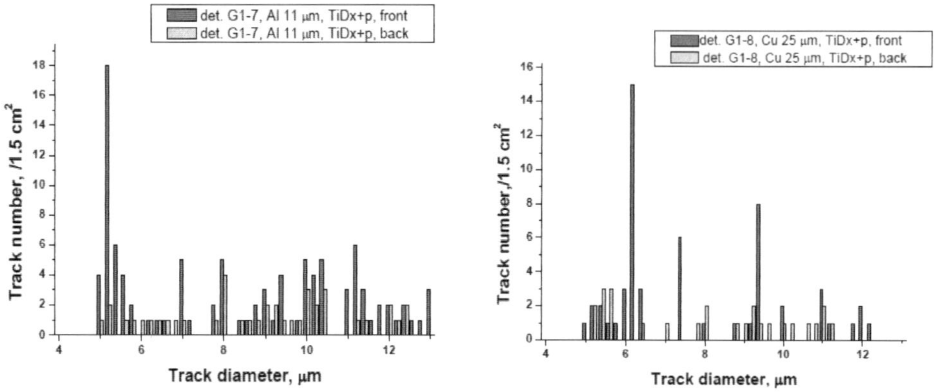

Figure 7. Diameter distribution of tracks on the detectors coated 11 μm Al (left) and 25 μm Cu (right), located above the sample Ti/TiO_2:D_x, irradiated by a H^+ beam with an energy of 23 keV. Comparison of readings front and back sides of the detectors indicates the presence of proton emission from the DD-reaction.

energy range of 10–25 keV to irradiate the Ti and Pd deuterated targets. Comparison of readings from the front and back sides of the CR-39 detectors placed over the target indicates the presence of proton emission from the DD-reaction – left peak in diameter distributions (see Fig. 7). Comparison of readings from the front and back sides of the CR-39 detectors under the target indicates the presence of neutron emission from the DD-reaction (Fig. 8).

Figure 9 shows the results of neutron flux measurements by the ^{3}He detector under the action of H^+ and Ne^+ beams on the Ti/TiO_2:D_x target. Background measurements were performed with similar beams on the Cu target. We can see that the beam exposure of the Ti/TiO_2:D_x target causes an increase in neutron detector indications over background values. Results obtained were similar in stimulating the target Pd/PdO:D_x beams of H^+ and Ne^+ [13,14].

To estimate the DD-reaction yields from targets during irradiation with the ion beam, we used a simplified model

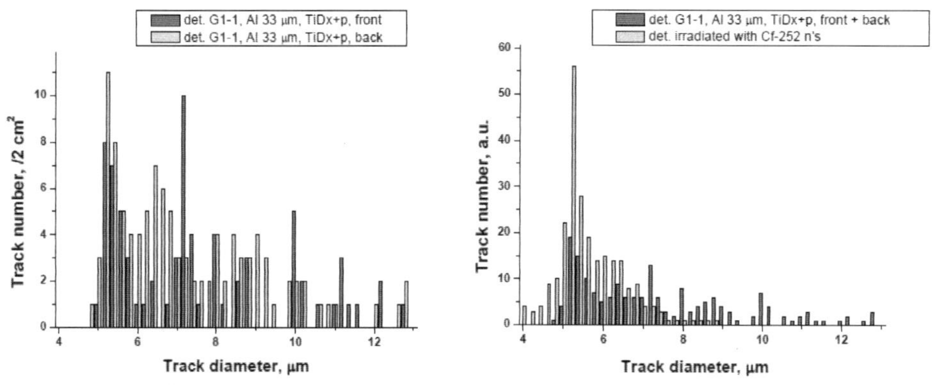

Figure 8. *Left* - the diameter distribution of the tracks on the front (red bars) and rear (green bars) side of the detector coated 33 mm Al, located below the sample Ti / TiO_2: D_x, irradiated by a proton beam with an energy of 23 keV. *Right* - Total tracks diameter distribution of the front and rear sides of the detector (red columns) compared with a detector irradiated by neutrons from the source Cf-252 (green bars).

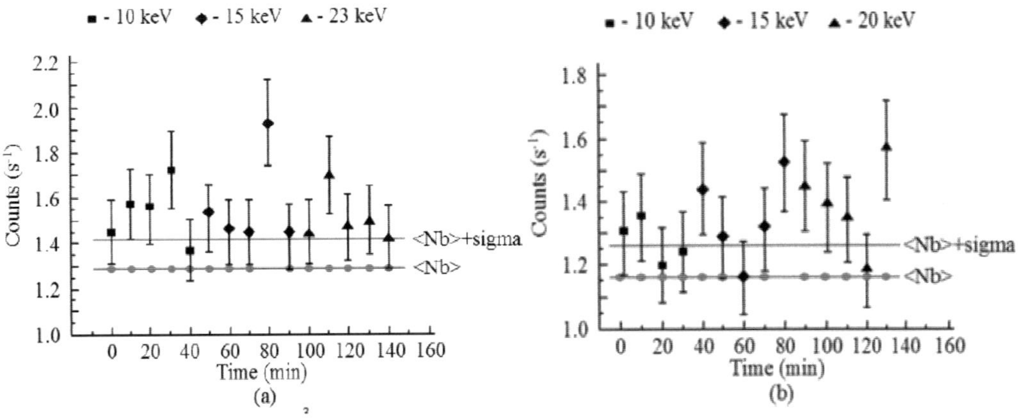

Figure 9. Counting rate of the ^{3}He-neutron detector (■, ♦, ▲). (a) Ti/TiO_2:D_x target 300 μm thick and H^+ beam (10, 15, and 23 keV), (b) Ti/TiO_2:D_x target 30 μm thick and Ne^+ beam (10, 15, and 20 keV). The average background ⟨Nb⟩ (red filled circles) was measured using the Cu target.

of the process, taking into account that deuterium desorption stimulated by radiation cause a deuteron current moving to the surface from the sample. This deuteron current and the deuterated surface can be considered the beam and target, respectively.

The screening potential U_e was estimated by the semi-empirical formula [17]

$$U_e = (T/T_0)^{-1/2}[a \ln(y) + b], \tag{2}$$

where $a = 145.3$ and $b = 71.2$ are numerical constants and $y = k \times y_0(J_d/J_0)$, here $k = \exp(-\varepsilon_d \Delta T/k_B T T_0)$, ε_d is the deuteron activation energy, y_0 = Me/D is the concentration ratio of metal and deuterium atoms in the target at $T_0 = 290$ K and $J_0 = 0.03$ mA/cm^2, and J_d is the deuteron current density from the target. The deuteron current was determined by the deuterium desorption rate from the sample.

- The screening potential for the Pd/PdO:D_x target, calculated using formula (2), is $U_e = 897$ eV.
- The screening potential for the Ti/TiO_2:D_x target, calculated using formula (2), is $U_e = 796$ eV.

3.3. CVD diamond target under the D^+ beam

In our previous investigations of DD-reaction in the crystal targets (Pd,Ti), anisotropy was observed: the neutron flux along the beam direction was higher than that in the transverse direction. This effect could be explained by presence of narrow channels in the samples, where the bulk of deuterium, trapped during the electrolysis, is concentrated. Particularly large anisotropy was observed using a polycrystalline textured CVD diamond samples.

We investigated the dependence of the DD-reaction yield from the incidence angle of the beam on the target. The relative yield of the DD reaction $Y_{dd} = n_n/(S \times I_d)$, where n_n is the longitudinal or transverse neutron flux, S the irradiated area of the target, and I_d is the ion beam current.

The measurements of the neutron flux in the beam direction are performed in dependence on the target angle, β, with respect to the beam axis. A significant anisotropy in neutron yield is observed, it was higher by a factor of 3 at β =0° compared to that at $\beta = \pm 45°$ NC-20 (80% of diamond, 20% of graphite) is composite material with

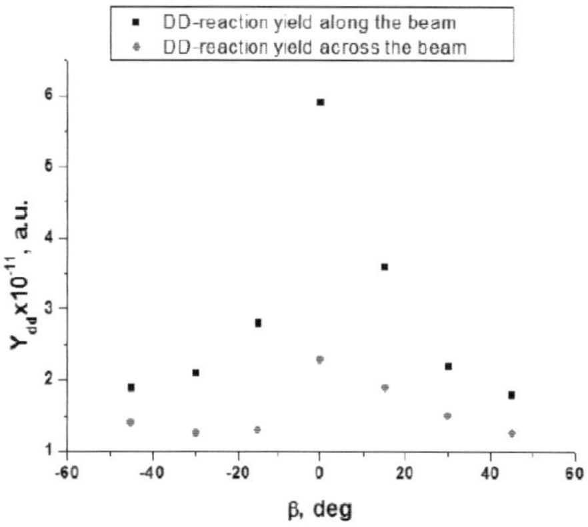

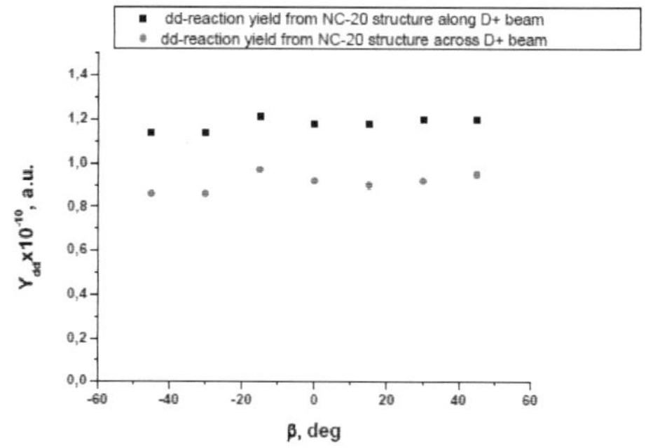

Figure 10. *Left* - The neutron yield obtained with the textured CVD-diamond sample as a function of the angle between the beam and the target plane norm, measured longitudinally (black filed squares) and transverse (red filled diamonds) directions with respect to the ion beam. Ion beam with the energy of $E_{\rm d} = 20$ keV and the current of 50–60 mA. *Right* – The neutron yield obtained with the diamond composite material (NC-20) as a function of the angle between the beam and the target plane norm, measured longitudinally (black filled squares) and transverse (red filled diamonds) directions with respect to the ion beam. NC-20 (80% of diamond, 20% of carbon) is composite material with isotropic structure. Ion beam with the energy of $E_{\rm d} = 25$ keV and the current of 20 mA.

isotropic structure. It was observed, that the neutron yield from isotropic NC-20 sample was practically independent of the sample orientation with respect to the beam (see Fig. 10).

Possible reasons for the increasing of DD-reaction yield in textured CVD diamond include the following:

- Screening effects of deuterium nuclei in the crystal structure.
- Collective processes associated with a high concentration of deuterium in certain directions.
- The effects of channeling, leading to an increase in the effective range of ions in the direction of the channel.

The results were presented in CHANNELING 2014 (Capri (Napoli), Italy, October 5–10, 2014) and published in [18].

4. Conclusions

The results of measurements of the DD-reaction yields from the Pd/PdO:D_x and the Ti/TiO_2:D_x heterostructures in the energy range of 10–25 keV show a significant effect of DD-reaction yield enhancement.

It was first shown that the impact of the H^+ and Ne^+ ion beams in the energy range of 10–25 keV at currents of 0.01–0.1 mA on the deuterated heterostructures results in the appreciable DD-reaction yield stimulation and enhancement.

It was observed that the crystalline structure and the orientation of the sample with respect to the beam has an impact on the neutron yield. The highest yield is recorded with the textured CVD-diamond target, oriented perpendicular to the beam.

Acknowledgements

This work was supported by the Russian Ministry of Education and Science (No. 16.518.11.7104 government contracts from 27 October 2011 and No. 14.518.11.7075 from18 June 2013.) The authors are very grateful to S.S. Gershtein and G.A. Mesyats for fruitful discussions.

References

[1] A.S. Roussetski, M.N. Negodaev and A.G. Lipson, *Proc. of ICCF15,* 5–9 Oct. 2009, Rome, Italy, p.182.

[2] F. Raiola, P. Migliardi, L. Gang et al., *Phys. Lett.* **B547** (2002) 193.

[3] H. Yuki, J. Kasagi, A.G. Lipson et al., *JETP Lett.* **68** (1998) 785.

[4] H. Yuki, T. Sato, J. Kasagi et al., *J. Phys. G: Nucl. Part. Phys.* **23** (1989) 1459.

[5] M. Aliotta, F. Raiola, G. Gyurky et al., *Nucl. Phys.* **A690** (2001) 790.

[6] K. Czerski, A. Hulke, A. Biller et al., *Europhys. Lett.* 54 (2001) 449.

[7] F. Raiola, P. Migliardi, G. Gyurky et al., *Euro. Physical J.* **A13** (2002) 377.

[8] A.G. Lipson, A.S. Roussetski, A.B. Karabut and G.H. Miley, *JETP* **100** (2005) 1175.

[9] H.S. Bosch and G.M. Halle, *Nucl. Fusion* **32** (1994) 611.

[10] I. P. Chernov, A. S. Rusetskii, D. N. Krasnov et al., *JETP* **112**(6) (2011) 952–960.

[11] A.V. Bagulya, O.D. Dalkarov, M.A. Negodaev, A.S. Rusetskii and A.P. Chubenko, *Bulletin Lebedev Physics Institute* **39** (9) (2012) 247–253.

[12] A.V. Bagulya, O.D. Dalkarov, M.A. Negodaev, A.S. Rusetskii and A.P. Chubenko, *Bulletin Lebedev Physics Institute* **39**(12) (2012) 325–329.

[13] A.V. Bagulya, O.D. Dalkarov, M.A. Negodaev, A.S. Rusetskii, A.P. Chubenko and A.L. Shchepetov, *Bulletin Lebedev Physics Institute* **40**(10) (2013) 282–284.

[14] A.V. Bagulya, O.D. Dalkarov, M.A. Negodaev, A.S. Rusetskii, A.P. Chubenko and A.L. Shchepetov, *Bulletin of the Lebedev Physics Institute* **40**(11) (2013) 305–309.

[15] V.G. Ralchenko, E. Pleuler, F.X. Lu, D.N. Sovyk, A.P. Bolshakov, S.B. Guo, W.Z.Tang, I.V. Gontar, A.A. Khomich, E.V. Zavedeev and V.I. Konov, *Diamond and Related Materials* **23** (2012) 172–177.

[16] J.F. Ziegler and J.P. Biersack, code SRIM 2003.

[17] A.G. Lipson, A.Yu. Tsivadze, B.F. Lyakhov, E.I. Saunin, D.M. Sakov, I.P. Chernov, Yu.P. Cherdantsev, Yu.I. Tyurin and A.S. Rusetskii, *Doklady Physics* **54**(4) (2009) 182–186.

[18] A.V. Bagulya, O.D. Dalkarov, M.A. Negodaev, A.S. Rusetskii et al., *Nucl Instr Methods Phys Res* B **355** (2015) 340–343.

J. Condensed Matter Nucl. Sci. 19 (2016) 274–280

Research Article

The Center to Study Anomalous Heat Effects [AHE] at Texas Tech University

Tara A. Scarborough* and Robert Duncan

Texas Tech University, Lubbock, TX 79409, USA

Michael C.H. McKubre

SRI International, Menlo Park, CA, USA

Vittorio Violante

ENEA, Unità Tecnica Fusione, C.R. Frascati, Frascati, Italy

Abstract

The Center for Emerging Energy Sciences at Texas Tech University (CEES) has been established to explore critical parameters in the observation of the anomalous heat effects (AHE). A large number of experiments report the production of heat from metal samples loaded with hydrogen or deuterium in amounts that are often thousands of times greater than the enthalpies of possible chemical reactions. The effect is anomalous because there is no agreed-to mechanism, and particle radiation rates are not reported at levels that are consistent with any known nuclear process.

Keywords: Calorimetry, Cryogenic, Electrochemistry, Heat-helium, Spectroscopy

1. Basis of Work

Heat-helium correlation was first documented by Miles [1,2] and co-workers in 1991 stating convincing evidence that their measured excess enthalpy correlated with D+D $\rightarrow {}^4$He + 23.4 MeV fusion reaction (by some unknown, and presumably many-body process). These same results have been reported by at least seven laboratories in three countries (including two of the present authors in three different experimental modes at SRI and ENEA).

While necessary improvements have been made to calorimetry, triggering and data acquisition, much is still left to be understood about the process and mechanisms of excess heat release. The intent of the work described here is not to verify nuclear processes, specifically, but to gain a fundamental understanding of the mechanisms at play. Extensive

*Corresponding author. E-mail: tara.scarborough@ttu.edu

© 2016 ISCMNS. All rights reserved. ISSN 2227-3123

materials characterization and eventual full reproducibility is the framework upon which the work performed at the AHE Center at Texas Tech. University (TTU) will be built upon, permitting direct correlation of excess heat production to materials properties that will point to the mechanism of this currently anomalous effect.

2. Strategy

Sufficient characterization of the metal foils both before and after their use in electrolytic cells is a critical path for correlation studies. While this characterization technique has yet to be standardized, the authors have embarked upon a path to define the necessary information to identify parameters that correlate with anomalous heat release.

Large scale, high resolution imaging is essential to investigate surface features of the cathodes. This will be achieved by use of atomic force microscope (AFM), scanning electron microscope (SEM), and confocal microscope (CM). All of these facilities are available at the Imaging Center at TTU which shares the same building as the AHE Center at Texas Tech.

To achieve nanometer resolution of areas $\sim 1 \times 4$ cm will require automated image acquisition and stitching software integrated with the SEM to register location. The registration will allow for the user to easily return to regions of interest for imaging at higher resolutions. Example foils included here were provided by the Violante Group at ENEA [3].

Contact with FEI regarding equipment upgrades for this project resulted in small demonstration in which the palladium foil was scanned and automatically stitched together as the scan progressed. The images below show resolution of this composite scan at several levels of magnification.

Additionally, plans to capture of the full spectrum of RF emission that have recently been observed to correlate with AHE production at TSEM, NRL, ENEA, and other laboratories is in progress. The effort described here will be led by the AHE Center at TTU and a collaborative agreement has been set in place with ENEA – Frascati (the Violante Group) to lead the further study of heat-helium correlations.

3. Electrolytic Cells

Based upon the pioneering work of Fleischmann and Pons, which has been further developed by McKubre [4], Violante [5], and other authors, correlation and rates of excess heat and helium production of deuterium and/or hydrogen loaded palladium foils will be investigated. While this work will continue at the established ENEA facility, the AHE Center at Texas Tech will establish working cells to begin testing characterization techniques, described in *Strategy,* in the next few months.

When satisfactory operation is achieved at Frascati, closely similar apparatus will be assembled in the AHE Center at TTU to check site-to-site repeatability. It is possible that a third laboratory will be selected in the future to further check repeatability.

We have purchased a PXIe system from National Instruments to aid in the data acquisition and processing necessary for calorimetric results. The system is equipped with an Intel Core i7, 12-bit oscilloscope/digitizers, DMMs, 26.5 GHz spectrum analyzer, simultaneous data acquisition (ADC/DAC + DIO + Timing/Triggering), and 5.7 TB RAID drive.

4. Exploding Wires

Considerable effort has been made to decrease data acquisition times, and therefore experimental setup and turnover times. As such, deuterium loading of palladium wires has emerged as a feasible alternative to foils. However, the process of loading the wires requires ultra-pure reagents in an ultra-clean environment, which introduces a different set of experimental requirements.

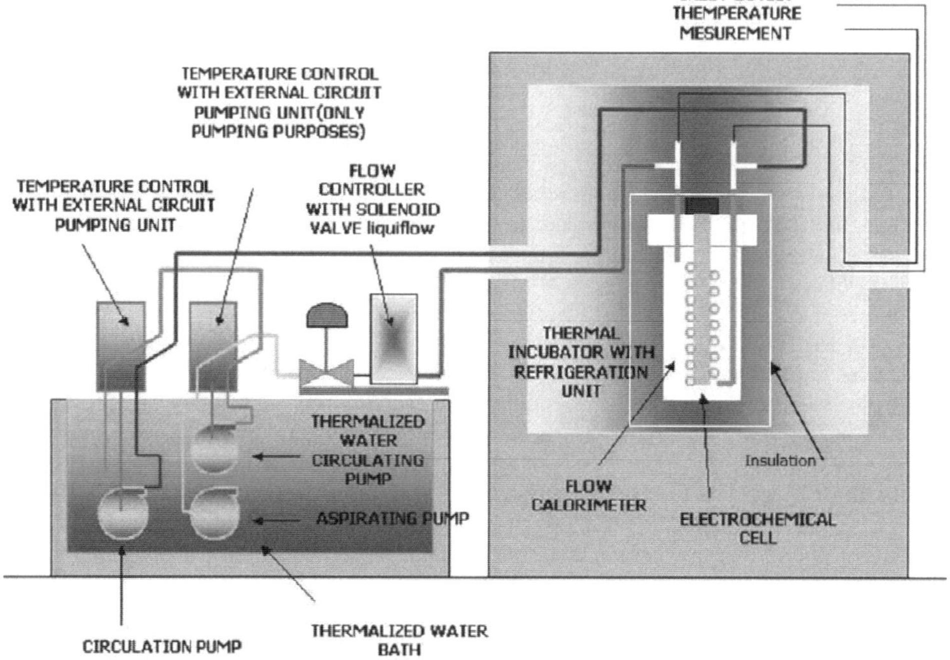

Figure 1. ENEA mass flow calorimeter.

The image below summarizes initial efforts made by Tanzella, Bao, and McKubre [6,7] to examine electrolytically loaded wires of various composition. The area below the line represents a region of no anomalous heat release, points above the line exhibit a significant percentage of excess heat. The AHE Center at TTU will be exploring the parameter space developed here in pursuit of correlations between the extent of heat release and helium production. Once operation is reached, site-to-site repeatability will be checked at Frascati.

5. Preliminary Results

During the last three months, we have investigated the necessary components for sufficient optical characterization of the electrolytic foils. Currently we have use of a Hitachi S4300 SEM which can be upgraded to provide surface feature resolution at the nanometer-scale, Energy Dispersive X-ray Spectroscopy (EDS) results for compositional information, and Forward Scatter Detector (FSD) imaging to detail the crystal structure orientation of the entire surface of each foil. Examples are shown below of the result of 33 stitched individual images with 30% overlap to guarantee feature matching. No smoothing has been applied.

Image defines sample as pure palladium with no impurities. Colors are associated with spectral markers in SEM image above. Red: La - K_{α}. The sample above shows a very strong preferred orientation with the $\langle 001 \rangle$ lattice orientation at the foil's surface.

Following preliminary tests with comparable models at Hitachi, EDAX, Oxford, FEI and Zeiss, TTU is in the process of installing a new Zeiss XB540 Crossbeam Focused Ion Beam FE SEM with Secondary Ion Mass Spectrometry

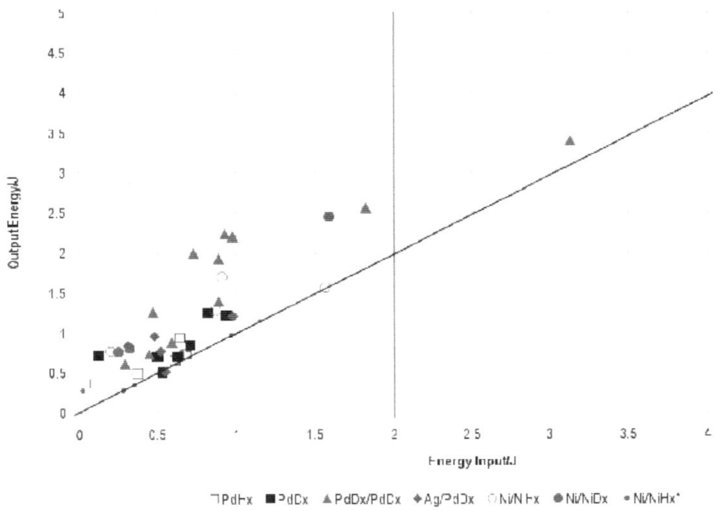

Figure 2. Calorimetrically determined output energy versus Joule input energy for a range of electrochemically loaded wires exercised electrically to destruction.

depth profiling (FIB-SIMS), material contrast with simultaneous Inlens SE and ESB imaging and Atlas 5 designed to create comprehensive multi-scale, multi-modal images with sample-centric correlative environment. The software capabilities include correlation of X-ray microscopy and FIB-SEM data to localize sub-surface features in 3D to target FIB sites that are not visible on the sample surface.

6. Mass Spectrometry

The AHE Center at TTU is in the process of establishing three separate methods of low-AMU mass detection for use in the previously mentioned experimental designs:

(1) *Leak Detection.* The Pfeiffer ASM 340 leak detector is calibrated for detection of H_2, ^3He, and ^4He, with

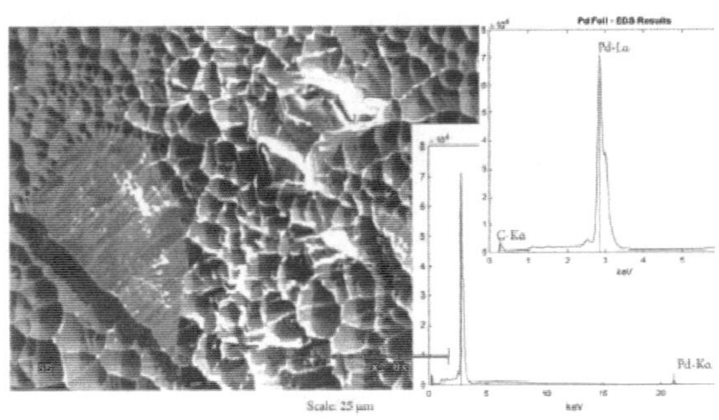

Figure 3. SEM and EDS results of a pure palladium electrode using Hitachi S4300 at TTU Imaging Center.

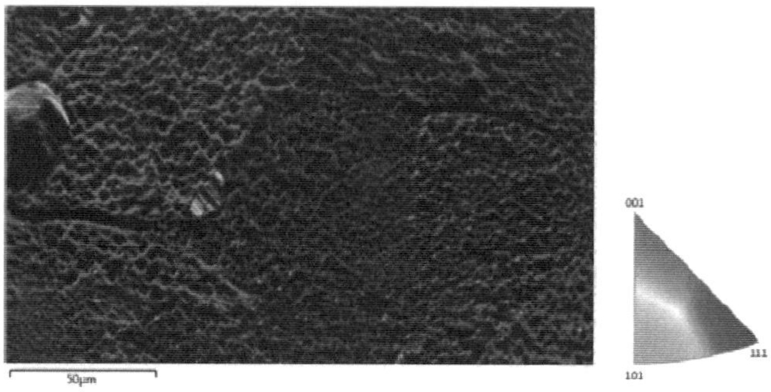

Figure 4. FSD and Euler Color Map overlay of palladium foil seen in Fig. 3 using the same Hitachi S4300 instrument.

a minimum detection rate of $< 5 \times 10^{-12}$ Torr l/s. This apparatus will be used to ensure that all seals are hermetic, but can also be utilized as a baseline calibration for background levels of the species listed above.

(2) *Gas Chomatography – Mass Spectrometer (GC-MS).* The JOEL GCMate is a compact double focusing mass spectrometer that incorporates a gas chromatograph before a traditional mass spectrometer that has a range that extends as low as m/z 1–3000. The figure to the right was obtained from JEOL by monitoring H_2^+ and D^+ from a mixture of D_2O and H_2O (mass difference of 0.00155 amu) with a resolving power of 3400. This instrument matches that used at ENEA – Frascati to facilitate in experimental repeatability once the AHE Center at TTU receives the mass flow calorimeter as described above.

(3) *Fourier Transform Ion Cyclotron Resonance (FT-ICR).* The Quantra Advance Mass Spectrometer utilizes a 1 Tesla magnet field ideal for confining small masses and lower voltages for small mass ion control. The piezo-electrically driven injection valve controls size and duration of sample sizes down to picoliter volumes while maintaining a consistent 10^{-10} Torr base pressure. Currently equipped with a 5 MHz waveform generator, the lowest detectable mass is 5.6 amu with a resolution of approximately 10,000, as it was originally designed for masses 2–1000, but utilizing the second harmonic helium and hydrogen can be trapped. Upgrading to a 40

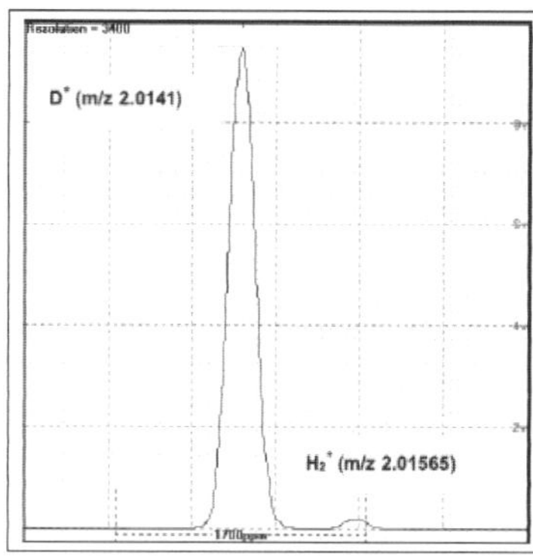

Figure 5. Illustrating the resolving power of the JEOL GC-MS, MS-071000A brochure available online.

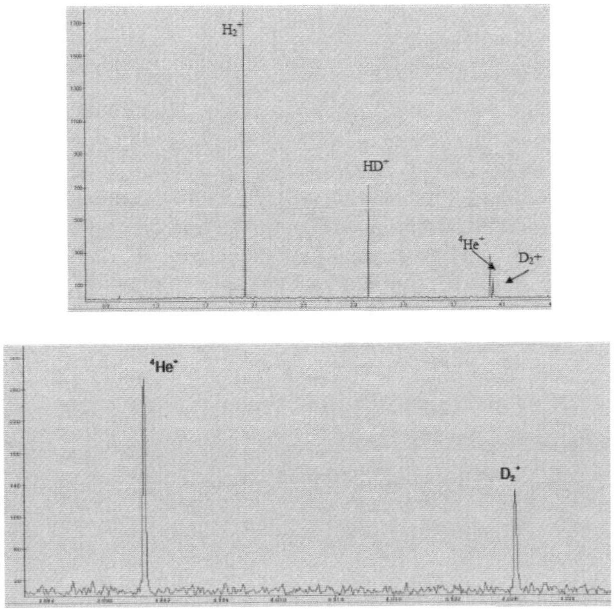

Figure 6. Plots of low mass spectrometer resolution provided by Siemens.

MHz generator would move detection limit to masses greater than 0.7 amu with a resolution of 400,000 for hydrogen and 200,000 for helium.

7. Custom Cryogenic Design

To initiate work on the exploding wire variations, Janis Research has assisted in the development of a custom research cryostat designed to operate with liquid argon (87 K), liquid nitrogen (77 K), or liquid helium (4 K).

The expected consumption rate of LAr is ~ 0.4 L/h with radiation baffles and bubble diffusers to ensure thermal stability of boil-off measurements before and during experimentation. Neutron and gamma ray transparent windows have been included in line with wires for emission detection with AmpTek X-123 Charge Drift Detectors.

Acknowledgements

Special thanks are acknowledged to: the Violante Group (ENEA – Frascati), Dean Davis (Siemens), Dr. Lin & Dr. Zhao (TTU Imaging Center), and Oxford Instruments & FEI – Houston. Undergraduate Laboratory Assistants: Thomas Hayes, Daniel Ballinger, and Onome Akene.

References

[1] E. Storms, *The Science of Low Energy Nuclear Reactions*, World Scientific, Singapore, 2007.

[2] M. Miles, B. F. Bush, G.S. Ostrom and J.J. Lagowski, Heat and helium measurements in cold fusion experiments, in *Proc. ICCF2*, Italy, 1991, p. 363.

[3] V. Violante, E. Castagna, S. Lecci, G. Pagano, M. Sansovini and F. Sarto, RF detection and anomalous heat prduction during electrochemical loading of deuterium in palladium, EAI, 2-3/2014.

[4] M.C.H. McKubre, F. L. Tanzella, P. Tripodi and P.L. Hagelstein, The emergence of a coherent explanation for anomalies observed in D/Pd and H/Pd systems: evidence for ^{4}He and ^{3}He production, in *Proc. ICCF8*, Italy, 2000.

[5] M. Apicella, E. Castagna, L. Capobianco, L. D'Aulerio, G. Mazzitelli, F.Sarto, A. Rosada, E. Santoro, V. Violante, M. McKubre, F. Tanzella and C. Sibilia, Some recent results at ENEA, *Proc. ICCF12*, Japan, 2005, p. 117.

[6] F.L. Tanzella, J. Bao, M.C.H. McKubre and P. L. Hagelstein, Triggered nuclear energy release from palladium deuteride, SRI Final Report to DTRA on Contract HDTRA1-08-0050, 2012.

[7] M.C.H. McKubre, J. Bao and F.L. Tanzella, Calorimetric studies of the destructive stimulation of palladium and nickel fine wires, in *Proc. ICCF17*, South Korea, 2012.

J. Condensed Matter Nucl. Sci. 19 (2016) 281–286

Research Article

Is the Abundance of Elements in Earth's Crust Correlated with LENR Transmutation Rates?

Felix Scholkmann*

Research Office for Complex Physical and Biological Systems (ROCoS), Mutschellenstr. 179, 8038 Zurich, Switzerland

David J. Nagel

The George Washington University, Washington DC, 20052, USA

Abstract

Nuclear transmutations are reported in many low-energy nuclear reaction (LENR) experiments. We showed in a previous study (Scholkmann and Nagel, *J. Condensed Matter Nucl. Sci.* **13** (2014) 485–494) that (i) the transmutation data of three independent experiments have a similar pattern and (ii) this pattern correlates with a model-based on the prediction of Widom and Larsen (WL). In the present study, we extended our analysis and investigated whether the abundance of elements in Earth's crust is correlated with either (i) the WL-prediction, or (ii) the three LENR transmutation data sets. The first analysis revealed that there is no statistically significant correlation between these variables. The second analysis showed a significant correlation, but the correlation only reflects the trend of the data and not the peak-like pattern. This result strengthens the interpretation that the observed peak-like pattern in the transmutation data sets does not originate from contamination. Further implications of our study are discussed and a recommendation is given for future transmutation experiments.

Keywords: Elements in Earth's crust, Low-energy nuclear reactions (LENR), Neutron scattering strength, Statistical analysis, Transmutation, Widom–Larsen theory

1. introduction

Over the last decades, many studies reported transmutations of elements induced by chemical energies (see for review [1–3]). These processes are taken to be the results of low-energy nuclear reactions (LENR).

In a previous study [4], we found that (i) the transmutation data of three independent experiments (Miley et al., [5–7], Mizuno et al. [8,9], Little and Puthoff [10]) show a similar pattern and (ii) this pattern correlates with a model-based prediction of Widom and Larsen (WL) [11] Both correlations were statistically significant.

*Corresponding author. E-mail: felix.scholkmann@gmail.com

© 2016 ISCMNS. All rights reserved. ISSN 2227-3123

There are two possible interpretations of these results: the correlations are results of LENR transmutation process, or they are artifacts originating from impurities of these used materials, which contain traces of elements present in nature.

The goal of the present study was to investigate this issue. In detail, our aim was to analyze whether the abundance of elements in Earth's crust is correlated with (i) WL-prediction, and/or with (ii) any of the three LENR transmutation data sets The results of this analysis will indicate if the observed pattern in the transmutation data is an artifact (originating from impurities of the probed material due to natural abundances of elements in Earth's crust) or not. To the best of our knowledge, such an analysis was not performed so far.

2. Material and Methods

2.1. Experimental data

As in the previous analysis [4], the following transmutation data sets from electrolysis experiment were selected from publications and used in the present analysis: data obtained by (i) Miley (*Miley data set*), (ii) Mizuno (*Mizuno data set*), and (iii) Little and Puthoff (*Little-Puthoff data set*). More detailed description of these data sets can be found in [4].

2.2. Data about Earth's elemental abundance

The data for the abundance of elements in Earth's crust was obtained by [12], which include the abundance for elements with atomic number (Z) ranging from 1 to 92 from several references.

2.3. Data analysis

Two types of analysis were performed: we analyzed whether the abundance of elements in Earth's crust is correlated with (i) the WL-prediction and/or (ii) the three LENR transmutation data sets (see Fig. 1).

For the first correlation analyses, the $f(A)$ function, with A the atomic mass number, given by the WL-prediction had first to be transformed to a $f(Z)$ function, since the abundance data of the elements in the Earth's crust are given according to their atomic number Z value. For more information about the $f(A)$ function and the WL-prediction please refer to [4,11]. The new $f(Z)$ function was derived by the following steps: (i) defining an approximate relation

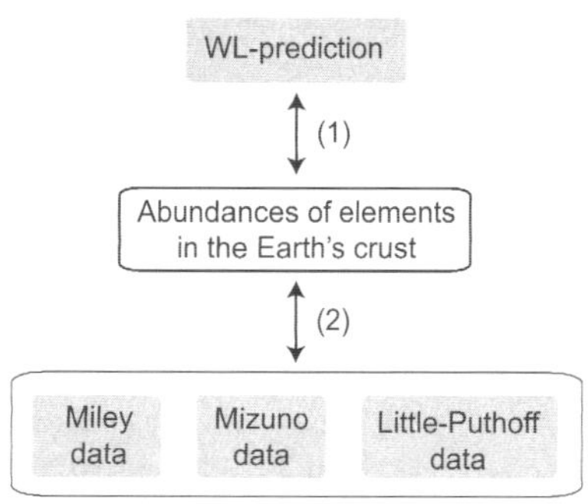

Figure 1. The two types of correlation analysis performed.

between A and Z as (a) $A = 1 \rightarrow Z = 1$ (hydrogen) and (b) $A = 238 \rightarrow Z = 92$ (uranium), (ii) finding the new values of the complex parameter z in Eq (1) of [4] by rescaling the $f(A)$ function such that the two constraints (a and b) were fulfilled. The derived z value is $4.89 + 0.5$ i. (iii) Manual adjustment of two values of the $f(Z)$ function to Z(5, 1.8) and $Z(12,$ 1.8) was necessary since the fitting procedure at these important points gave a poor result The peak-like structure of the obtained $f(Z)$ function is identical to that of the $f(A)$ function, except the y-axis scaling (Z instead of A).

2.3.1. *Comparison of the Earth's elemental abundance with the WL-prediction*

The data set for the abundances of elements in Earth's crust was normalized to the range [0,1] to simplify the next data processing steps. The $f(Z)$ function was then fitted to the normalized abundances of elements in Earth's crust by using a classical least squares optimization method. The fitting was performed by multiplying the $f(Z)$ function with a scaling factor (θ), which then was changed until the sum of squared errors between the $f(Z)$ function and the data was minimized. Then the data sets of the normalized abundances of elements in Earth's crust was segmented into intervals corresponding to the peaks and troughs of the $f(Z)$ function. Thus, ten intervals were determined. The borders of the intervals were chosen to cover the peak and troughs position of the $f(Z)$ function optimally. The borders for the peak intervals were (given in Z): 4–6, 12–13, 23–27, 42–48, and 69–79. The borders for the trough intervals were (given in Z): 7–9, 15–21, 29–39, 50–68, and 80–92. To test whether the mean values of the new data set in the peak and trough intervals are significantly different statistically, a t-test was used for each interval pair. To account for the multiple comparison situation a Bonferroni-correction was used, leading to the new level of significance: $p<0.05/N=0.01$, with $N=5$ for the five t-tests performed. Thus, a p-value of <0.01 was used to indicate a statistically significant difference. The data processing cascade for this analysis is visualized in Fig. 2(a).

2.3.2. *Comparison of the earth's elemental abundance with LENR transmutation data sets*

For this analysis, the correlation between the $f(Z)$ function and each of the three LENR data sets as well as the combined data sets were computed. For each data set, the θ value was determined separately. The correlation strength was quantified by calculating the Pearson correlation coefficient (r), ranging from –1 to 1 where 0 indicates that no linear correlation between the variables exists. On the other hand, a value of –1 or 1 implies a perfect negative or positive, respectively, correlation. In order to test if the correlation is due to chance, the statistically significance (p-value) of the correlation was calculated. A p-value of < 0.05 was chosen as a threshold value for statistical significance. A limitation of this analysis is that it only provides an indication of the trends if the variables agree A correlation of the fine-structure cannot be quantified with this analysis.

3. Results

3.1. Comparison of the Earth's elemental abundance with the WL-prediction

The analysis gave the following p-values corresponding to the significance of the difference between the values in the intervals 1–10: (i) interval 1 vs 2: $p = 0.2600$, (ii) interval 3 vs 4: $p = 0.0296$, (iii) interval 5 vs. 6: $p = 0.0522$, (iv) interval 7 vs. 8: $p = 0.0063$ (v) interval 9 vs. 10: $p = 0.1486$ (see Fig. 2) Thus, only the difference between interval 7 and 8 was statistically significant ($p < 0.05/5 = 0.01$). Since the sign of the difference is opposite to the expectation (i.e., the WL-prediction gives a peak at this interval instead of a trough), this single significant correlation is not a proof for a general correlation between the Earth's elemental abundance and the WL-prediction.

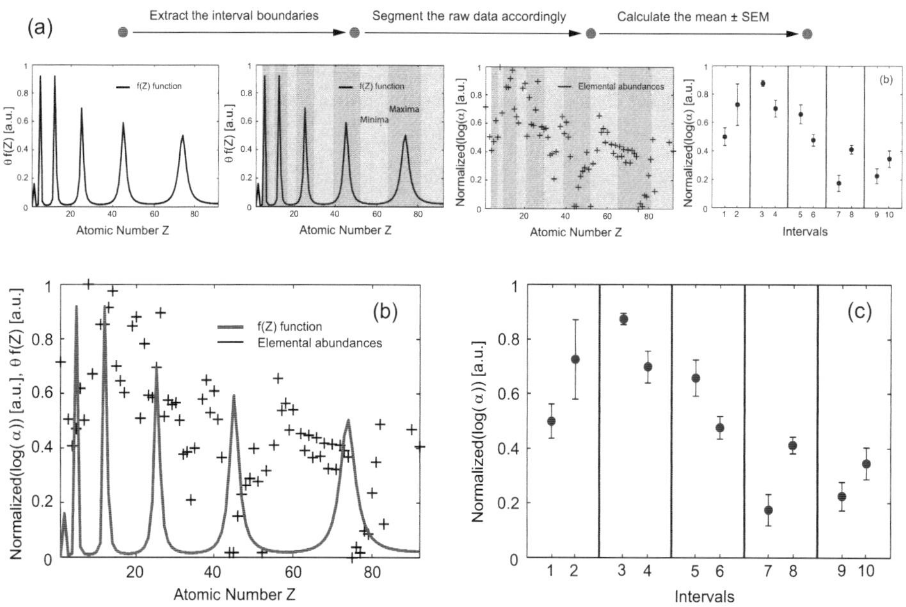

Figure 2. (a) Shows the signals processing cascade for the first type of data analysis performed. (b) Shows the comparison between the abundances of elements in Earth's crust (α) and the $f(Z)$ function based on the WL-prediction. θ: scaling coefficient ($\theta = 0.5102$, estimated using least squares fitting). (c) Interval-wise comparison between the abundance data and the WL-prediction. Interval definition: (i) intervals with maxima: (1): 4–6, (3): 12–13, (5): 23-27, (7): 42–48, (9): 69–79; (ii) intervals with minima: (2) 7–9, (4) 15–21, (6) 29-39, (8) 50–68, (10) 80–92. The interval boundaries are defined according to the $f(Z)$ function. Data are given as the mean $\pm$ standard error of the mean (SEM) .

3.2. Comparison of the Earth's elemental abundance with LENR transmutation data sets

The analysis showed a statistically significant correlation between these data sets ($p < 0.5$). For the obtained $p$values and the values of the correlation coefficient (r) please refer to Fig. 3.

4. Discussion and Conclusion

The first analysis (*comparison of the Earth's elemental abundance with the WL-prediction*) showed clearly that the peak-pattern of the WL-prediction does not correlate with the Earth's elemental abundance. The correlation was not statistically significant, apart from the difference between interval 7 and 8. However, as already explained, since the sign of the difference is opposite to the expectation (i.e., the WL-prediction), this single significant correlation is not a proof for a general correlation between the Earth's elemental abundance and the WL-prediction. In short, this test does not find a proof that there is an association between the Earth's elemental abundance and the WL-prediction.

The second analysis (*comparison of the Earth's elemental abundance with LENR transmutation data sets*) revealed that, on the one hand, the LENR transmutation data are significantly correlated with the Earth's elemental abundance data. However, on the other hand, this correlation is caused by the global trend of the data sets and not of their fine structure (i.e., the peak-pattern). What could be the cause for the same global trend (i.e., higher abundance values at low Z values and vice versa) in the data? The global (decreasing) trend in the Earth's elemental abundance data can be explained by the cosmic nucleosynthesis processes that created them [13]. An explanation for the same decreasing

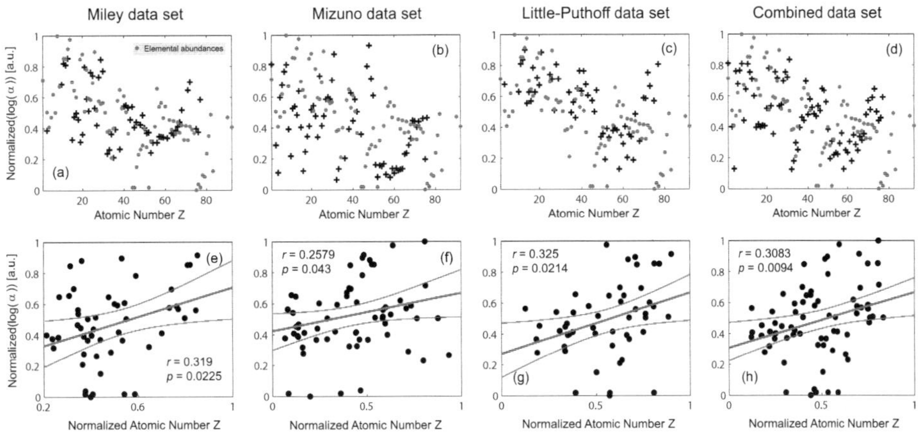

Figure 3. Results of the correlation analysis. Shown are the abundances of the Earth's elements and the three different LENR transmutation data sets (a–c, e–g) as well a combination of all these data sets (d,h). Subfigures e–h show the scatter plots of the Earth's elemental abundances vs. the LENR transmutation data sets. r: correlation coefficient, p: p-value corresponding to the significance of the correlation. In figures e–h the normalized atomic number Z is given which is defined as $Z \in [0\ 1]$.

trend present in the LENR transmutation data sets is missing, as far as we know. We can speculate that production of heavier elements requires sequential LENR. However, there is neither experimental nor theoretical evidence for that possibility at present.

How should our analysis be interpreted regarding the question of whether or not the observed and correlated patterns in the three LENR transmutation data sets are simply an artifact? The present analysis seems to strengthen the opinion that the experimental patterns cannot be explained by the distribution of the Earth's elemental abundance. However, there remain two significant concerns. We consider them each, and conclude they are not likely.

First, our analysis cannot rule out the possibility that contamination did play a role in the LENR transmutation experiments analyzed. In general, there might be two sources of contamination: (i) dirt in the laboratories that has an elemental abundance different from that of the Earth's crust, and (ii) other materials that are in contact with the experimental materials and also have specific elemental distributions. However, it is highly unlikely that the distributions of contaminations in all three of the laboratories would have been as similar as reported, whatever the origin of possible contaminations.

Second, there is concern over the sequence of events needed to link elements from the outdoor environment to those found in the three LENR experiments. The transport path involves three steps: (a) from the environments into the laboratories, (b) from the laboratories into the experiments, and (c) deposition within the experiments onto the surfaces that were later analyzed. We believe that it is very unlikely for the starting distribution of environmental contaminations and the selectivity of all three steps to be so identical that very similar final reported distributions of elements resulted. Even less likely is the possibility that the starting environmental distributions were really different, but the selectivity in the three transport steps happened to lead to the same end results. A simpler conclusion is that the three LENR experiments, despite their great differences, all yielded similar results because of the still unknown nature of LENR.

Since the analyzed experimental data in the three experiments were obtained years ago, and no samples of external or indoor dirt and other materials involved in the experiments were obtained, analyses of these elemental distributions are not possible. Thus, for future LENR transmutation experiments, we recommend collecting samples from different materials that are in the outdoor and laboratory environments and in the materials used for the experiments. This will allow subsequent data analyses as discussed in our previous study [4] and in this paper. Extensive consideration of other desirable aspects of future transmutation experiments is available in [4].

References

[1] E. Storms, *The Science Of Low Energy Nuclear Reaction* , World Scientific, Singapore, 2007

[2] M. Srinivasan, G.H. Miley and S. Storms, *Low Energy Nuclear Reactions: Transmutations*, in *Nuclear Energy Encyclopedia: Science, Technology and Applications*, S.B. Krivit, J.H. Lehr and T.B. Kingery (Eds.), Wiley, New York, 2011, pp. 503–540.

[3] D. Nagel, The status of 'cold fusion', *Radiation Phys. Chem.* **51**(4–6) (1998) 653–668.

[4] F. Scholkmann and D. Nagel, Statistical analysis of transmutation data from LENR experiments and comparison with a model-based prediction of Widom and Larsen, *J. Condensed Matter Nucl. Sci.* **13** (2014) 485–494.

[5] G.H. Miley, *Reaction product and heat correlation for LENRs*. 2000. Lerici (La Spezia), Italy.

[6] G.H. Miley, Possible evidence of anomalous energy effects in H/D-loaded solids – Low energy nuclear reactions (LENR), *J. New Energy* **2**(3–4) (1997) 6–13.

[7] G.H. Miley and J.A. Patterson, Nuclear transmutations in thin-film nickel coatings undergoing electrolysis, *J. New Energy* **1**(3) (1996) 5-3.

[8] T. Mizuno, Isotopic changes of elements caused by various conditions of electrolysis, in *The 237th ACS National Meeting*, 2009, Salt Lake City, UT, USA.

[9] T. Mizuno, T. Ohmori and M. Enyo, Isotopic changes of the reaction products induced by cathodic electrolysis in Pd, *J. New Energy* **1**(3) (1996) 31–45.

[10] S. Little and H.E. Puthoff, Search for Evidence of Nuclear Transmutations in the CETI RIFEX Kit. http://www.earthtech.org/experiments/rifex/rifex.pdf, 1998.

[11] A. Widomand and L. Larsen, *Nuclear Abundances in Metallic Hydride Electrodes of Electrolytic Chemical Cells.* arXiv:cond-mat/0505026, 2006.

[12] Jefferson-Lab., *It's Elemental — The Periodic Table of Elements,* 2013; Available from: http://en.wikipedia.org/wiki/Abundance_of_elements_in_Earth%27s_crust.

[13] E.M. Burbidge et al., Synthesis of the elements in stars, *Rev. Mod. Phys.* **29**(4) (1957) 547–650.

J. Condensed Matter Nucl. Sci. 19 (2016) 287–297

Research Article

Impact of Electrical Avalanche through a ZrO_2–NiD Nanostructured CF/LANR Component on its Incremental Excess Power Gain

Mitchell R. Swartz* and Gayle Verner
Nanortech Inc., Wellesley Hills, MA 02481-0001, USA

Peter L. Hagelstein
Research Laboratory of Electronics, Massachusetts Institute of Technology, Cambridge, MA 02139, USA

Abstract

Cold fusion nanomaterials, in general, and NANOR®-type LANR components (derived from them), in particular, have two distinct regions of performance on each side of the electrical avalanche. This had lead to the identification of three (3) distinct regions of their electrical operation. We now report that the optimal power gain of NANOR®-type cold fusion components is found far below the breakdown voltage and that the power gain decreases continuously as the electrical avalanche threshold is approached. Beyond the region of electrical avalanche, the previously active preloaded LANR quantum electronic components then give a thermal output similar to a standard ohmic control (a carbon composition resistor). Therefore, use of this technique of driving an active CF/LANR nanomaterial component into, and beyond, their avalanche threshold, provides verification of the excess heat an additional way, which confirms that the calorimetry was calibrated. Also, this investigation indicates where, on the input power axis, to drive them for a maximum effective use. We also report that deuterium can fuel nanomaterial ZrO_2–Ni systems, consistent with the previously report involving aqueous CF/LANR systems by Swartz et al. (ICCF-9).

Keywords: Deuterated nickel, Nanomaterials, Nanostructured ZrO_2, ZrO_2–NiD

1. Introduction

Active cold fusion dry nanomaterials, both unloaded and preloaded with D or H, have three regions of electrical operation, as discussed below in detail. We report the continuous measurement (and loss) of excess heat production from an active CF/LANR component during and after its electrical breakdown, namely the, avalanche state. This transformation of active CF/LANR components from active to inactive state will be shown to be not only critical to successfully understanding and controlling these energy producing reactions, but also, providing a significant and giant step forward.

*Corresponding author. E-mail: mica@theworld.com

© 2016 ISCMNS. All rights reserved. ISSN 2227-3123

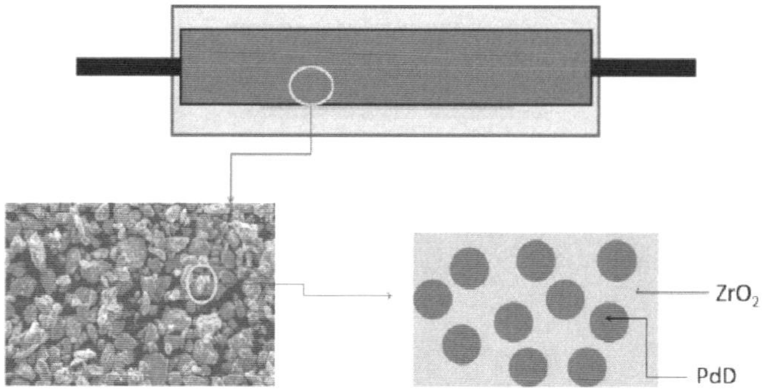

Figure 1. A schematic and actual view of an early NANOR®-type component. (*Top*) Shown is the two terminal component with the cylindrical shaped core. (*Bottom left*) A scanning micrograph shows the nanomaterial in the core. (*Bottom right*) A schematic illustration of the electrically insulating zirconia within which lie islands of deuterated alloy.

2. Background. A new generation of CF components

Dry ZrO_2–PdNiD NANOR®-type cold fusion or lattice assisted nuclear reaction (CF/LANR) components containing activated nanocomposite ZrO_2–PdNiD and ZrO_2–PdD which are preloaded with deuterons (thereby avoiding requisite loading each time) are capable of producing excess heat with significant energy gain [1,2] over long periods of time with reasonable reproducibility and control. These self-contained, two-terminal nanocomposite ZrO_2–PdNiD CF/LANR components feature new composition, structure, and superior handling properties (enabling portability and transportability) compared to aqueous and/or gas-loaded cold fusion (LANR) systems. They represent a new generation of LANR (CF) quantum electronic devices which are active and demonstrate significant improvement over all of their predecessors both regarding in sustained activity, improved reproducibility, and a large incremental power gain $(P_{\text{output}}/P_{\text{input}})$.

The initial NANOR®components are smaller than 2 cm length, and with 30–200 mg of active LANR material. For these NANOR®-type LANR components, the fuel for the nanostructured material is deuterium which is contained in the core volume (or chamber), a hermetically sealed enclosure.

The enclosure is tightly fit because both contamination is a potential problem and because of the potential toxicity, from the nanomaterial. Early NANOR®-type CF/LANR components were made of nano-scale regions of PdD, PdNiD or NiD in ZrO_2 [3,4]

Later, NANOR®-type LANR devices were loaded with additional D for greater loadings (ratio of D to Pd) which are estimated at 100–130% or more, although additional shallow traps are not ruled out. Thus, the deuterons are tightly packed ("highly loaded") within the nanostructured lattices [4]. Although the correct terminology is ZrO_2–$PdNiD_{1.3}$, for simplicity this subscript is left off.

Lattice assisted nuclear reactions (LANR or cold fusion) use such highly loaded hydrogen alloys to create heat and other products [5]. They do this by enabling deuterium fusion to form an excited *de novo* helium nucleus ($^4He^*$) at near room temperature, under difficult-to-achieve conditions. The product is believed to be *de novo* ^4He based on previous CF/LANR studies in aqueous systems, such as by Miles [5]. The "excess heat" observed is energy derived from the subsequent coherent de-excitation of the $^4He^*$ to its ground state via an internal conversion using the lattice phonons. Here, the excess heat is stimulated by the applied electric current; and the deuterium loading is separated from the excess heat operation.

Usually, in the past, successful LANR required engineering of multiple factors including loading, adequate confinement time, loading rate, and prehistory (with careful avoidance of contamination and materials and operational protocols which quench performance). Today, dry, preloaded NANOR®-type technology makes LANR reactions more accessible. What is different here is that most importantly, the components are pre-loaded so that LANR activation of the desired cold fusion reactions is, for the first time, separated from loading. In every other system known, Fleischmann and Pons, Arata, Miles, and the others, the loading has been tied to activation at room temperature [6].

2.1. States of CF/LANR electrical drive

It is a long, expensive, arduous effort to prepare these preloaded nanocomposite CF/LANR components, but by contrast, in these pre-loaded NANOR®-type CF/LANR components, the desired reactions are driven, and activated, by an applied electrical current. The development of more reproducible nanostructured CF/LANR components has not been easy, and has directly been linked to improved materials, with complete avoidance of low-threshold electrical breakdown states with their electrical avalanches and their CF/LANR quenching tendencies [1,2,4].

However, until now we have never completely measured the heat production from the active component on both sides the electrical breakdown (avalanche) state. This is important because although we have reported both the observed excess heat on one side of the electrical avalanche and the quenching of that excess heat on the other side, until now we have never observed both with continuity (observable in Fig. 1). This transformation of active CF/LANR components from active to inactive states has been critical to successfully controlling CF/LANR, and is critical to understanding how to engineer these systems.

2.2. Preliminary early results, including five month open demonstration at MIT

Early NANOR®-type components have been deliberately engineered to be small and low power (Excess power gain, P_{xs}, of hundreds of milliwatts) to enable them to be carefully evaluated for energy gain, including during, and after, the January, 2012 IAP MIT Course on CF/LANR. This public demonstration ran from Jan. 30, 2012 through mid-May 2012 [2]. A Series 6 NANOR®-type CF/LANR component provided long term CF/LANR activity and was monitored by parallel diagnostics including calorimetry, input-power-normalized delta-T, and focused heat flow measurement (such as Omega HFS Thin Film) and calibration with an ohmic (thermal) control located next to the NANOR®. The NANOR and the thermal control were at the center of much larger thermal mass in the calorimeter discussed in more detail in [1,2].

The NANOR®-type preloaded LANR device openly demonstrated clear cut energy gain. Over weeks, the NANOR®-type preloaded LANR component openly demonstrated energy gain (COP) which ranged generally from 5 to 16 (e.g. 14.1 (~1412%) while the MIT IAP course was ongoing, [11]) confirmed by three methods and time integration. It was driven for more than a year with careful evaluation for energy gain under a variety of conditions. The excess energy gain compared to driving input energy exceeded 20 times (sometimes more). This NANOR®-type preloaded LANR system had a much higher energy gain compared to the 2003 CF/LANR demonstration unit using an aqueous high impedance system with a Phusor®-type electrode (energy gain 14.1 in 2012 vs an energy gain $\sim$2.7 in 2003). It also had an improved controlling/driving system which provided a reliable low power, high-efficiency, energy production component for demonstration and teaching purposes of size smaller than a centimeter, with an active site weight of less than 50 mg. Although small in size, this NANOR®-type preloaded LANR device is actually not *de minimus* because the LANR excess power density is more than 19,500 W/kg of nanostructured material [1] and the carbon footprint is zero.

3. Experimental

3.1. Materials - Impact of E-field on nanomaterial

Nanostructured materials are important in LANR and are also produced in codeposition structures, observed producing non-thermal near infrared emissions when active, and exhibit typical CF/LANR excess heat correlated with the size of the Pd–D nanostructures [4,5]. However, the development of more reproducible nanostructured CF/LANR components has not been easy, and has directly been linked to improved materials, with complete avoidance of low-threshold electrical breakdown states with their CF/LANR quenching tendencies. The problem is that there exist three operating regions and one involves electrical breakdown (electrical avalanche transconduction).

Avalanche behavior with three regions were first observed with ZrO_2–NiH NANOR®-type component [4]. Control of these breakdown states and quenching tendencies has been critical and has also required surmounting the extremely high electrical resistances (as high as hundreds of gigohms) of these nanomaterials [1,3] and their complicated polarization/transconduction phenomena including an electrical current "avalanche (transconduction electrical breakdown) effect" which has a critical negative role on excess heat generation because it quenches the desired reactions even as the input power increases. The very high DC electrical impedances can suddenly drop, as the voltage across the sample (transsample voltage) is increased to as low as $\sim$ 24 V. It can be shown theoretically that this sudden reduction can be attributed to an "avalanche effect" that is typical of the current–voltage behavior that occurs in Zener diode, but perhaps better understood as follows.

The nanostructured material is a composite distribution of nanostructured ferromagnetic "islands" separated among a vast dielectric zirconia "ocean". The dielectric zirconia embeds uncountable numbers of nanostructured metal ternary alloy islands. The high resistance occurs because the zirconia dielectric matrix is insulating at low voltage and it keeps the nanoscale metal islands electrically separated and prevents the aggregation of the islands. Each nanostructured island acts as a short circuit elements during electrical discharge. One hypothesis of the excess heat is that these "islands" allow deuterons to form a hyperdense state in each island, where the deuterons thereafter are able to be sufficiently close together to fuse and form $^4He^*$, by some pathway not known involving paired deuterons or possibly more.

There have been important implications from the fact that NANOR®-type component current and voltage characteristic shows a breakdown effect. For example, some electronics destroyed when voltage exceeded NANOR®-type component breakdown voltage [7]. Therefore, an effort was made to develop more robust electronics capable of driving the NANOR®-type component through breakdown and on both sides

3.2. Methods - Electrical activation of NANOR®-type CF/LANR component

The LANR preloaded, stabilized NANORs were driven by a high DC voltage circuit up to 1000+ V rail voltage. The duty cycle was split into a rest period with no input to anything, followed by a period with input power going to a control portion, and then to the CF/LANR component. That was followed by another rest period and control portion. The input power was delivered by a carefully controlled electrical DC pulse. The control was an ohmic resistor which was used to thermally calibrate the calorimeter by providing a series of well known input power, and by using a precise amount of time, energy-measured pulses [1,2]. Data acquisition was taken from voltage, current, temperatures at multiple sites of the core, around the heat flow sensor, and outside of the calorimeter. Data acquisition sampling was at data rates of 0.20–1 Hz, with 24+ bit resolution; voltage accuracy $0.015^{+/-0.005}$ V, temperature accuracy $< 0.6°C$). The noise floor (power) of the calorimeter is in the range of $\sim$1–30 mW. The noise power of the Keithley current sources is generally $\sim$10 nW. The implication of this is that the excess power generated (if it is present) must exceed the noise floor of the calorimeter in order to be observable. It also means that the driving of the calorimeter below that noise floor can on occasion result in false positives without further techniques to rule this out [8].

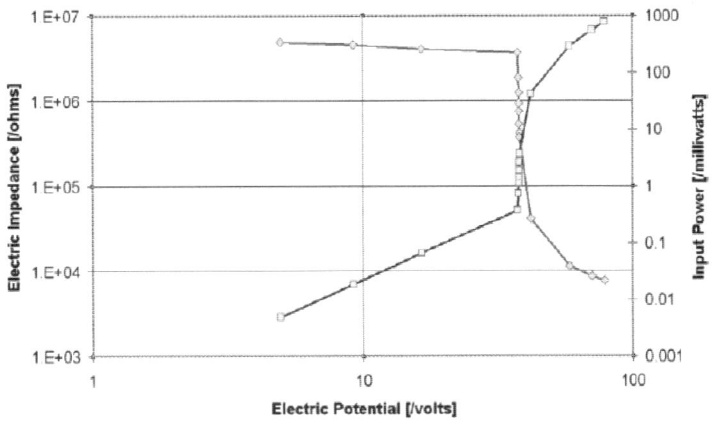

Figure 2. Impedance and Electrical current as a function of applied electric potential (volts). This figure shows the response, and particularly the Avalanche behavior with three regions, as first observed with ZrO_2–NiH NANOR®-type component. Three regions are seen; consisting of the initial plateau of electrical impedance, then the rapid fall off, and finally a sloped region at higher voltage [4,1].

Input power is defined as $V \times I$. There is no thermoneutral correction in denominator. Therefore, the observed incremental power gain is actually a lower limit. The result is heat measurement of this preloaded NANOR®-type LANR three (3) ways ending in calorimetry, input-power-normalized delta-T (dT/P_{in}), and input power normalized heat (HF/P_{in}) [11]. These three methods of verification are pooled to derive very useful information, including the energy produced ("excess energy") and sample activity. The instantaneous power gain (power amplification factor (non-dimensional)) is defined as P_{out}/P_{in}. As discussed above, the energy is calibrated by at least one electrical joule control (ohmic resistor) used frequently, and with time integration for additional energy validation. The excess energy, when present, is defined as (P_{output} - P_{input})$\times$ time. The amount of output energy (and therefore, both power, and energy, gain) is determined from the heat released producing a temperature rise, which is then compared to the input energy. The output of the component compared to the output of the precisely driven ohmic control.

4. Results

4.1. Interesting transconduction/XSH results and NiD can yields XSH

The new results are interesting. First, we continue to see improvements in loading and fabrication with each new generation of NANOR®-type components with respect to excess power gain and excess heat (XSH) At the time this manuscript was initially submitted, the Series 7 NANOR®-type components were the best so far with respect to incremental power gain; under select conditions very high (>1000) power gains observed. These components have demonstrated the first evidence of Mode B behavior [9], where an activation energy is not required beyond the input. Second, the new NANOR data reveals (as was shown for aqueous CF/LANR systems at ICCF-9) that deuterium can fuel ZrO_2–Ni systems, too [10].

4.2. Imaging a CF/LANR component through the electrical avalanche

Third, as shown in detail below, it is now possible to take electrical and calorimetry data systematically in the different regions of the electrical avalanche, and on both sides of that avalanche. Figures 3–6 show the examination of

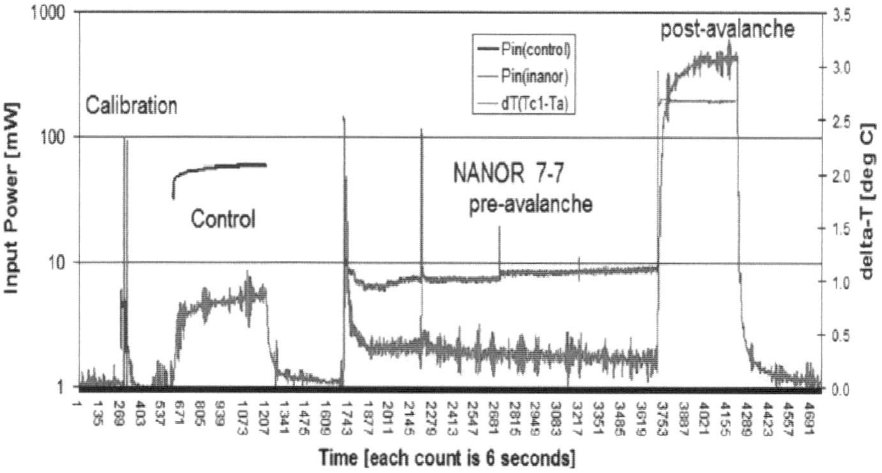

Figure 3. Temperature difference induced by a ZrO_2–NiD NANOR®-type component and ohmic control, both electrically driven, showing both excess heat and avalanche behavior.

this ZrO_2–NiD NANOR®-type component looking for possible excess heat on both sides of the electrical avalanche, and from several points of view. The figures include both raw data (Fig. 3) and derived information from the runs (Figs. 4–6). They demonstrate conclusively that LANR excess energy can be achieved, and it is heralded by input power normalized incremental temperature (delta-T) changes shown in Fig. 4, and verified and confirmed by calorimetry (Fig. 5).

Figure 3 is a set of curves which plot the differential incremental increase in temperature (°C) for the case with no input ("Background"), for the case of an ohmic thermal control (labeled 'control') at the same location, and for the NANOR®-type LANR component, both pre- and post-avalanche. The curves in Fig. 3 plot the raw data as the incremental temperature rise and the applied input electrical power. The applied electrical power is switched between an ohmic (thermal) control and the self-contained quantum electronic Series 7 NANOR®-type component CFLANR component. The graph presents these three curves and demonstrates the temperature rise in response to several different levels of electrical input power administered to both the ohmic control and then to the NANOR. For each of those electrical input powers, shown is the input power and then induced incremental output temperature rise.

There is shown an initial calibration pulses at the very beginning of the sequence, which is used for accuracy and precision checks of voltages and currents. The x-axis represents time, and each count represents 6 s. The y-axis on the left side represents electrical input power in watts. The y-axis on the right side represents the amount of temperature rise (differential temperature increase) in response to the electrical input power (degrees Centigrade). The graph shows first the response to the calibration pulse, then the response of the ohmic control, and then the response of the NANOR. A very large rise in power input and output (delta- T) result after the electrical avalanche at about count 3700.

Compare the output for NANOR®-type LANR component to the thermal (ohmic) control in Fig. 3. Two distinct regions of performance are seen on each side of the avalanche which is labeled in Fig. 4.

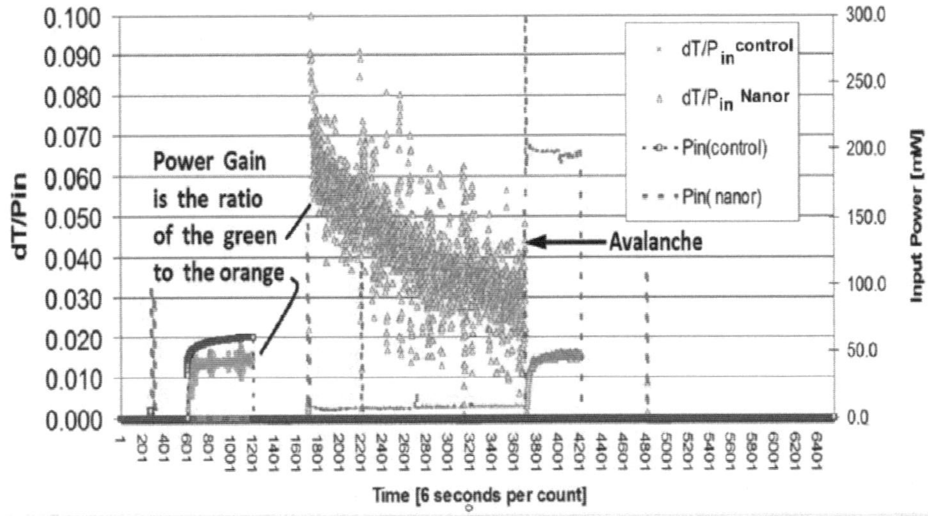

Figure 4. These curves show the dT/P_{in} ratios, which enable determination of the power gain, both in the excess heat region and after the avalanche behavior, for a ZrO_2–NiD NANOR®-type component. The region of the Avalanche is labeled.

4.3. On other side of the avalanche the power gain is essentially unity

To determine the effectiveness of the heat source, Fig. 4 is set of curves which presents the results of the same experiment, but which plots the temperature rise (delta-T in °C) of the preloaded NANOR®-type LANR component and the ohmic control with both normalized to input electrical power. This derived value, delta-T/P_{in}, is important because it enables semiquantitative determination of the incremental power gain.

Figure 4 shows the differential temperature rise normalized to input electrical power for the preloaded NANOR®-type LANR component, and for the case with no input power ("Background" but not labeled in the figure), and for the case of input to the ohmic thermal control, located at the core. The x-axis represents time, and each count represents 6 s. The y-axis on the left side represents electrical input power in watts. Each of the outputs are read off on the right-hand side. The y-axis on the right side represents the amount of temperature rise (differential temperature increase) normalized (that is, divided by) to the electrical input power. The units of this axis are in °C/W. A calibration pulse, used for accuracy and precision check of voltage and current measurement, is again also shown at the beginning and end (not labeled) of the run.

Because these curves plot the temperature rise normalized to input electrical power as a function of time, the ratios can be used to estimate incremental power gain. That is done here by taking the ratio of the response of the NANOR (green) to the ohmic control (orange). Compare the delta-T output normalized to input power for preloaded NANOR®-type LANR component to the thermal (ohmic) control, delta-T/P_{in}.

By comparing that ratio, note the active preloaded LANR quantum electronic component again clearly shows significant improvement in thermal output, here input-power-normalized compared to a standard ohmic control (a carbon composition resistor). Observe that despite lower input electrical power to the NANOR®, the temperature rise normalized to input electrical power observed in the core was higher than expected, as compared to the ohmic control. The graph therefore shows quite clearly a demonstrated active over-unity thermal output power from the NANOR®-type cold fusion (LANR) component, before the electrical avalanche.

Attention is directed to the fact that Fig. 4 clearly demonstrates a larger, significant improvement in differential thermal output (incremental increase in temperature in °C) for the preloaded NANOR®-type quantum electronic component compared to the standard ohmic control (a carbon composition resistor), before the electrical avalanche.

Figure 4 thus heralds the significant incremental power gain, and therefore excess energy achieved, by this Series 7 NANOR® type of LANR component. The ratio discussed above indicates that the input power normalized delta-measurements suggests strongly the presence of excess heat. Quantitatively, the amount of this differential temperature increase divided by the input electrical power for the preloaded NANOR®-type component compared to the control heralds great utility, and so to the degree that these components can be improved for total power out, this is a possible future efficient heat production source.

Figure 5 is a set of curves which present the results of the same experiment, but which plots the electrical power input and the thermal output power both for the two terminal NANOR®-type component Series 7 component, the ohmic control, two control pulses, and the background (no electrical input to either components). The curves present the electrical input power at several different input electrical power levels. Evaluation of the calorimetric response of both the ohmic control and the NANOR®-type component also includes presenting electrical energy input and the thermal heat (energy) output.

The figure shows the input, and the calorimetry, of preloaded NANOR along with that for the ohmic thermal control used to calibrate the system. Those calibration pulses, used for accuracy and precisions checks of voltages and currents and time, are also shown. The input to the thermal ohmic control followed by the preloaded NANOR®-type component is shown, as are the calibrated calorimetric outputs for both. The x-axis represents time, and each count represents six (6) seconds. The$y$-axis on the left side represents electrical input power in watts. The y-axis on the right side represents the amount of energy released. The units of this axis are in joules. Each of the power and energy outputs are read off of the left and the right-hand sides, respectively. The latter curves (on the right-hand side axis) represent time integration to determine total energy. They thus rule out energy storage, chemical sources of the induced heat, possible phase changes, and other sources of possible false positives.

Compare the output of the NANOR®-type LANR component to the thermal (ohmic) control. As can be seen, this semiquantitative calorimetry, itself calibrated by thermal waveform reconstruction, was consistent with excess heat being produced only during energy transfer to the active NANOR®-type LANR component. Notice that the active preloaded LANR quantum electronic component clearly shows significant improvement in thermal output compared to a standard ohmic control (a carbon composition resistor). The graph shows quite clearly demonstrated over unity thermal output power from the demonstration-power-level NANOR-type cold fusion (LANR) component, at least during the pre-avalanche period. It is clear that Fig. 5 demonstrates the excess heat from this component versus an ohmic control.

4.4. Optimum power gain found well below breakdown voltage

Figure 6 is a curve which presents the results of the same experiment, and which plots the incremental power gain of this two terminal NANOR®-type component Series 7 component and the ohmic control. There is obvious that the incremental power gain for the CF/LANR component until the avalanche, at which time (count ~3700) the component has no energy gain, but has a response similar to an ohmic resistor. This saliently demonstrates that outside of the CF/LANR active state, the component acts as any other ohmic resistor, and thus confirms that the calorimetry was calibrated, and verifies the excess heat yet an additional way.

4.5. Optimum power gain found well below breakdown voltage

In this experiment, we have for the first time obtained calorimetric data while scanning the applied electrical voltage in a series of levels, through the avalanche and then continuing on the other side of the avalanche (at even high applied

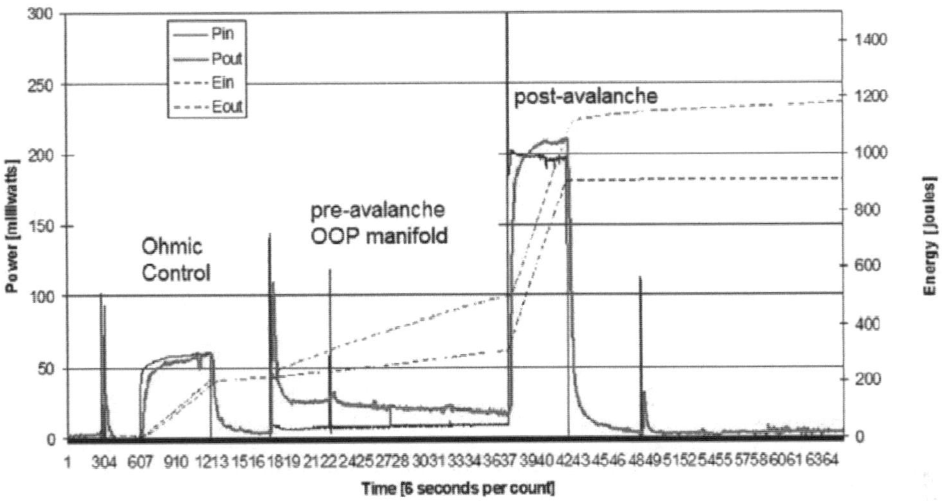

Figure 5. Four curves showing the input power and thermal power output, and their time integrals showing both excess heat and avalanche behavior, in a ZrO_2–NiD NANOR®-type component.

voltages). Thus, we have serniquantitatively measured this drop in power gain on the way to the electrical avalanche, and the total loss of that power gain during the electrical avalanche which occurs even though higher voltage is applied.

Several things have been made clear from this experimentation. First, the optimal power gain of NANOR®-type cold fusion components is found far below the breakdown voltage. This investigation has revealed that the power gain decreases continuously as the electrical avalanche threshold is approached. Second, in fact, as the voltage which produces the electrical avalanche is approached, the power gain, appears to monotonically decrease. Third, it is important to recognize the importance of our discovery of the fact that on the far side of the electrical avalanche. the power gain of the NANOR®-type component is essentially unity. Therefore, these discoveries have important imp1ications, First, it indicates where to drive these systems for at maximum effective use. Second, it indicates a key (present) limitation of these over-unity CF/LANR components.

4.6. On other side of the avalanche, the NANOR®-type component acts "dead", providing yet an additional control

This investigation has revealed that, despite driving at higher input electrical power, on the other side of the electrical avalanche, these NANOR®-type components act as little more than electrical resistors which are conventional, not over-unity, and therefore are functionally "dead" with respect to producing excess heat. Although this appears limiting in some ways, this phenomena does provide yet an additional control to check calorimetry beyond the measurements involving simply using a simple ohmic, thermal control. Therefore, driving a component into this region thus adds an additional verification of the actual excess heat which is developed in these studies.

4.7. NiD can yield XSH

This result also demonstrates clear excess heat (XSH) can be obtained from a Ni–D nanomaterial system. This is consistent with the previously reported impact of D to the high impedance aqueous CF/LANR system [10].

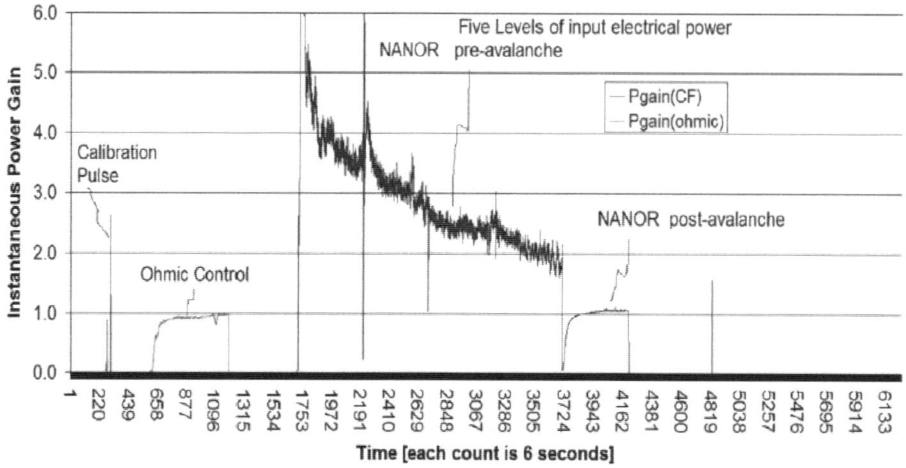

Figure 6. This graph presents the power gain of the ZrO_2–NiD NANOR®-type component CF/LANR component as a function of time, and shows that the excess heat which is generated decreases with high input power, eventually reverting to normal, ordinary, resistor-like operation after the electrical avalanche.

Swartz et al. presented at ICCF-9, the addition of D does NOT poison the reaction in the aqueous Ni system (under low drive conditions) but actually increases the incremental power gain. Only in the long run and at higher input power does D damage the Ni system [10].

5. Conclusion

5.1. Future of clean, high performance energy production components

It is clear that these preloaded nanostructured NANOR®-type CF/LANR quantum electronic components are quite useful [1,11]. They have shown significant improvement over their predecessors, including the highly successful metamaterial PHUSOR®-type of LANR component. They comprise an effective very clean, highly efficient, energy production system which has worked successfully as a test bed for cold fusion experiments for years, and has enabled the way to higher instantaneous power gain, total energy gain, imaging [12], emissions [13], open demonstrations [2,11], and a better understanding of the impact of applied magnetic fields [7], electrical transconduction [1,4], and things that quench the desired reactions (as presented at ICCF17, ICCF18, and ICCF19). Therefore these dry, preloaded, ready-to-be-activated, NANOR®-type LANR components/systems/materials, if the input power levels can be increased, may be in the future of clean efficient energy production including in preassembled IC components and systems.

Acknowledgements

The authors would like to thank Jeff Tolleson, Alex Frank, Alan Weinberg, Richard Goldbaum, Allen Swartz, Brian Ahern, Jeff Driscoll, and Charles Entenmann for their suggestions and support. This effort was supported by JET Energy Inc. and New Energy Foundation. NANOR® and PHUSOR® are registered trademarks of JET Energy

Incorporated. NANOR®-technology, and PHUSOR®-technology are protected by U.S. Patents D596724, D413659 and several other patents pending.

References

[1] M.R. Swartz, G. Verner et al., Energy gain from preloaded ZrO_2–PdNi–D nanostructured CF/LANR quantum electronic components, *J. Condensed Matter Nucl. Sci.* **13** (2014) 528. www.iscmns.org/CMNS/JCMNS-Vol13.pdf.

[2] M.R. Swartz and P.I. Hagelstein, Demonstration of energy gain from a preloaded ZrO_2–PdD nanostructured CF/LANR quantum electronic device at MIT, *J. Condensed Matter Nucl. Sci.* **13** (2014) 516. www.iscmns.org/CMNS/JCMNS-Vol13.pdf.

[3] Y. Arata and Y.C. Zhang, Observation of anomalous heat release and helium-4 production from highly deuterated palladium fine particles, *Jpn. J. Appl. Phys.* **38** (7A), Part 2 (1999) L774–L776.

[4] M.R. Swartz, LANR nanostructures and metamaterials driven at their optimal operating point, *J. Condensed Matter Nucl. Sci.* **6** (2012) 149; *LANR/LENR Sourcebook*, Vol. 3, October 21, 2011; www.iscmns.org/CMNS/JCMNS-Vol6.pdf.

[5] M. Swartz, Survey of the observed excess energy and emissions in lattice assisted nuclear reactions, *J. Scientific Exploration* **23**(4) (2009) 419–436.

[6] M. Miles et al., Correlation of excess power and helium production during D_2O and H_2O electrolysis using palladium cathodes, *J. Electroanal. Chem.* **346** (1993) 99–117.

[7] M. Swartz, G. Verner et al., Amplification and restoration of energy gain using fractionated magnetic fields on ZrO_2–PdD nanostructured components, *J. Condensed Matter Nucl. Sci.* **15** (2015) 66; www.iscmns.org/CMNS/JCMNS-Vol15.pdf.

[8] M. Swartz, Patterns of failure in cold fusion experiments, *Proc. of the 33rd Intersociety Engineering Conference on Energy Conversion*, IECEC-98-I229, Colorado Springs, CO, August 2–6, 1998.

[9] M.C.H, McKubre, F.L.Tanzella, V. Violante, What is need in LENR/FPE studies? *J. Condensed Matter Nucl. Sci.* **8** (2012) 187–197.

[10] M. Swartz, G.M. Verner, A.H. Frank, The impact of heavy water (D_2O) on nickel-light water cold fusion systems, *Proc. 9th Int. Conf. on Cold Fusion (Condensed Matter Nuclear Science)*, Beijing, China, Xing Z. Li (Ed.), May 2002, pp. 335–342.

[11] M. Swartz, G. Verner, J. Tolleson and P.L. Hagelstein, Dry, preloaded NANOR®-type CF/LANR components, *Current Science* **108** (4) (2015) 595. http://www.currentscience.ac.in/Volumes/108/04/0595.pdf.

[12] M. Swartz, G. Verner et al., Imaging of an active NANOR®-type LANR component using CR-39, *J. Condensed Matter Nucl. Sci.* **15** (2015) 81; www.iscmns.org/CMNS/JCMNS-Vol15.pdf.

[13] M. Swartz, Incremental high energy emission from a ZrO_2–PdD nanostructured quantum electronic component CF/LANR, *J. Condensed Matter Nucl. Sci.* **15** (2015) 92; www.iscmns.org/CMNS/JCMNS-Vol15.pdf.

J. Condensed Matter Nucl. Sci. 19 (2016) 298–315

Research Article

Fundamental of Rate Theory for CMNS

Akito Takahashi*,†
Technova Inc., 1-1-1 Uchisaiwai-cho, Chiyoda-ku, Tokyo 100-0011, Japan

Abstract

The condensed matter nuclear reactions (CMNR) are thought to happen for trapped H(D) particles within some chemical (electromagnetic) potential well with finite lifetime. As the lifetime is much longer than the collision time of two-body interaction of free particles, the CMNR reaction rates are significantly (of the order of 19–20 in magnitude) enhanced if we compare with estimated reactions rates by the two-body collision formula. The basis of CMNR rate theory is reviewed in this paper by extracting essence of the TSC theory tools developed until now. Derivation of Fermi's golden rule with nuclear optical potential, rate formulas by Born–Oppenheimer wave function separation, estimation of bracket integral of inter-nuclear strong interaction rate, estimation of time dependent barrier penetration probability by the HMEQPET method for dynamic D(H)-cluster condensation/collapse process, and DD fusion power levels as functions of inter-nuclear d–d distance and effective existing (life) time are given. A DD fusion power level of 10 kW/mol-dd-pairs is possible for a 1 pm inter-nuclear d–d distance with 10 ato-seconds lifetime. The level of 2.8 nano-mol 4D/TSC formations/s may release 10 kW neutron-free heat-power with ^{4}He ash.

Keywords: Condensed matter nuclear reaction, Enhanced fusion rate, Fermi's golden rule, Finite lifetime, 1 pm d–d distance, Time-dependent barrier factor, Trapped D(H) particles, 10 kW heat power

1. Introduction

This paper reviews the fundamental aspect of rate theory for the development of cold fusion condensed matter nuclear reaction (CMNR) theories [1–15]. The basic nuclear reaction rate formulas for trapped state D(H) particles in condensed matter and some examples on estimated reaction rates and heat-power levels are shown.

Despite the importance of reaction rate estimation in modeled condensed matter nuclear science (CMNS) theories, only a few authors have treated nuclear reaction rates properly (see many theory papers in list [1]). Some theories have borrowed rate formulas of two body collision process which is for the case of nuclear reactions by the random free particle motion as in plasma and gas phase or beam–target interactions. Intrinsically the CMNS nuclear reactions (CMNRs) should happen between trapped particles of protium and deuterium (H or D) in negative potential wells organized by the ordering of condensed matter such as periodic lattice, mesoscopic nano-particle and surface deformed/fractal

*E-mail: akito@sutv.zaq.ne.jp
†Also at: Osaka University, 2-1 Yamadaoka, Suita, Osaka 565-0071, Japan.

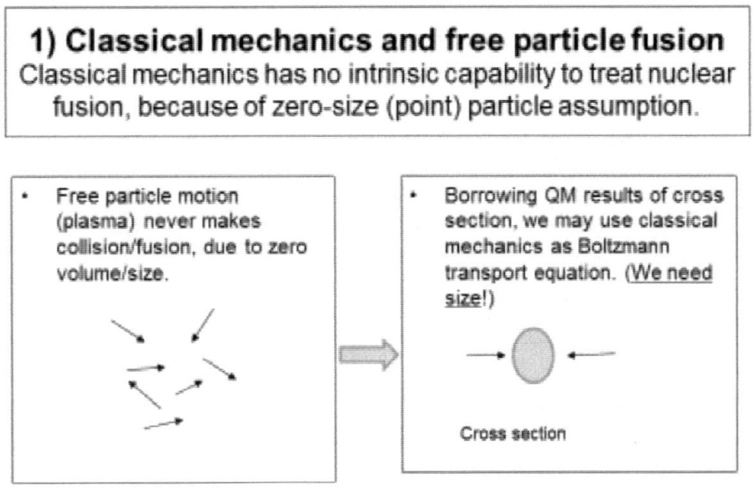

Figure 1. A simple image of two-body collision and cross section for random.

conditions. Such trapped H(D)-particles should have finite lifetime or existing time in negative potential wells and are keeping mutual inter-nuclear distances for finite time-intervals before nuclear fusion reactions. Therefore, the application of collision cross section formulas which treat instantaneous interactions (within about 1.0 × $^{-23}$ s collision time, for instance) is not proper for our CMNS nuclear reactions which may happen within much longer time interval of dynamic (or time-dependent pseudo-trapped state) and hence rate of CMNR is drastically enhanced if we compare with the estimated reaction rate by the two-body collision formulas. We need therefore to use formulas based on the Fermi's first golden rule for rate estimation, due to the finite lifetime of trapped particles [4]. This paper recalls the procedure and formulas for the fundamental of rate theory for CMNS, as has been used in the TSC theory development [2–15]. The outline is as follows:

(1) Why two-body collision theory cannot be used for nuclear reactions of CMNS is explained. It is shown that the collision rate theory underestimates drastically (on the 19–20 orders of magnitude) fusion rates of trapped deuterons in condensed matter, in comparison with more rigorous rates by the Fermi's first golden rule for trapped deuterons.
(2) The quantum mechanical (QM) formulation of rate calculation by the Fermi's first golden rule is given with nuclear optical potential and chemical Coulomb potential. The scaling of imaginary part component of nuclear optical potential is revisited [8].
(3) Formulas with electro-magnetic (EM) trapping potential well and nuclear interaction potential are derived by the adiabatic approximation, namely the Born–Oppenheimer approach [4–8].
(4) Collapsing condensation of D(H)-cluster is needed to have observable nuclear reaction rates by overcoming Coulombic repulsion, namely drastically enhancing barrier penetration probability of trapped particles in condensed matter. Time-dependent analysis is needed here. The HMEQPET method calculation of barrier factors as a function of d–d inter-nuclear distance is re-shown after references [4–7].
(5) It is shown that inter-nuclear d–d distance of D-cluster/molecule should be smaller than 1.0 pm to have d–d fusion rate with experimentally claimed heat-power levels. This issue is explicitly given first time in this paper,

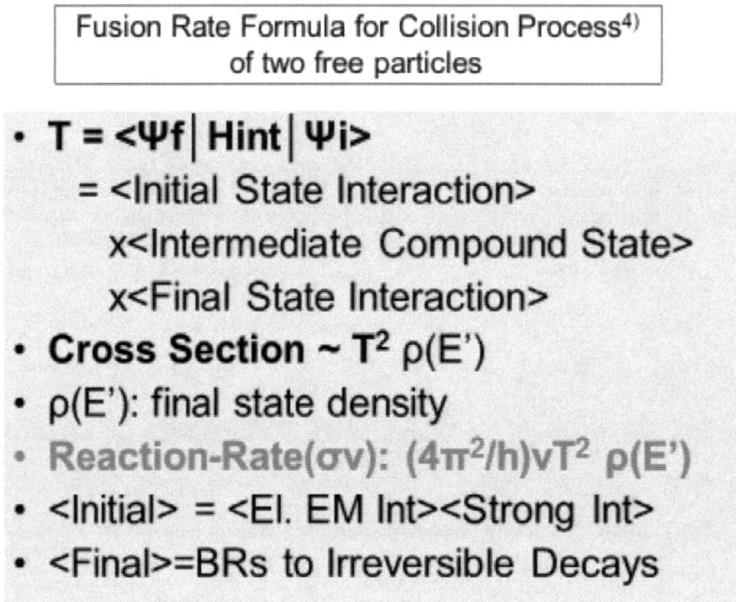

Figure 2. A summary slide of two-body collision cross sections and transition matrix.

though already it has been given implicitly in [5,6].

(6) Some example model/calculations are given for D(H)-cluster condensation/collapse processes and final nuclear products (for details see [4–7]).

This study only treats the initial state interaction of three steps (initial, intermediate, and final states) of nuclear reaction process, since the fusion reaction rates for explaining the observed anomalous heat levels of cold fusion (CMNR) experiments are to be determined by the initial state interactions. The studies on the intermediate and final states interactions (reaction products) are given, for the TSC theory, in [8–10,13]. For modeling the sites of D(H)-cluster, such as 4D/TS, in condensed matters are given in [4,12,14,15]. The overall background of theoretical modeling for condensed cluster fusion is given in [4] and very briefly in [3,7].

As the motivation of this study was for a tutorial review of fundamental tools used in the TSC theory development [2–15], description of mathematical derivations of used formulas is not repeated in the text description and instead the slide-type figures are used to explain the process of quantitative analyses in easily visible way. To understand mathematical derivations of used physical formulas, please refer the original papers [5,6,8,11,13].

2. Brief Review of Two-body Collision Rate Formulas

The some detail explanation of QM derivation of cross section and reaction rate for two-body free particle collision process is summarized in [8]. A very basic image of two-body collision in random motion of free particles is shown in Fig. 1.

As shown on the left-hand side part of Fig. 1, no collision ever happens in random motion of Newtonian classical mechanics particles, since the particles sizes are regarded as zero, namely points. To cause any collision in classical

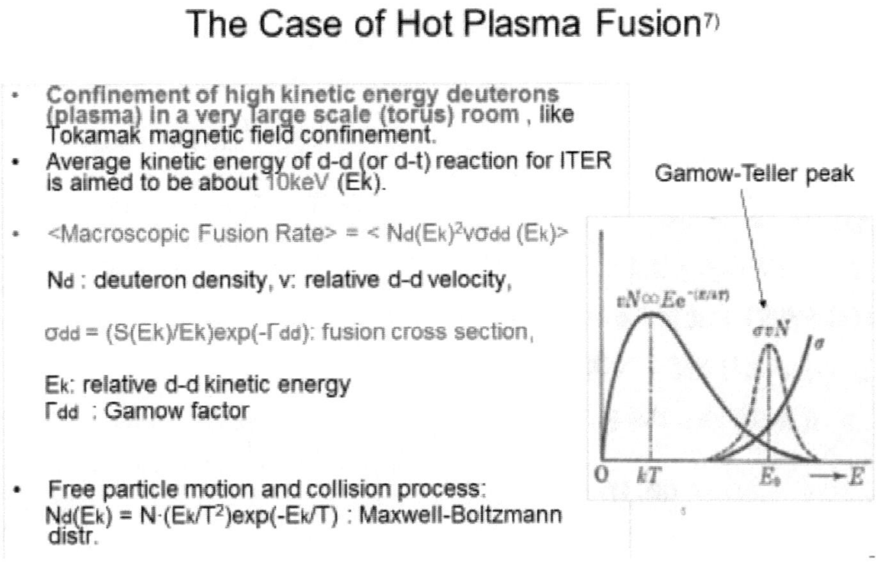

Figure 3. Fusion reaction rate formulas for hot plasma fusion.

mechanics, we need to assume finite sizes of particles. Borrowing the simple image of nucleus as closely packed nucleons (neutrons and protons for virtual instantaneous quantum chromodynamics (QCD) oscillating states inside nucleus by exchanging quarks with gluons) to be $R_0A^{1/3}$ fm, we can estimate the geometrical cross section of collision in the view of classical mechanics particle motion. Here R_0 is 1.2 fm and A is the mass number (sum of nucleon numbers, $A = N + Z, N$ for neutrons and Z for protons).

We know that the geometrical cross sections do not agree with the measured data (mostly by accelerator beam with target) and we have had to accumulate evaluated data with comparison of measured data and QM cross sections formulated by the squared collision amplitude of scattered particle QM wave function (see a simple derivation of QM cross section formulas in [8]). Usually, the QM collision cross section becomes much larger than the classical geometrical cross section mainly due to the spread wave nature of QM particle wave functions. The second important aspect is that the classical geometrical cross section does not have capability to treat nuclear fusion process which should make the sub-atomic (inter-nuclear) level transition of structure in quark–gluon interaction. The sub-atomic level transition is made by the strong force interaction or the weak force interaction, resultantly to cause mass deficit (defect) that contributes to the nuclear excited state of intermediate compound state and the released particle kinetic energies after the final state strong/strong, strong/electro-magnetic or strong/weak force-driven transitions [10]. As the fact, we need to use fully QM formalisms to know the reaction rates of 'hither-to-unknown' nuclear reactions in condensed matter.

In Figs. 2 and 3, we show brief summary of QM two-body cross section formulas and fusion rate estimation for deuteron electron plasma. Cross section is given to be proportional to the square of transition matrix **T** for the free particle collision process (Fig. 2). The averaged fusion rate of free-gas like motion of D-gas and D-plasma is given by effective Gamow peak for the Maxwellian average of two-body d–d cross section, which is a product of astrophysical S-factor (intrinsic inter-nuclear strong interaction factor) and barrier factor (QM tunneling probability for Coulombic

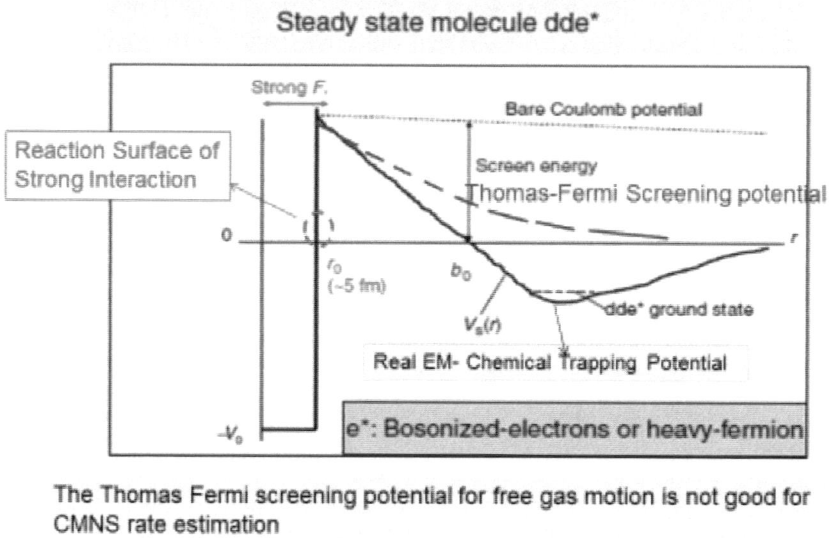

Figure 4. Comparison of charge-screening effects to cause nuclear fusion enhancement between the bare Coulomb potential, the Thomas–Fermi potential and the pseudo-potential of d–d pair chemical trapping in condensed matter. To estimate fusion rate, we estimate Gamow integral of barrier factor from b_0 point to fusion interaction surface distance (about 5 fm of d–d distance).

interaction between deuterons) [3]. The screening effect (decreasing repulsive force between deuterons) by existing negative charge of electrons in gas and plasma state is usually estimated by using the Thomas–Fermi free gas model. Since the screened potential by using the Thomas–Fermi model does not have negative well of particle trapping, its application to the D(H) particles trapped in some kind of chemical potential of condensed matter becomes a very rough approximation, as we show in the following of this paper.

Figure 4 shows the image of charge screening effect comparing the bare Coulomb potential, the Thomas–Fermi free gas potential and the pseudo-chemical potential of dynamic condensed matter.

In intuition, we may have understanding that the pseudo-potential of D(H) particle trapping in (dynamic conditions of, as explaining later) condensed matter may give much more enhanced screening effect than the Thomas–Fermi model of screened potential effect.

3. Rate Formulas for Trapped D(H) Particles in Condensed Matter

A simple image of QM state of d–d pair trapped in a chemical potential well is illustrated in Fig. 5.

A simple case of d–d pair trapped in a chemical potential is a D_2^+ ion, namely a d–e–d three body system, as illustrated by the left-hand side figure of Fig. 5. The chemical trapping potential is Vs1(1,1) as given in [5,6,8,12]. For treating the very fast dynamic process of TSC condensation/collapse motion, time variation of TSC trapping potentials [5,6,12] were approximated (substituted) with effective-mass variation of Vs1(m,Z) pseudo-potentials [6]. Here the Vs1(m,Z) pseudo-potential is defined as the d–d (or p–p) pair trapping potential of electronic quasi-particle expansion theory (EQPET) [8] dde* molecule with effective mass m and effective charge Z. The selection of (m,Z) is not necessarily a set of integers, but continuous numerical values can be used to follow the continuous variation

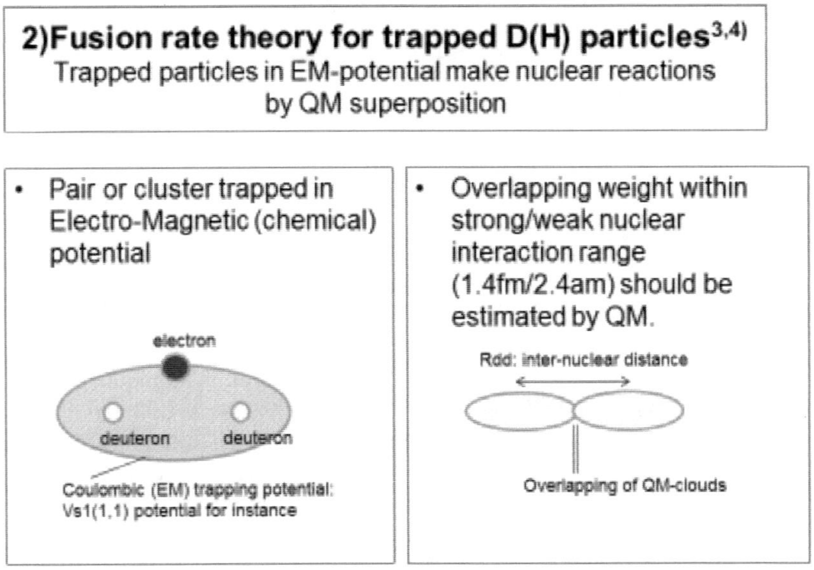

Figure 5. A simple image of trapped d–d pair in QM chemical potential well (*left figure*) and a very small domain of strong interaction for overlapping QM particle density functions for d–d pair (*right figure*).

of elapsed time of condensation. Examples of Vs1(m,Z) pseudo-potentials used for time-dependent condensation motion is shown in Section 4. The Vs1(1,1) potential is the real one of D_2^+ molecule which is stable in nature; that means its infinite lifetime at ground state (R_{dd}(gs) is 138 pm with electron mean kinetic energy 13.6 eV [5]). However, the pseudo-d–d trapping potential for dde* pseudo-molecule is a kind of pure mathematical tool to describe an adiabatically discretized time-dependent trapping potential with very small time mesh [5,6]; this means the very fast time variation of TSC-like cluster condensation motion can be numerically simulated by using Vs1(m,Z) potential with the empirically related substitution from time-step to discretized-m with as practically small as possible time mesh [5,6]. In every discretized time-mesh at every time-step, we need to calculate the weight of superposed wave functions of two deuterons (as shown on the right-hand side part of Fig. 5) that is the time-dependent barrier penetration probability for getting to the interaction surface (R_{dd} = ca. 5 fm) of strong nuclear interaction domain.

We now have to be cautious about the fact that the superposition of two deuteron wave functions does not mean fusion reaction yet. To cause a fusion reaction event, there should occur the transition of sub-atomic (inter-nuclear) structure of superposed two deuteron's admixture; namely the fusion transition from the d+d pair to the intermediate transit compound particle ^{4}He*(Ex) should occur. The interaction Hamiltonian for this transition is written by the component proportional to the imaginary part $W(r)$ of nuclear optical potential [8] $V(r)$+ i$W(r)$. To derive a QM averaged fusion transition rate, we shall treat the balance equation of particle QM density function $\rho(r) = \Psi\Psi^*$ with the forward wave function Ψ and the adjoint (backward) wave function Ψ^* of particle-pair, as shown in Fig. 6.

The right-hand most term of the bottom equation of particle QM density balance corresponds to the rate of nuclear fusion [4,8]. The fusion rate formula is therefore become the top most equation of Fig. 7.

Our real fusion rate calculation has to be done time dependently by using every substituted EQPET pseudo-potential Vs1(m,Z) at every small time step. In Fig. 8, we illustrate an image of time-dependent pseudo-potential

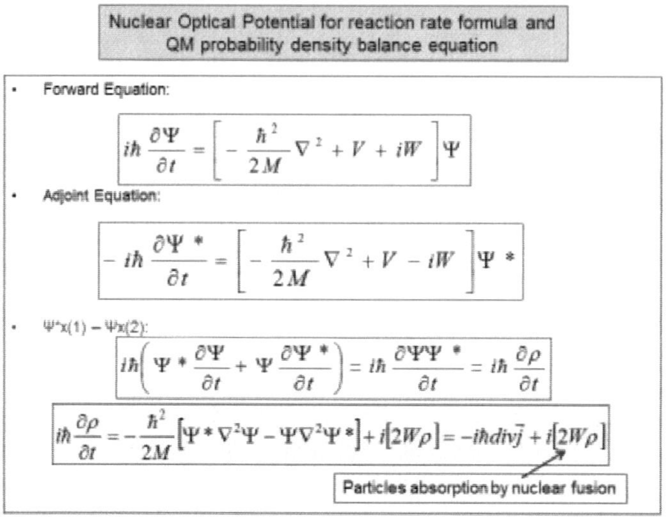

Figure 6. Derivation of the balance equation of particle-pair QM density function by using the forward and the backward Schroedinger equations with nuclear optical potentials.

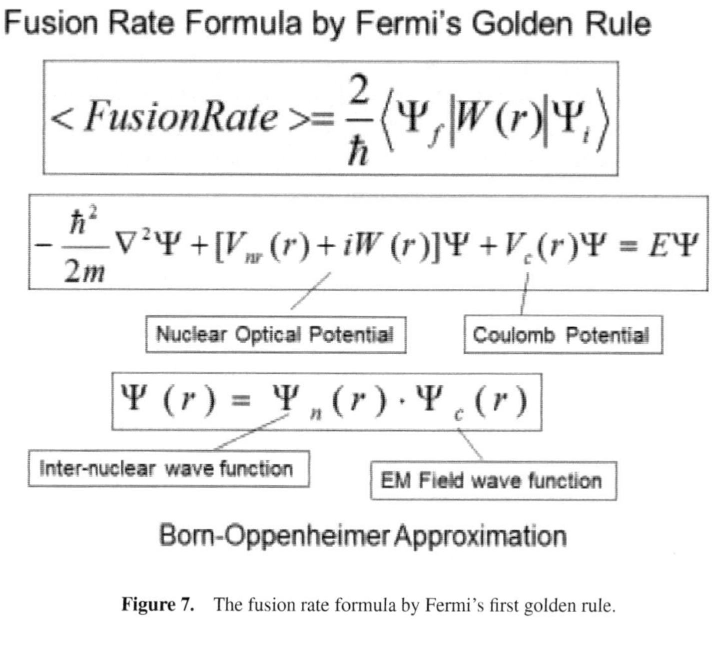

Figure 7. The fusion rate formula by Fermi's first golden rule.

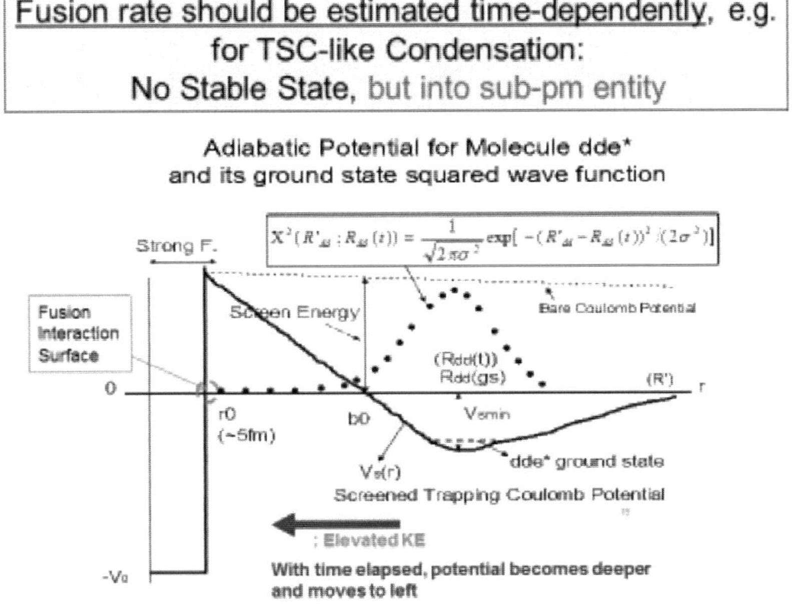

Figure 8. Illustration of time-dependent adiabatically approximated pseudo-potential of d–d pair trapping.

for EQPET molecule dde*(m,Z). In numerical calculations for ground state eigen-value search of dde*(m, Z) with Vs1(m, Z) potential, we applied the variational principle [6,11] with Gaussian d–d pair wave function (this is approximation for wave function form). As the elapse of condensing time of TSC-like symmetric D(H)-cluster becomes to approach the collapse time, pseudo-potential at the discretized small time step becomes deeper and virtual deuteron kinetic energy becomes larger, by reflecting in the Heisenberg Uncertainty Principle (HUP) [11].

In the HMEQPET method, we adiabatically treat a quasi-steady state of EQPET pseudo-particle dde* within a very small discretized time interval; step-dependent time-steps of 0.05–0.0000001 fs were used in actual calculations by the QM Langevin calculation code [6,12]. We needed to choose as very small time mesh as 0.0000001 fs when condensation motion approaches very closely to the collapse time t_c. By applying the adiabatic HMEQPET method for simulation, however, we make it possible to treat only steady state Schroedinger equations for every dde*$(m,2)$ pseudo-molecule.

A steady state Schroedinger equation, with two components of nuclear optical potential and chemical (or EM) d–d pair trapping pseudo-potential, can be again adiabatically solved by the Born–Oppenheimer approximation, as shown in Fig. 9.

Then, the fusion rate formula is converted to be more practical one as shown in Fig. 10. We calculate the inter-nuclear fusion rate, namely the bracket integral of the imaginary part $W(r)$ of nuclear optical potential, being independent of outer chemical EM (Coulombic) field. The second term of bracket integral corresponds to the so-called barrier penetration factor (or QM tunneling probability), for which we can approximately use the WKB Gamow integral treatment [8]. We should be cautious to the domain of bracket integrals for both cases. The domain of bracket integral must be within the domain of strong (or weak) nuclear interaction range as approximately given with the $4\pi R_0^2\lambda_\pi$, where

Inter-Nuclear QM Schroedinger Equation:

$$-\frac{\hbar^2}{2m}\nabla^2\Psi_n(r)+[V_{nr}(r)+iW(r)]\Psi_n(r)=E_n\Psi_n(r)$$

Outer-Nuclear QM Schroedinger Equation for Electro-Magnetic Field:

$$-\frac{\hbar^2}{2m}\nabla^2\Psi_c(r)+V_c(r)\Psi_c(r)=E_c\Psi_c(r)$$

Figure 9. Born–Oppenheimer approximation is applied for separately calculating the inter-nuclear wave function and the outer-nuclear (chemical or EM-field) wave function of d–d pair.

λ_π is Compton wave length of pion (for strong interaction, 1.4 fm) or weak W-boson (weak interaction). Estimation of inter-nuclear fusion rates for 2D, DT, 3D, and 4D fusion was empirically done [6,8]. Estimation of the barrier factors is given in Section 4.

4. Effect of Lifetime of Trapped D(H)-particles in Dynamic Pseudo-potentials

According to our past studies on dynamic motions of D(H)-molecules and clusters [4,5,12], we have recognized that feasibly enhanced DD fusion rates as large as experimentally claimed anomalous heat-powers do not exist for

Fusion Rate Formula by Fermi's First Golden Rule with Born-Oppenheimer Approximation

Inter-nuclear fusion rate   Barrier Factor

$$< FusionRate >=\frac{2}{\hbar}\langle\Psi_{nf}|W(r)|\Psi_{ni}\rangle_{Vn}\cdot\langle\Psi_{cf}|\Psi_{ci}\rangle_{Vn}$$

$$Vn \approx 4\pi R_n^{\,2}\lambda_\pi$$: Effective Volume of Nuclear Strong (Weak) Interaction Domain

λ_π : Compton wave length of pion (1.4 fm) (weak boson: 2.5 am)

R_n : Radius of Interaction surface of strong (weak) force exchange

Figure 10. Practical fusion rate formula by Born–Oppenheimer approximation for the Fermi's golden rule.

<W> value Estimation

- Using Tn ~ (PEF)[5] in S-value analysis:

Cluster	<W> (MeV)
DD	0.008
DT	0.115
3D	1.93
4D	62.0

$<W> = \langle \psi nf | W(R) | \psi ni \rangle$

Inter-nuclear wave functions are governed by
the real part of optical potential V(R),
having Woods-Saxon type well shape
that makes inter-nuclear wave functions
very localized near around V(R).
Therefore, <W> becomes approximated by
the surface sticking at R=R0.

Figure 11. Estimated bracket integral of $W(r)$ for two-body and multi-body fusions.

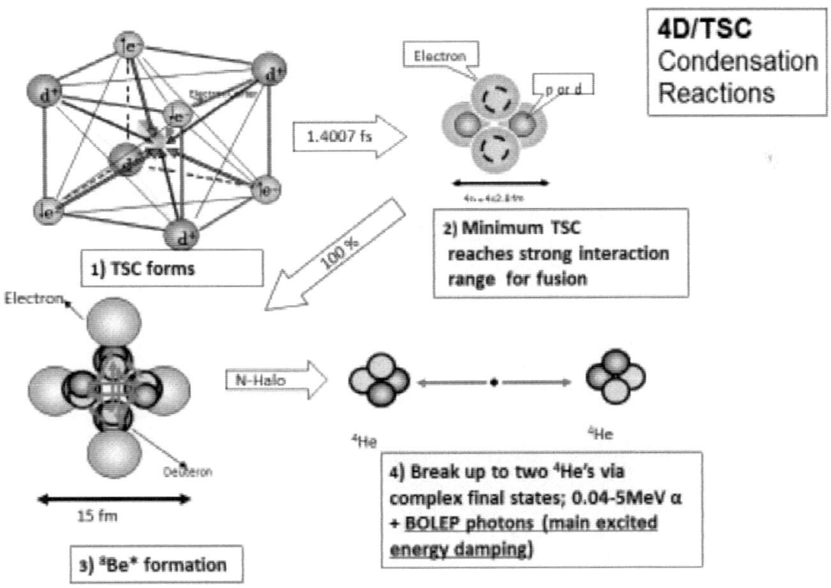

Figure 12. The four steps of 4D/TSC condensation; (1) TS formation, (2) collapse, (3) ^{4}Be* intermediate compound, and (4) the final state decay processes to produce low energy helium and BOLEP photons.

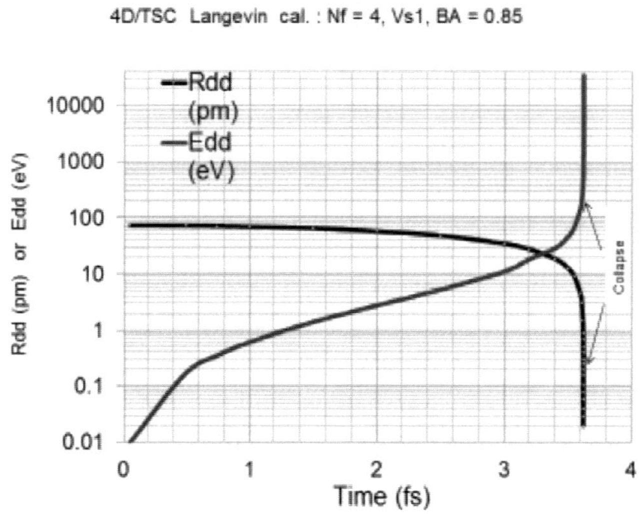

Figure 13. 4D/TSC condensation/collapse motion calculated by the QM Langevin code [12].

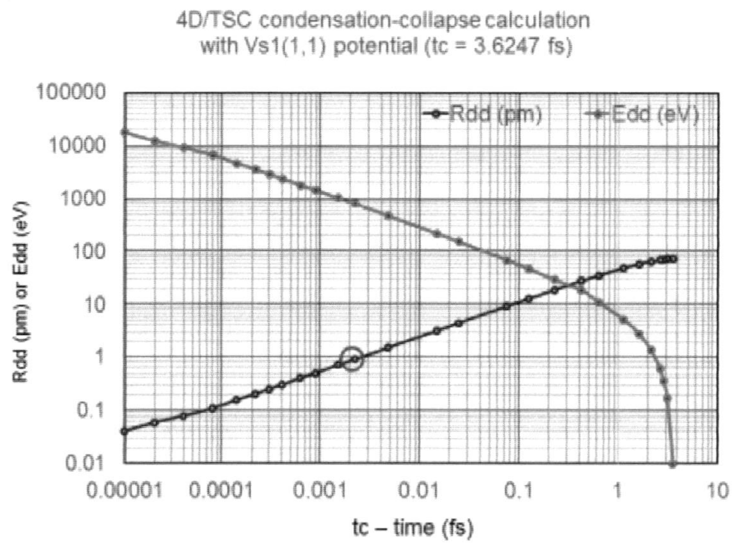

Figure 14. The log–log plot of 4D/TSC condensation as a function of reversed time direction from the collapsed time t_c (3.6247 fs for this case) to the starting point of 4D/TS formation time ($t = 0$). When R_{dd} reaches at 35 fm, relative deuteron kinetic energy of d–d pair is increased to the level of ca. 20 keV.

R_{dd}=R_{gs} (pm)	P_{2d} ; 2D Barrier Factor	P_{4d}; 4D Barrier Factor
0.0206	4.44E-2	1.98E-3
0.0412	1.06E-2	1.12E-4
0.103	1.43E-3	2.05E-6
0.206	3.35E-5	1.12E-9
0.412	9.40E-7	2.16E-13
0.805 (μ dd)	1.00E-9	1.00E-18
1.03	9.69E-11	9.40E-21
2.06	6.89E-15	4.75E-29
4.12	9.38E-21	8.79E-41
10.3	2.16E-32	4.67E-64
21.8 (dde*(2,2)	1.30E-46	1.69E-92
74.1 (D_2 molecule)	1.00E-85	1.00E-170

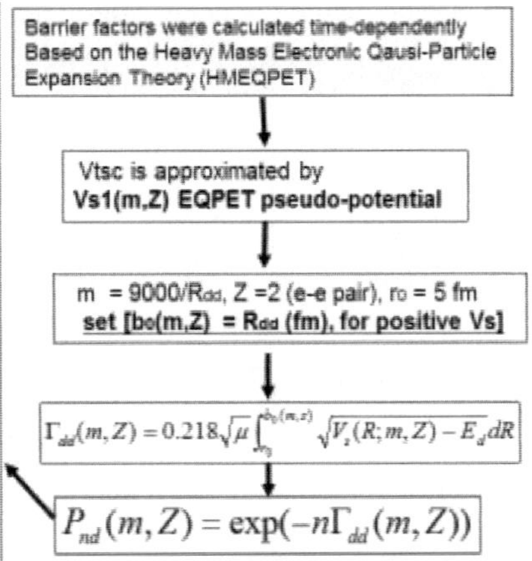

Figure 15. The HMEQPET method of calculating time-dependent barrier factors as a function of R_{dd} or condensation time.

steady molecules of D_2^+, D_2, D_3^+ or their transient time-dependent motion of formation. However, we have found that three-dimensionally symmetric (or orthogonally coupled electron wave function and deuterons wave function in the three-dimensional space) D(H) cluster such as 4D(H)/TSC, 6D(H)/RDC, and 8D(H)/RDC may have no steady ground states but may make ultimate-condensation/collapse-type one-through motion time dependently [5,6,12]. The case of 4D/TSC condensation/collapse motion in four sub-divided steps is copied in Fig. 12. Here, TSC is the tetrahedral symmetric condensate and the RDC is the Rhombic dodecahedron condensate, respectively.

The corresponding QM Langevin code calculation with Vs1(1,1) potential for every d–d pair in 4D/TSC is shown in Fig. 13. To see the detail behavior near to the collapse time, the log-log plot of inter-nuclear d–d distance R_{dd} and mutual kinetic energy of d–d pair is given in Fig.14, by reversing the flow of time.

We have estimated already that 100% 4D fusion will take place until the condensed R_{dd} reaches at 20 fm [6]. However, the following discussions will be made for arguing the possibly enhanced d–d fusion rate by this kind of dynamic condensation/collapse motion, for understanding general feature of fusion rate enhancement as a function of quasi-lifetime (effective existing time interval) of d–d pair and also as a function of second parameter of d–d distance R_{dd}.

Just for example, we put a marker for $R_{\mathrm{dd}} = 1$ pm region (see red circle in Fig. 14). The condensation motion is changing continuously. However, according to our HMEQPET adiabatic analysis with very small discretized time meshes, we see the effective existing time of d–d pair with diminished size of 1 pm R_{dd} distance has about 2 as (ato second = 1.0×10^{-18} s) "pseudo-lifetime" of dde*(m,2) pseudo-EQPET molecule. What we want to know now is the rough measure of DD fusion power per one molar d–d pairs ($6.023 \times 10^{+23}$ pairs). To know this, we need to estimate the value of barrier factor for the $R_{\mathrm{dd}} = 1$ pm adiabatic state. The calculation process of time-dependent barrier factor is shown in Fig. 15.

Implement of Gamow integral calculation with a pseudo-potential Vs1(m, Z) needs a fix-up technique to erase negative values of potential well region inside the square root of Gamow integral. Before showing that, some examples of used pseudo-potentials are shown in Fig. 16. A dde* adiabatic pseudo-molecule with the ground state d–d distance 100 pm with –36 eV negative well is given by the Vs1(0.4, 2) EQPET potential. A dde* adiabatic pseudo-molecule with the ground state d–d distance 10 pm with –394 eV negative well is given by the Vs1(4,2) EQPET potential. A

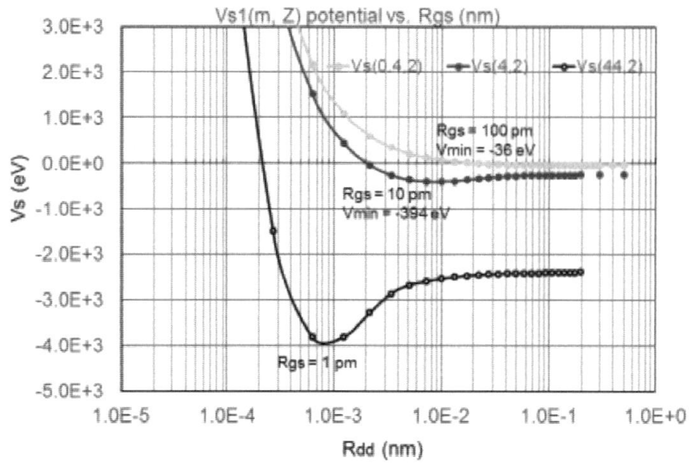

Figure 16. Some examples of used pseudo-potentials to approximate real dynamic cluster trapping potentials as $V_{\rm tsc}$.

dde* adiabatic pseudo-molecule with the ground state d–d distance 1 pm with –4000 eV negative well is given by the Vs1(44,2) EQPET potential. Here Z =2 is assumed and fixed. As discussed in detail in [8], anti-parallel spin coupling with reversed momenta for two electrons locating on a square domain of a D_2 molecule-like face of 4D(H)/TSC under condensation/collapse motion were thought to make a kind of "transient Cooper pair" on each square face of TSC cube. The pseudo-electronic quasi-particle e*(2,2) is of approximate "bosonized state" of electrons in the Platonic symmetry of condensate [8,6].

To erase negative values in the square root of Gamow integral, we have introduced the fix-up technique as illustrated in Fig. 17.

Instead of carrying out Gamow integration with the deep negative well of Vs1(44,2) pseudo-potential for $R_{\rm dd}$ = 1 pm dynamic pseudo-molecule state, we use Vs1(9,2) pseudo-potential to implement Gamow integral calculation from the b_0 point (potential zero crossing point) of 1 pm to the strong interaction surface range about 5 fm. We have one-to-one empirical relation for the $R_{\rm dd}$(gs) to b_0-corresponding potential for time-varying m value [6].

Calculated barrier factors by the HMEQPET method are plotted in Fig. 18 as a function of d–d distance.

The estimation of DD fusion power level as a binary-variable function of $R_{\rm dd}$ and lifetime was done by a calculation procedure as shown by the slide of Fig. 19. Illustration of DD fusion power levels vs. parameters of lifetime and d–d distance by graph is given in Fig. 20.

We see that with $R_{\rm dd}$ = 1 pm and 10 as (1.0×10^{-17} s) lifetime the DD fusion rate reaches 10 kW/mol-dd level and we meet the similar conditions of experimentally claimed anomalous excess heat–power.

However, according to the author's CMNR/Cold-Fusion studies until now, DD fusion cannot explain the claimed ^{4}He production per 24 MeV energy release without hard gamma rays and neutrons [3–11].

The author has been asserting that 4D/TSC fusion (see Fig. 12) may be the real source of ^{4}He/23.8 MeV nuclear events in CMNR experiments using deuterium loading systems. In this case, as shown in Fig. 21, 10 kW heat–power level may be attained by the tetrahedral symmetry (TS) state formation rate of 2.8 nano-mol per second in the condensed matter. The rate theory for TS state formation is of pure solid state, surface physics, and chemical problems, and is yet to be studied [4], although we have proposed some models of TS formation on catalytic surface sub-nano-hole (SNH) [7,14,15], lattice defects (vacancies) and non-linear QM oscillation of deuterons trapped in a

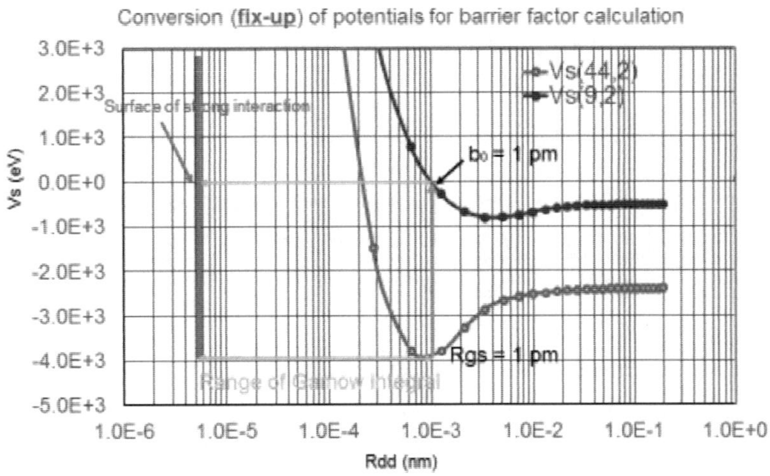

Figure 17. Fix-up technique to erase negative values in the root of Gamow integral.

global mesoscopic potential [15] of nano-metal-particle.

Finally, the calculated fusion rates for steady dde* molecules, including muonic dd molecule and 4D/TSC-final-collapsed state in Table 1. Evidently, the collision theory drastically underestimates fusion rates of trapped DD pairs

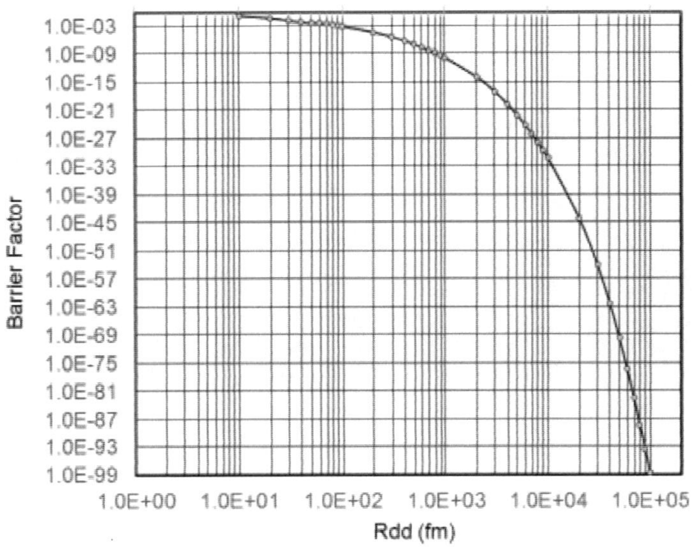

Figure 18. Calculated barrier factors as a function of inter-nuclear d-d distance of pseudo dde* molecule.

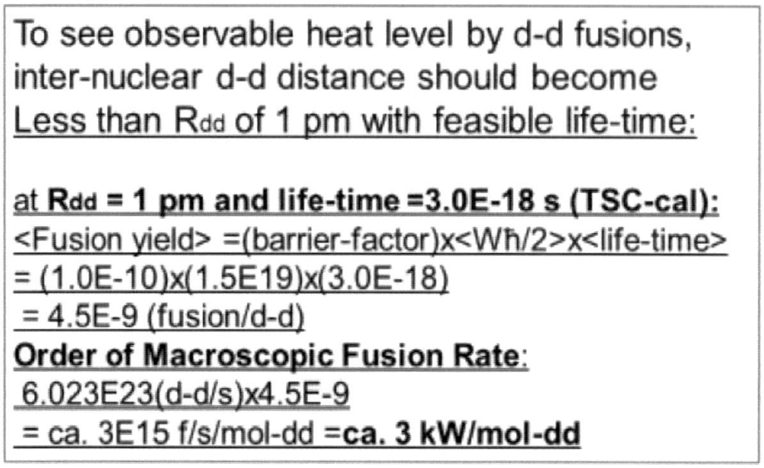

Figure 19. Slide to show how to calculate heat power of DD fusion with parameters of life-time and d-d distance The estimation of DD fusion power level as a binary-variable function of Rdd and life-time was done by a calculation procedure as shown by the slide of Fig.19. Illustration of DD fusion power levels vs. parameters of life time and d-d distance by graph is given in Fig.20.

in negative chemical potential well which is reasonable for condensed matter nuclear reactions. The underestimation

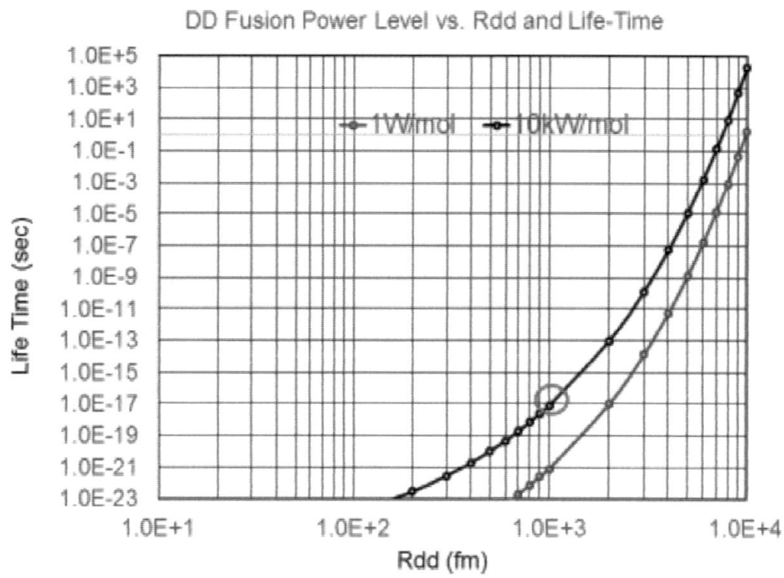

Figure 20. Scaling curve of heat-power level per one molar d-d pairs as functions of life-time and d-d distance.

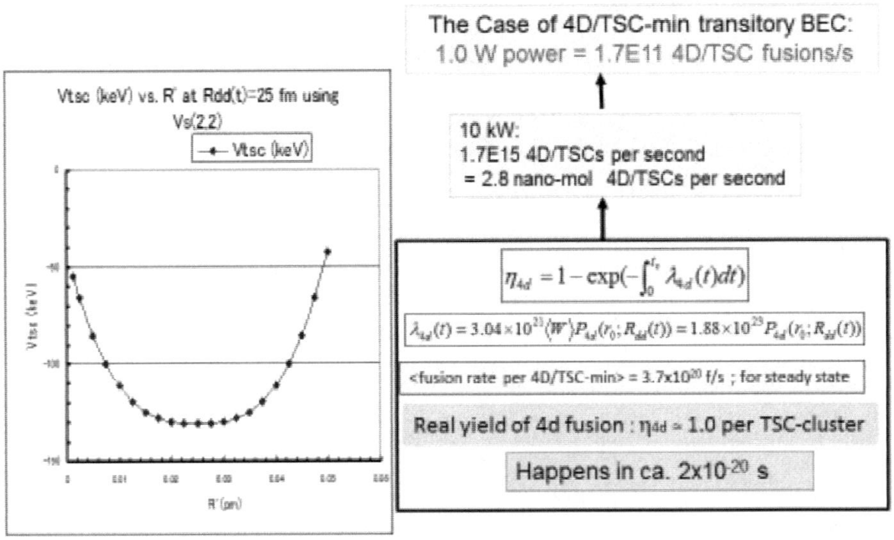

Figure 21. Heat power level of 4D/TSC fusion reaches 10 kW level by 2.8 nano-mol TS formation per second.

is very large order as 19–20 orders of magnitude. When we make corrections with ground state oscillation frequency (ca. 1.0×10^{14} Hz for D_2 molecule), the underestimation is still large as ca. six orders of magnitude.

We have already pointed out that a dynamically diminished size ca. 1 pm of d–d pair with lifetime of ca. 10 ato s will meet the feasible condition that can explain the excess-heat-power levels of experimental claims by many authors (see papers by Fleischmann–Pons, M. Miles, M. McKubre, V. Violante, Y. Arata, and so on in the list of [1]).

Collision Rate Formula UNDERESTIMATES fusion rate of steady molecule/cluster

Cluster	Rdd = Rgs (pm)	Barrier Factor	Steady Cluster d-d Fusion Rate (f/s)	Steady Cluster 4d Fusion rate (f/s)	Fusion Rate for d-d collision formula (f/s)
D2	74.1	1.0E-85	2.4E-66		3.6E-86 (4.0E-72)*
dde*(2,2)	21.8	1.3E-46	3.2E-27		1.0E-46 (1.0E-31)*
ddμ	0.805	1.0E-9	2.4E+10		1.5E-9 (1.0E+8)*
4D/TSC-minimum	0.021	1.98E-3		3.7E+20	

* Frequency of d-d pair oscillation by QM-Langevin calculation was considered. 25

Table 1. Calculated fusion rates by trapped state DD pair, compared with those by two-body collision theory.

Looking at Fig. 20 again, we see that the $R_{dd} = 10$ pm state of diminished d–d size requires a lifetime of ca. 1.0 $\times$ 10^{+4} s, namely 2.78 h for the quasi-steady (isomeric) dde* molecule. Such an isomeric d–d quasi-molecule state has never been observed in condensed matter science (CMS), since the transition from the ca. $R_{dd} = 100$ pm state of usually considered d–d pair state in PdD-like lattice to the $R_{dd} = 10$ pm state with long lifetime will emit ca. 360 eV photons to be observable, according to referring pseudo-ground state levels of trapped d–d pair in Vs1(0.4,2) and Vs1(4,2) pseudo-potentials as shown in Fig. 16. Of course, deuterium gas, namely D_2 molecule with R_{dd} =74.1 pm has no possibility to observe any DD fusion event in our lifetime (at longest 110 years), because its fusion rate is so small as (2.4×10^{-66}) $\times$ ($6.023 \times 10^{+23}$) $= 1.44 \times 10^{-42}$ f/s/mol. The muonic DD fusion has a great possibility in principle, as 100% d–d fusion yield takes place in ca. 100 ps after a muonic dd molecule is formed [6,12], but intense generation of muonic flux by high energy accelerator is not economical. Until now, the author has found a plausible/feasible mechanisms of high power level CMNR events only in the dynamic condensation/collapse motion of such D(H)-clusters as TSC and RDC [4]. We have shown by the QM Langevin code simulations [12] that condensation/collapse times for such polyhedral symmetric condensates (PSC) took several femto seconds only. From Fig. 20, by seeing the 1.0×10^{-15} s lifetime level, we may conclude that the maximum inter-nuclear d–d distance to observe the experimentally claimed anomalous heat power level is ca. 1 pm under dynamic/transient condition of condensed matter.

5. Conclusions

For the CMNR as known as cold fusion which has been studied in emerging CNMS, we need to take life-time of D(H)s in chemical trapping potential well into account properly to formulate reaction rate formulas.

We cannot confidently use formulas for two-body collision theory for CMNR, as they drastically underestimate fusion rate as we have shown in this work. The underestimation is as huge as 19–20 orders of magnitudes. We understand that, because of much longer life-time (or existing time) of diminished size d–d pair trapped dynamically/transiently in condensed matter, a d–d pair can have very enhanced chance to cause stochastic DD fusion.

Fermi's first golden rule should be used for rate formulas for CMNR. We have shown a summary review of reaction rate formulas for CMNR based on the Fermi's golden rule. For applying the Fermi's golden rule to dynamic/time-dependent D(H)-cluster condensation/collapse motion, we have developed some mathematical tools such as the QM-Langevin code for analysis of time-dependent condensation motion and the HMEQPET code for time-dependent QM tunneling probability estimation with eigen-value search of pseudo-potential trapping (the variational principle is applied).

Feasibly enhanced cluster fusions may happen only for collapse states of dynamic condensation: 4D/TSC, 6D/RDC, 8D/RDC for examples, as studied in our recent works.

Therefore, we need time-dependent fusion rate calculation to estimate real fusion yield per D(H)-condensing-cluster formation. We have done some elaborations for the deuterium (D) related CMNRs based on the strong nuclear reaction mechanisms, and extension (not described in this paper) for protium (H) related CMNRs.

If inter-nuclear d–d distance for condensate becomes less than 1.0 pm, we can expect reaction rates with 'observed heat-power levels'. The inter-nuclear d–d distance of 10 pm state of d–d pair in condensed matter lattice dynamics is not short enough to observe experimentally claimed excess-heat power level.

References

[1] See paper compilation of ISCMNS: http://www.iscmns.org/library.htm

[2] Preprints of the following papers are down-loadable from http://vixra.org/author/akito_takahashi

[3] Akito Takahashi, Development status of condensed cluster fusion theory, *Current Science* **108** (4) (2015) 514–515.

[4] Akito Takahashi, Background for condensed cluster fusion, *Proc. JCF15*, to be published in *JCFRS* http://jcfrs.org/proc_jcf.html (see also [2]).

[5] Akito Takahashi, The basics of deuteron cluster dynamics as shown by Langevin equation, *Amer. Chem. Soc. LENRSB (Low Energy Nuclear Reaction Source Book)* **2** (2009) 193–217.

[6] Akito Takahashi and Norio Yabuuchi, Study on 4D/TSC condensation motion by non-linear Langevin equation, *Amer. Chem. Soc. LENRSB* **1** (2008) 57–83.

[7] Akito Takahashi, Physics of cold fusion by TSC theory, *J. Physical Sci. Application* **3**(3) (2013) 191–198.

[8] Akito Takahashi, Fusion rates of bosonized condensates, *JCMNS* **1** (2007) 62–85.

[9] A. Takahashi and D. Rocha, Nucleon halo model of ^{8}Be*, *Proc. JCF-13*, 2013, pp.10–31. http://jcfrs.org/file/jcf13-proceedings.pdf.

[10] Akito Takahashi, Nuclear products of cold fusion by TSC theory, *Proc. ICCF18*, to be published in *JCMNS* (see also http://vixra.org/abs/1309.0072).

[11] Akito Takahashi, Kinetic reaction energy of cold fusion, *JCF-12-7, Proc. JCF-12*, pp. 67-76 (2013), http://jcfrs.org/file/jcf13-proceedings.pdf.

[12] A. Takahashi and D. Rocha, D(H)-cluster Langevin code and some calculated results, *Proc. JCF14*, 2014, pp. 113–140 (see also preprint at http://vixra.org/abs/1401.0202).

[13] Akito Takahashi, 4H/TSC Fusion by simultaneous weak and strong interactions, *JCF12-14, Proc. JCF-12*, 2013, pp. 115–122.

[14] A. Takahashi, R. Seto, Y. Fujita, A. Kitamura, Y. Sasaki, Y. Miyoshi and A. Taniike, Role of PdO surface coating in CNME D(H)-gas loading experiments, *JCMNS* **4** (2010) 17–33.

[15] A. Takahashi, A. Kitamura, Y. Miyoshi, H, Sakoh, A. Taniike, R. Seto and Y. Fujita, Mesoscopic catalyst and D-cluster fusion, *Proc. JCF11*, 2011, pp. 47–52.

J. Condensed Matter Nucl. Sci. 19 (2016) 316–318

Research Article

Theoretical Study of the Transmutation Reactions

T. Toimela*

Vaasa University of Applied Sciences, Wolffintie 30, 65200 Vaasa, Finland

Abstract

Transmutation reactions are studied from a theoretical point of view. An idea is proposed to explain the variations in the transmutation ability of different elements, especially the relative inertness of palladium compared to the other elements. Proposals are made in order to verify experimentally this explanation and to enhance the transmutation signal.

Keywords: LENR, Surface, Theory, Transmutation, WKB approximation

The Condensed Matter Nuclear Science has been studied for the last 26 years since the original work by Fleischmann and Pons [1]. While the cold fusion of two deuterium atom may be more important from the technological point of view, the transmutation reactions are scientifically more interesting. The existence of low energy transmutation reactions is intrinsically even more surprising than the cold (low energy) deuterium fusion, because the low tunneling probability (the Gamow factor) is supposed to suppress very effectively the transition rate with the increasing nuclear charge.

In this note, transmutation reactions will be studied theoretically. In what follows we shall assume that there is some kind of mechanism, which greatly enhances the apparently small tunneling probability. We will not make any restrictive assumptions about the actual mechanism, except just assuming that the mechanism *itself* is independent of the nuclear charge Z.

The reported experiments show a very large variety in the transmutation products [2,3]. Not only the products, but also the methods and materials by which transmutation had been found, have a large variety [4,5]. Among all the published transmutation studies the most prominent are those by Iwamura et al. [6,7]. The interpretation of the experimental results allowed them to propose the existence of the following reactions:

$$4\,\mathrm{D} + {}^{133}_{55}\mathrm{Cs} \rightarrow {}^{141}_{59}\mathrm{Pr} + \ 50.49\ \mathrm{MeV} \tag{1}$$

and

$$6\,\mathrm{D} + {}^{138}_{56}\mathrm{Ba} \rightarrow {}^{150}_{62}\mathrm{Sm} + \ 67.61\ \mathrm{MeV}. \tag{2}$$

* E-mail: tuomo.toimela@vamk.fi

It is worth noting that the corresponding reaction of palladium (e.g. ${}^{108}_{46}\mathrm{Pd}$)

$$4\,\mathrm{D} + {}^{108}_{46}\mathrm{Pd} \rightarrow {}^{116}_{50}\mathrm{Sn} + 54.55\ \ \mathrm{MeV} \tag{3}$$

has not been found, although palladium is present far more extensively than the small amount of the deposited cesium or barium. In some other transmutation studies evidence has been found also of transmutation reactions of palladium itself. However, it can be taken granted that the transition rates for the transmutation reactions of palladium are several orders of magnitude smaller than those of cesium or barium.

Intrinsically one may assume that the rates of the transmutation reactions depend prominently on the nuclear charge Z (through the tunneling probability) and the energy released Q. In both of these respects the palladium reaction should be more copious than for example the corresponding cesium reaction.

If we consider a bare nucleus, the transition rate of the deuterium absorption can be obtained by the semi-classical WKB approximation, giving the leading behavior of the probability:

$$P_1 \sim \mathrm{e}^{-\int_0^r 2\sqrt{\frac{2\alpha\mu c Z}{\hbar x}}\,\mathrm{d}x} \sim \mathrm{e}^{-4\sqrt{2\alpha\mu c Z r/\hbar}}, \tag{4}$$

where r is the equilibrium distance of the deuteron from the nucleus, μ is the reduced mass of the deuteron and the heavy nucleus, and α is the electromagnetic fine structure constant ($\approx 1/137$). Equation (4) is the static counterpart of the Gamow factor that corresponds to a collision of a charged deuterium projectile and a target nucleus. If we consider a transition, where four deuterons are absorbed simultaneously, Eq. (4) is replaced by

$$P_4 \sim \mathrm{e}^{-16\sqrt{2\alpha\mu c Z r/\hbar}}. \tag{5}$$

Taking into account the modification caused by the electron screening, Eq. (5) is replaced by

$$P_4 \sim \mathrm{e}^{-8\int_0^r \sqrt{2\mu V(x)/\hbar^2}\,\mathrm{d}x} \sim \mathrm{e}^{-8\int_0^r \sqrt{\frac{2\alpha\mu c}{\hbar}\times\frac{Z-\frac{4\pi x^3}{3}\rho_0-\sum_i\int_0^x \mathrm{d}^3x\,\rho_\mathrm{i}(x)}{x}}\,\mathrm{d}x}, \tag{6}$$

where ρ_0 is the number density of the free electron gas in metal (including the electrons of the loaded deuterium atoms). $\rho_\mathrm{i}(x)$ are the electron probability functions that in the first approximation can be assumed to be the electron probability functions of the hydrogen atom with the appropriate effective nuclear charge for different energy levels [8,9]. The inclusion of the charge screening by the electrons does not change the fact that the transition probability depends very effectively (i.e. exponentially) on the distance of the deuterons from the nucleus.

These tunneling probabilities shown in Eqs. (5) and (6) are of course incredible small numbers. However, we shall assume that there is some kind of yet unknown CMNS mechanism enhancing these extremely small numbers, and we shall explore here the differences between different elements.

Actually the dependence on the distance r is more important than the dependence on the nuclear charge in Eq. (6). If in some configuration the deuteron is lying (or be pushed) more close to the nucleus, the transition rate increases substantially, which may then allow the enhancing mechanism to compensate the extremely low tunneling probability. These shorter distances can be achieved if there are atoms or molecules, which are either neutral or have the effective charge less than the elementary charge, in the surface region. There is a simple situation, where one can assume this to occur. While the palladium atoms in the lattice are ions as they have given an electron in the surrounding electrons gas, the base metal atoms deposited on the surface can be assumed to be mostly oxidated, and are then part of neutral entities. Especially this is true for alkali metals or alkaline earth metals. Hence the deuterons lie more close to those base metal atoms than the palladium atoms and therefore the base metal atoms transmute more easily.

Few percent reduction of the distance in Eq. (6) increases the probability factor by many orders of magnitude. Of course, these probability factors remain still extremely small numbers, because it is not possible reduce the distances between the deuterons and the absorbing nucleus by chemical means enough to cause the nuclear reaction directly.

However, the increase of the probability factor by the reduction of the distance may be enough that the enhancing mechanism can compensate that small factor.

Of course, the oxidation and the whole surface chemistry are far more complicated than the oversimplified picture above. However, the above idea may give a reasonable picture on the relative transmutation ability of different elements qualitatively.

If this idea is correct, one can assume sodium to be an element that can be transmuted more easily than cesium or barium for three reasons. Firstly the energy released in the corresponding reaction

$$4\,\mathrm{D} + {}^{23}_{11}\mathrm{Na} \rightarrow {}^{31}_{15}\mathrm{P} + 67.45\ \mathrm{MeV} \tag{7}$$

is larger compared to Eq. (1). Also the nuclear charge of the sodium is very much smaller. Sodium is also chemically very active and will be oxidated very rapidly when exposed to the air. Therefore, it would be profitable to check experimentally, whether sodium will be transmuted in the gas permeation experiment along Eq. (7), or by some other (more rapid) transmutation reaction.

Another application of this idea is to embed just below the surface some nonmetal element, which will stay as neutral (or almost neutral) atoms in the palladium lattice. Experiments should be done to see whether transmutation signal can be increased in such a manner. However, one has to ensure that the concentration of the deposited or embedded element remain rather low so that those impurities do not prevent the concentration of the deuterium in the nuclear active environment (NAE) to exceed the critical value needed.

The proposed experiments would give more insight for the further theoretical studies and they can be used to restrict possible theoretical explanations (mechanism and NAE).

References

[1] M. Fleiscmann, S. Pons and M. Hawkins, *J. Electroanal. Chem.* **261** (1989) 301.

[2] G. Miley, G. Narne, M. Patterson, J. Nix, D. Cravens and H. Hora, in *ASTI Workshop on Anomalies in Hydrogen/Deuterium Loaded Metal* , W. Colllis (Ed.), Italian Phys. Soc., Italy, 1997.

[3] T. Mizuno, *Nuclear Transmutation: The Reality of Old Fusion*, Infinite Energy Press, NH, USA, 1998.

[4] E. Storms, *The Science of Low Energy Reaction*, 2007, World Scientific, Singapore, and references cited therein.

[5] M. Srinivasan, *J. Condensed Matter Nucl. Sci.* **13** (2014) 495 and references cited therein.

[6] Y. Iwamura, M. Sakano and T. Itoh, *Jpn. J. Appl. Phys.* **41** (2002) 4642.

[7] Y. Iwamura, T. Itoh and S. Tsuruga, *J. Condensed Matter Nucl. Sci.* **13** (2014) 242.

[8] E. Clemeti and D.L. Raimondi, *J. Chem. Phys.* **38** (1963) 2686.

[9] E. Elementi, D.L. Raimondi and W.P. Reinhardt, *J. Chem. Phys.* **47** (1967) 1300, also in www.webelements.com.

J. Condensed Matter Nucl. Sci. 19 (2016) 319–324

Research Article

Heat Production and RF Detection during Cathodic Polarization of Palladium in 0.1 M LiOD

Vittorio Violante*, E. Castagna, S. Lecci, G. Pagano, M. Sansovini and F. Sarto

ENEA, Unità Tecnica Fusione, C. R. Frascati, Via Enrico Fermi, 45 - 00044 Frascati (Rome), Italy

Abstract

The study of the Fleischmann and Pons Effect (FPE), i.e. the production of excess power production during electrochemical loading of deuterium in palladium (in the past labeled as cold fusion) has had of noticeable multidisciplinary character, involving as solid state and nuclear physics, material science, electrochemistry and others. Correlation between the material status and the occurrence of the effect as well as some changing of the electrochemical interface status have been observed. Although during the last two decades in several calorimetric experiments the effect was observed to be well above the measurement uncertainties, the mechanism producing the excess power is not completely understood. The lack of reproducibility has been so far responsible for the absence of a clear explanation of the phenomenon based on measurement chains conceived to enhance the spectrum of information required to define the effect. Recently, In order to improve this aspect of this research, a specific work has been carried out in order to investigate if the excess power was produced concurrently with the emission of Radio Frequency from the active cathode. Suitable probes and triggering included into the RF measurement chain allowed to reveal RF signal emission during electrochemical loading of Pd samples and a correlation between the heat production, in active samples, and RF signals emission was found. The preliminary results are highlighting the importance of performing studies on the electrodynamic effects that may be involved in the phenomenon and in the design of appropriate instruments conceived to investigate unexplored regions of metal hydrogen systems.

Keywords: Anomalous heat effect in metal hydride, Electrochemistry, Electromagnetic stimulation, Fleischmann and Pons effect, Metal hydrides and deuterides, Palladium, Palladium alloys, Radiofrequency

1. introduction

The excess power effect was observed by using both Pd and Pd90Rh10 cathodes [1]. Some specific features were identified for the active electrodes with both materials [2–4]. The active palladium electrodes had mostly (100) orientation of the poly-crystals, enhanced hydrogen (deuterium) diffusivity, average grain size around 100 μm and a well-defined surface morphology identified by means of the power spectral density function (PSD).

Active Pd90Rh10 cathodes have a high diffusivity and a surface morphology, in terms of PSD [4], very similar to the one belonging to the active Pd samples. The differences are a mixed orientation of the crystals and a lower

*Corresponding author. E-mail: vittorio.violante@enea.it

© 2016 ISCMNS. All rights reserved. ISSN 2227-3123

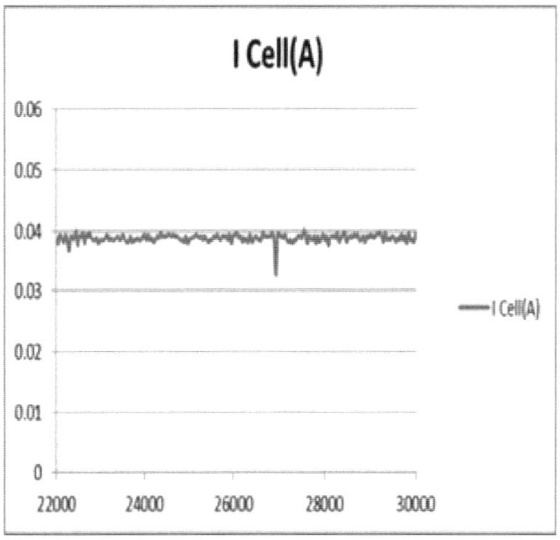

Figure 1. Current ripple during excess of power at low data acquisition rate.

grain size, typically some ten's of microns. Doping with contaminants has been performed for both Pd and Pd90Rh10 cathodes in order to achieve the material characteristics belonging to samples producing excess of power.

The current ripple observed during galvanostatic loading of deuterium in Pd cathodes lead to consider some RF emission from active cathodes during excess power.

Figures 1 and 2 are showing the current evolution during the excess, Fig. 1 is showing a ripple at low da1a acquisition rate (sampling interval 40 s), the ripple in this case was ascribed to an under sampled signal. Figure 2 is showing again a current and voltage a clear current and voltage oscillations turn out.

2. Experimental Results

This evidence supported the hypothesis that a radio frequency signal was produced during the event and a first experimental campaign was performed [5] and signals up to some hundreds MHz were detected.

An experimental campaign was conceived in order to investigate an higher frequency range up to some tents of GHz [6]. During this campaign also a magneto-acoustic trigger was used and the RF signal was detected by using a down-converter, located into the electrolyte, and connected to a National Instrument RF spectrometer.

Both the magnetic trigger and the down converter were given by an anonymous electronic company. During the magnetic stimulus a clear RF signal was revealed by the RF spectrometer and the electrolyte temperature started to grow up of several °C in about 20 min. The trigger was disconnected as soon as the temperature increasing was observed .

Figure 3 shows the evolution of the cell temperature during the excess, Fig. 4a shows the observed RF signal during the event and Fig. 4b shows the typical background RF signal for not active samples.

The calorimetric approach followed during this experimental campaign was based on a simple isoperibolic temperature calibration. During this experimental campaign active and not active lots of palladium cathodes have been identified and active lots were obtained by using a specific metallurgical procedure.

A further experimental campaign was carried out by using a Peltier calorimeter and one of the electrodes belonging to the lots that gave active behavior during the experimental campaign above described showed a clear excess power. The calorimetric behavior is shown in Fig. 5.

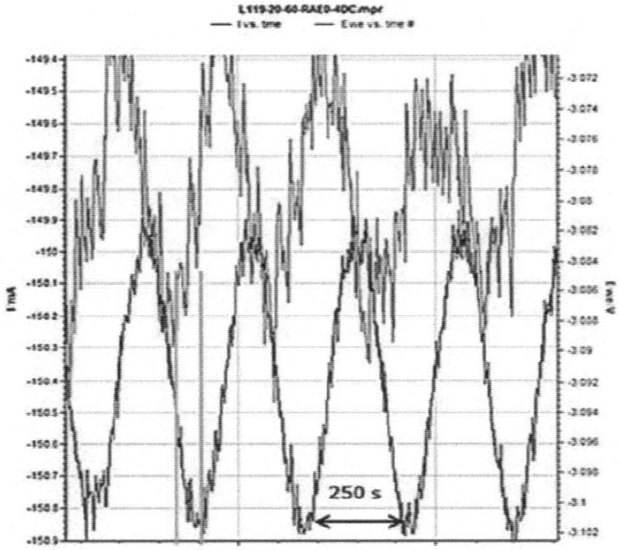

Figure 2. Current ripple at high data acquisition rate.

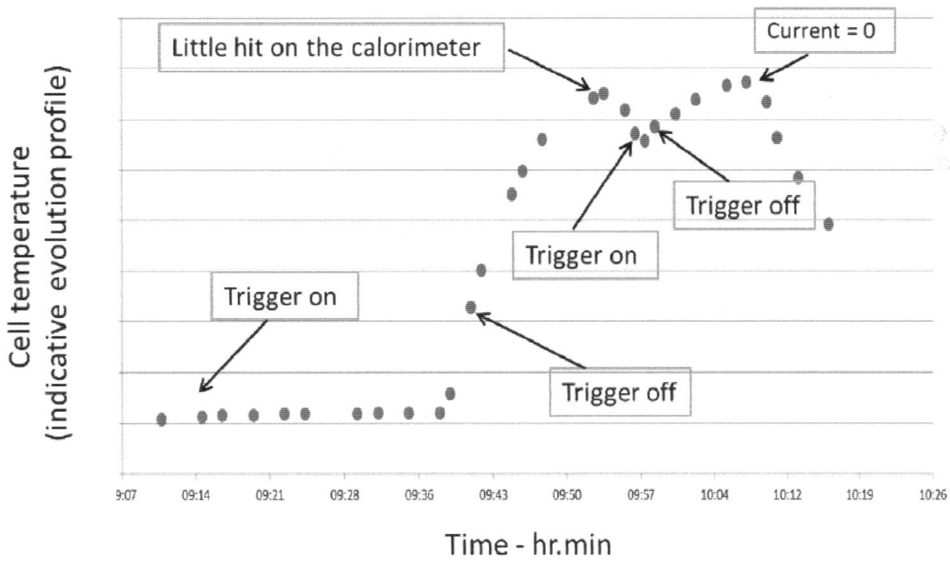

Figure 3. Indicative temperature evolution of the cell wall.

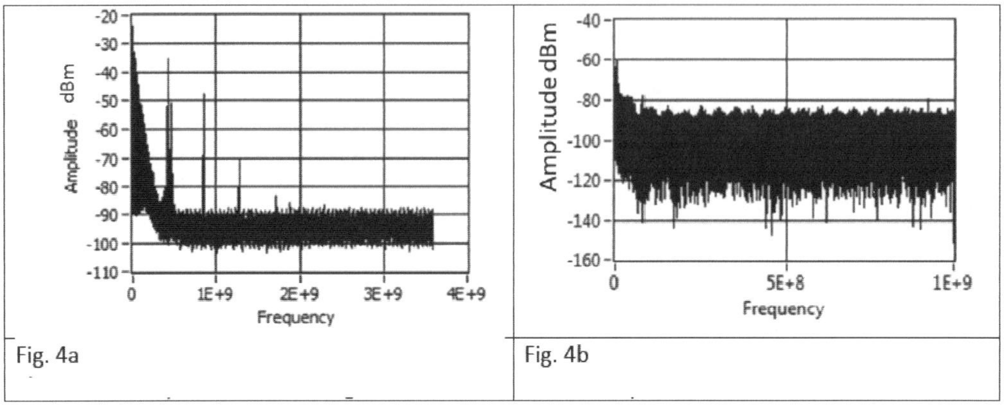

Figure 4. (a) Spectral RF signal for active sample during electrolyte temperature increasing. (b) Typical Background RF signal for not active samples.

The Peltier calorimeter was realized by using a thermally insulated aluminum pipe containing the electrochemical cell. The Peltier cell was located between the bottom of the pipe (not thermally insulated) and an heat sink. The calibration was based on the voltage signal given by the Peltier cell and the following relationship was used to measure

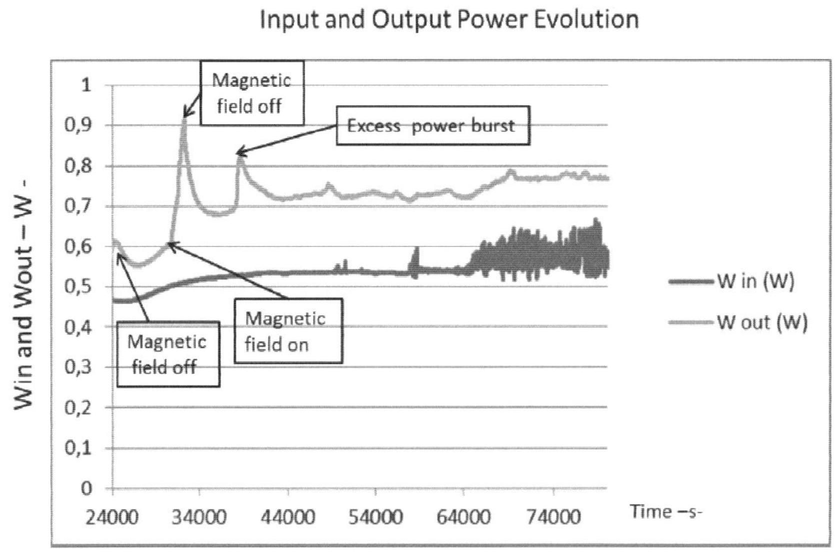

Figure 5. Input and output power evolution during excess power measured by using a Peltier calorimeter. Spikes in input power starting at ~50,000 s and increasing in amplitude and frequency at ~65,000 s were due to voltage fluctuation. This effect is not reflecting in the output power because of the calorimeter time constant, much larger than the fluctuation frequency.

the cell output power:

$$P_{\text{out}} = a\,\Delta V + b,$$

where ΔV is the Peltier cell voltage, a and b are coefficients given by the calibration performed with light water (LiOH). The input power in DC current mode was just the scalar product between the current and voltage since the cell used was a closed one.

We may observe in Fig. 5 that after applying the magneto-acoustic trigger the output power (W_{out}) was exceeding the input power (W_{in}) and the excess was about 0.1 W when the input power was 0.5 W. The trigger was applied again for several minutes and then removed. After this operation an excess power bust occurred. The plot shows a maximum value of the difference between the output and input power of about 0.3 when the input power was 0.55 W. The burst survived for a time too short compared to the calorimeter transient time and clearly it was much larger than 0.3 W. After this event the excess survived at an average value of 0.2 W and after several hours the experiment was switched off and the electrode removed in order to analyze the sample.

Table 1 shows a synoptic of the experiment performed and we may see that only samples belonging to lots 122 and 137 (produced by using the same preparation protocol to have the same metallurgy) were active.

3. Conclusion

RF signal emission has been observed during the excess power production but such a signal has been obtained, with the same amplitude observed during excess power, also when the excess power was not observed. The detection limit of

Table 1. Synoptic of the experimental results. If not specified anodes are of stainless steel coated with platinum.

Electrode	Lot	Electrolyte	Current mA	Notes	RF	Heat
Steel	NA	H_2O	53	NA	No	No
Steel	NA	H_2O	53	NA	No	No
L124(10–50)	Pd90/Rh10 (Al)	H_2O			No	No
L127(20–130)	H.M. + Pt	H_2O			No	No
L128(105–145)	H.M. + Pt	H_2O			No	No
L128(145–185)	H.M. + Pt	H_2O			No	No
L58(165–200)	Johnson Matthey MM29560	H_2O			No	No
L93(110–150)	Alfa Aesar 307622	H_2O			No	No
L95(50–90)	Alfa Aesar 307622	H_2O			No	No
L121(90–130)	Pd90/Rh10	H_2O			No	No
L125(170–206)	H.M. 2108539 + Pt	H_2O			No	No
L122(120–160)	Pd+Doping	H_2O	24		Yes	Yes
L122(120–160)	Pd+Doping	H_2O	53		Yes	Yes
L122(160–200)	Pd+Doping	H_2O	53		Yes	Yes
L122(160–200)	Pd+Doping	H_2O	53	Charge 107 mA	Yes	Yes
L122(236–272)	Pd+Doping	H_2O	53		No	No
L122(308–366)	Pd+Doping	H_2O	53		Yes	No
L137(100–140)	Pd+Doping	H_2O	53		No	No
137(180–216)	Pd+Doping	H_2O	107	Anode Pt	No	No
L137(60–100)	Pd+Doping	H_2O	107	Anode Pt	Yes	No
L137(60–100)	Pd+Doping	H_2O	107	Anode Pt	Yes	Yes
L137(300–340)RAEM	Pd+Doping	H_2O	107	Anode Pt	Not measured	YES
L137(380–420)RAEM	Pd+Doping	H_2O	107	Anode Pt	Yes	No

the isoperibolic calorimeter, used during RF measurements, was in the order of 100 mW. The excess power observed during RF emission was well above such a detection limit .

This is indicative that RF emission is not the effect of the excess but perhaps the cause. Electrochemical instability is observed in coincidence with the onset of the effect. Active electrodes have been obtained by doping a rough palladium originally inactive. Such a result should be considered as preliminary since a more significant statistical basis of data is required. A further effort in this direction could improve the control of the effect.

References

[1] V. Violante et al., Material science on Pd–D system to study the occurrence of excess of power, *Proc. ICCF-14*, Washington DC 10–15/8/2008, Vol. 2, pp. 429–436.

[2] V. Violante et al., Evolution and progress in material science for studying the Fleischmann and Pons effect (FPE), *Proc. ICCF-15*, Oct. 5–9, 2009, Rome, Italy, p. 1.

[3] V. Violante et al., The study of the Fleischmann and Pons effect through the materials science development, *Proc. the XVI Int. Conf. on Condensed Matter Nuclear Science*, Chennai, India, Feb. 6–11, 2011, p. 313.

[4] V. Violante, E. Castagna, S. Lecci, F. Sarto, M. Sansovini, T. D. Makris, A. Torre, D. Knies, D. Kidwell, K. Grabowski, D. Dominguez, G. Hubler, R. Duncan, A. El Boher, O. Aziz, M. McKubre and A. La Gatta, Excess of power during electrochemical loading: materials, electrochemical conditions and techniques, *Proc. ICCF18 Int. Conf. on Condensed Matter Nuclear Science*, 20–25 July 2013, Columbia, MU, USA.

[5] D. Dominguez, D.A. Kidwell, G.K. Hubler, S.F Cheng, M.A. Imam, K.S. Grabowski, D.L. Knies, L.De Chiaro, A.E.Moser, J.H. He and V. Violante, *17th Conf. on Condensed Matter Nuclear Scienc*e, Daejeon, South Korea, 12–17 August 2012. http://lenr-canr.org/acrobat/DominguezDanomalousr.pdf.

[6] V. Violante, E. Castagna, S. Lecci, G. Pagano, M. Sansovini and F. Sarto, EAI – Energia Ambiente Innovazione is edited by ENEA, Italian National Agency for New Technologies, Energ and Sustainable Economic Development, 2-3/2014, p. 6; DOI: 10.12910/EAI2014-62.